D0999991

WHAT'S GOTTEN INTO YOU

WHAT'S GOTTEN INTO YOU

The Story of
Your Body's Atoms,
from the Big Bang
Through Last Night's Dinner

Dan Levitt

HARPER

An Imprint of HarperCollins*Publishers*

WHAT'S GOTTEN INTO YOU. Copyright © 2023 by Dan Levitt. All rights reserved. Printed in the United States of America. No part of this book may be used or reproduced in any manner whatsoever without written permission except in the case of brief quotations embodied in critical articles and reviews. For information, address HarperCollins Publishers, 195 Broadway, New York, NY 10007.

HarperCollins books may be purchased for educational, business, or sales promotional use. For information, please email the Special Markets Department at SPsales@harpercollins.com.

FIRST EDITION

Designed by Bonni Leon-Berman

Library of Congress Cataloging-in-Publication Data has been applied for.

ISBN 978-0-06-325118-2

23 24 25 26 27 LBC 5 4 3 2 1

To Ariadne, Zoe, Eli,
and my parents, Lore and Dave

We are an example of what hydrogen atoms can do, given fifteen billion years of cosmic evolution.

—*Carl Sagan*

CONTENTS

Introduction

$1,942.29 IN THE BANK

This book was inspired by a warning. When my teenage daughter decided to become a vegetarian, I recalled a caution I had received from my grandfather who once owned a small hunting lodge and, at ninety, still loved wild game. He told me it was unhealthy not to eat meat. I also recalled that Henry David Thoreau once received a similar admonition from a farmer. "You cannot live on vegetable food solely, for it furnishes nothing to make bones with." As the farmer spoke, Thoreau noted wryly, he stood behind sturdy oxen who pulled "him and his lumbering plow along in spite of every obstacle" with "vegetable-made bones."

So when my daughter told me of her intentions, I was sympathetic. I knew her bones were not in grave danger if she became a vegetarian. But I began to wonder—what exactly *are* we made of? I'd studied science in college and spent years making science films. I thought I knew a fair amount. Yet I realized that I knew less about what *I* was composed of than I knew about what lay inside my computer or my car. So I began asking questions. What *is* inside us? That seems easy: muscle, organs, and bones. What are these made of? Cells, molecules, and atoms. Go on . . . what are *they* made of? Hmm. That's a lot harder. Where did they come from? I'm not sure. And how do we know any of this? Not a clue.

My questions sparked a flurry of googling, reading, and conversations with remarkably patient scientists. I soon found myself hooked by an epic tale. I discovered that our atoms have seen so much that, if they could speak, they'd probably never stop talking. Their history begins at the beginning of time. And their dramatic odyssey across billions of years was only revealed, I discovered, by many surprising, even shocking, discoveries.

As time went on, my investigation evolved into something deeper. Themes and connections emerged, and I discovered that taking the long view also revealed the majesty of our existence in a way I'd never before appreciated. Carl Sagan once famously said we are made of star stuff.

This is the improbable story of how it happened.

The story begins with the truly bizarre discovery that all matter—everything around and within us—has an ultimate birthday: the day the universe was born. We'll see how, on our atoms' long strange journeys toward us, they were transformed into stars, helped assemble our planet, and then hunkered down while the newborn Earth endured unimaginable catastrophes. Once things calmed down, those dead atoms, in stunning recombinations, brought forth life, terraformed Earth, and created plants—which at last made our own existence possible.

Finally, we'll see how our bodies transform the food on our dinner plates into us. Mind you, our own interiors are so vast that it's difficult to fathom how bewilderingly complex we truly are. You are an ever-changing mosaic, a colony of thirty trillion cells, each made of over a hundred trillion atoms that are dancing in riotous vibrating shuffles. You contain a billion times more atoms than all the grains of sand in the earth's deserts. If you weigh 150 pounds, you carry around enough carbon to make 25 pounds of charcoal, enough salt to fill a saltshaker, enough chlorine to disinfect several backyard swimming pools, and enough iron to make a three-inch nail. If you sorted all your elements, you'd find that about sixty on the periodic table are represented inside you. If you could sell them, you'd pocket a cool $1,942.29 (exact estimates vary by weight and market price).

In pursuing this strange tale, I came to understand that the mere fact that we can reconstruct the history of our atoms across billions of years is as deeply surprising as if we were able to piece together the history of a typhoon from the shadows of a few raindrops. But the clues were all around us; they had simply lain long hidden. They were to be found in the fleeting traces of invisible particles that rained down from space, in the fact that each element emits a unique set of wavelengths of light, in the unexpected timetables of comets' returns to Earth, in the discovery that life flourishes in the bone-crushing depths of the dark ocean floor.

The stories of the scientists in these pages are as compelling as the discoveries they unearthed. Behind their unexpected findings were investigations marked by fierce rivalries, obsession, heartbreak, flashes of insight, and flukes of blind luck. Again and again, clues were overlooked until someone was willing to entertain ideas that everyone else "knew" were wrong.

I had no idea when I began that I would also learn about the unconscious workings of our brains that cloud our thinking. We all succumb to unconscious assumptions, called cognitive biases or thinking traps, that influence how we see the world. The negative bias, for instance, leads us to pay more attention to unpleasant events and helps explain why our media focuses on bad news. In these pages we will see that a small number of particular thinking traps repeatedly prevented whole communities of scientists from recognizing great breakthroughs—even when confronted by overwhelming evidence. We'll encounter six biases so frequently that I've given them nicknames:

- The "Too Weird to Be True" bias
- The "If Our Current Tools Haven't Detected It, It Doesn't Exist" bias
- The "As an Expert, I've Lost Sight of How Much Is Still Unknown" bias
- The "You Look For and See the Evidence That Matches Your Existing Theory" bias

- The "World's Greatest Expert Must Be Correct" bias
- The "Because It Seems Most Likely, It Must Be True" bias

In a world rife with biases, the scientists who made great breakthroughs were tremendously brave to go against the grain. Some are well known. Most are not: like two women, one a Jewish physicist, the other, well, a Nazi, who collaborated to search for subatomic particles; the Austrian empress's personal physician who discovered photosynthesis; the hard-of-hearing chemist who discovered DNA over eighty years before Watson and Crick found its structure; or the renegade biochemist, ridiculed as a fraud and prophet of doom, who revolutionized our understanding of our cells. It seems progress in science often leaps out of dark corners and unexpected places.

Let's begin this strange history of your body and mine, then, with a man who always wore black.

PART I

THE
JOURNEY
BEGINS

From a Big Bang to a Rocky Home

In which we make the shocking discovery that all the
particles within us were born in a single instant, discover
their surprising nature, puzzle out the outlandish ways in
which they became our elements, and learn how, improbably,
a monstrous cloud of dust with our atoms scattered within it
created a planet hospitable to life.

1

HAPPY BIRTHDAY TO EVERYONE

The Priest Who Discovered the Beginning of Time

All great truths begin as blasphemies.
—*George Bernard Shaw*

On a cold, uncharacteristically dry London day in September 1931, a short, stocky man with slicked-back hair, a piercing gaze, and a hell of a lot of nerve walked along Storey's Gate Street. He entered Central Hall, Westminster, a large assembly place near Westminster Abbey. It's hard to imagine that this man, a thirty-seven-year-old Belgian professor of physics, did not feel some trepidation. The soaring dome of the Great Hall imposed grandeur on the proceedings: a celebration of the hundredth anniversary of the British Association for the Advancement of Science. Many of the world's most eminent physicists were among the audience of two thousand to whom Georges Lemaître was about to present a theory that bordered on the crackpot.

Lemaître was not just a physicist and a mathematician, but a Catholic priest as well, and he was to speak in a session on a topic that physicists had just begun to grapple with: the evolution of the universe. Dressed in his black clerical garb and white collar, as if prepared to take confessions, he stepped to the podium and presented an idea that veered perilously close to theology. He had discovered, he claimed, a moment when the entire universe exploded out of a tiny "primeval atom."

Many other speakers presented ideas that were also electrifying, to say the least. The famed astronomer James Jeans suggested that the universe's days were numbered. The mathematician Ernest Barnes (himself an Anglican bishop) speculated that the universe was so vast, it must contain many inhabited worlds, and there must exist beings on some of them "who are immeasurably beyond our mental level." But Lemaître's theory was the strangest of all. He claimed that physics could address almost the very instant of creation.

Few of the great and good in the hall took his remarks seriously. Those who were not simply mystified were deeply skeptical. Virtually all the physicists and astronomers in the audience believed that the universe had always existed. Lemaître's claim that it had not seemed preposterous.

Although they failed to realize it at the time, his insight would lead to one of the greatest achievements in all of science—the mindblowing discovery that there was a single moment when the most elementary particles within all visible matter, including you and me, sprang into existence.

Lemaître's search for truth began years earlier—in the bloodsoaked trenches of the First World War. He had been a student at the Catholic University of Leuven intending to take up the sensible career of a coal-mine engineer when, on the morning of August 4, 1914, German troops poured across the Belgian border, plunging Europe into war. Instead of taking a planned cycling trip, both Lemaître and his brother promptly enlisted and hiked four days to join a volunteer unit at the front. Within two weeks they were fighting, with outdated single-loading rifles.

As an infantryman, Lemaître had the misfortune to witness the first successful poison gas attack in war. The Kaiser's army, acting on the brainstorm of the chemist Fritz Haber (whom we'll meet again), released chlorine gas along the front. It dissolved the lungs of unsuspecting Allied soldiers, sending them shrieking from the battlefield. "The madness of it would never fade from his memory," Lemaître's colleague recalled. Later, in the artillery, he was mired in ghastly exchanges of explosive shells. Family legend says his scientific bent kept him from getting a promotion because he couldn't keep from correcting the ballistic calculations of his superior. He lacked, it seemed, the attitude expected of an officer.

However, Lemaître had brought physics books with him, and in the intervals of trench warfare, while waiting for shells to fly, he somehow found the concentration to read the work of the French physicist Henri Poincaré and ponder the ultimate nature of reality. In the squalor of wood-and-dirt trenches, Lemaître mused about a big question: What was the universe ultimately made of? For the young man, who came from a deeply religious family, physics and prayer both brought solace.

Lemaître emerged from the war a decorated veteran, his brother, an officer. But the war had seared Lemaître's soul. When peace arrived after four years, a practical career in engineering no longer seemed so important. Instead, he was torn between his two loves: religion and science. Returning to university in Belgium, he sped through a master's degree in mathematics and physics. These were exciting times. At the University of Berlin, a brash physicist named Albert Einstein had just confounded his colleagues with a radical, profoundly unsettling theory that the mass of an object actually warps the space and time around it. Lemaître was captivated. Yet, when he graduated, he abruptly changed direction. He enrolled in a seminary. "There were two ways of arriving at the truth," he would say later. "I decided to follow them both." Once ordained, Lemaître took a vow of poverty. He joined a small priestly association, Les Amis de Jésus, which emphasized the continued development of piety. Then he promptly returned to physics. Some of his more progressive professors at university, following the teaching of Saint Thomas Aquinas, taught that the Bible could not be a literal guide to science, just as science could not offer a guide to religion.

With his cardinal's blessing, Lemaître headed to Cambridge University to study with Arthur Eddington, soon to become *Sir* Arthur Eddington, thanks to a celebrated discovery he'd made four years earlier. Anticipating a solar eclipse, the astronomer had organized expeditions to the coasts of West Africa and Brazil, and brought back photographic evidence that Einstein was right. As unlikely as it seemed, light curved as it traveled around the Sun. His observations were proof that mass warps space and time, and transformed both men into celebrities. When Lemaître arrived to work on relativity, Eddington found his new student "wonderfully quick and clear-sighted." He was so impressive that Eddington recommended that, after his year in England, he study at Harvard with Eddington's friend, Harlow Shapley, the first astronomer to measure the size of our galaxy.

Lemaître arrived in Cambridge, Massachusetts, in 1924, just as new observations were roiling astronomy. Two years earlier, most scientists had believed that the entire universe consisted of the Milky Way and a few other galaxies. That was it, because that was all they could see. But in 1922, at California's Mount Wilson Observatory, Edwin Hubble shocked them. Peering through the world's most powerful telescope, he discovered that the universe was vastly larger. It contained an incredible number of other galaxies, all of which, the *New York Times* proclaimed, were "'island universes' similar to our own." It was humbling to realize that we lived in a small corner of the universe, and electrifying to know that there was so much more to learn about it. It was as if a bank had informed astronomers, "Sorry, we made a mistake. You don't have 500 dollars in your account, you have 500 trillion dollars."

Lemaître immersed himself in the raging debates of astronomers who were struggling to make sense of this immense new universe. Oddly, the latest measurements taken by a few of them seemed to suggest that the newfound galaxies were not standing still. They were moving away. Even more puzzling, the most distant galaxies were moving away faster than those closer to us.

Lemaître was intrigued. He returned to Belgium to begin teaching at his old university and there he dove deep into Einstein's equations to see if they might predict this strange state of affairs. He finally emerged with a disconcerting solution. It suggested not just that the galaxies were moving apart, but that the universe itself was actually growing larger. This was bizarre, to say the least, perhaps one of the strangest ideas ever proposed in science. Furthermore, he argued, the galaxies were not actually moving away from one another in space. Instead, the space between them was actually expanding and pulling the galaxies apart, like raisins in a loaf of rising bread.

Feeling triumphant, the unknown physics professor swiftly published his finding in French in a little-known Belgian periodical. It was not, perhaps, his smartest move. His paper was completely ignored. He sent it to his old teacher, Eddington; no response. He sent it to Einstein and to the famous cosmologist Willem de Sitter. Nothing.

Finally in 1927, a frustrated Lemaître had a chance to sound out Einstein directly. As he strolled through the alleys of Brussels's Parc Léopold at the famed Solvay Conference, a meeting of the world's greatest physicists, he was introduced to Einstein by Auguste Piccard (the inspiration for Professor Calculus in the comic series *The Adventures of Tintin*). This was Lemaître's opportunity to meet the world's greatest living scientist. Einstein's general theory of relativity—a set of ten equations that modestly described the interaction of space, time, and gravity—was reshaping our understanding of the universe. Nothing would have pleased Lemaître more than Einstein's approval. To the older man, however, Lemaître was simply an obscure Belgian priest whose paper had attracted no notice at all.

Einstein's reaction to his theory was to the point: he hated it.

Deep in his soul, Einstein believed that the universe had to be static. His powerful intuition, which had guided him astonishingly well in the past, told him that some simple order must lie beneath the chaos of the material world. He could not believe that the universe itself was expanding. That just didn't seem right, and he had little interest in suggestions to the contrary. It was simply too weird to be true. "Your calculations are correct," he told Lemaître as they walked through the park, "but your physical insight is abominable." Becoming more polite, he explained that, a few years earlier, he had rejected a similar calculation by the Russian mathematician Alexander Friedmann. In fact, Einstein hated the idea so much that he had introduced a fudge factor he called the "cosmological constant" into his equations to keep his universe static. When Einstein and Piccard got into a taxi, Lemaître tagged along. He tried to tell Einstein about new observations, of which Einstein seemed unaware, that implied that the galaxies were flying apart at puzzling speeds. Einstein brushed him off and began speaking with Piccard in German.

Two years later, Edwin Hubble published new observations from Mount Wilson. He was still capitalizing on the unprecedented light-collecting power of an eight-foot-diameter telescope mirror, over two feet wider than any other. His data confirmed that the most distant galaxies were indeed moving away from us faster than the closer ones. Now Eddington himself reexamined Einstein's equations and found that, despite Einstein's disbelief, they implied that the universe *was* expanding. Soon after, Eddington was embarrassed to find that he had read, but forgotten about, the paper Lemaître had sent to him two years earlier that reached the same conclusion. Eddington swiftly arranged for Lemaître's article to be published in English in the *Monthly Notices of the Royal Astronomical Society*. Now, Einstein had to pay attention to Lemaître's theory. Although in most textbooks Hubble gets credit for proposing the expanding universe, Lemaître had discovered it first.*

---

* The Russian physicist Alexander Friedmann saw the possibility in Einstein's equations of an expanding or contracting universe, but tragically, he died in 1925. Lemaître discovered the idea independently and was the first to recognize that astronomical observations supported an expanding universe.

Undeterred by the great man's rebuff, meanwhile, Lemaître had dug even deeper into Einstein's equations and taken a more daring leap. Rewinding time in his head, he reasoned that, if the universe is expanding now, sometime back it must have been smaller, and before that, smaller still. Taking his logic to its seemingly absurd conclusion, he proposed that, at one point, it was so small and dense that every single galaxy in existence today—not to mention every fundamental particle within all matter—was nestled inside what he called a single "primeval atom."

In 1931, at a meeting of the British Association for the Advancement of Science, Lemaître presented his brainstorm. At its origin, he claimed, the whole universe was produced by the "disintegration" of this tiny atom. That may not have been the most poetic way of putting it, but Lemaître did know how to turn a phrase. He would later write, "The evolution of the universe can be likened to a display of fireworks that has just ended: some few wisps, ashes and smoke. Standing on a well-chilled cinder, we see the fading of the suns and try to recall the vanished brilliance of the origin of the worlds."

Lemaître had been mulling over some version of this theory even longer. His friend, Bart Jan Bok, recalled Lemaître saying at Harvard, "Bart, I've had a funny idea. Maybe the whole universe started out of a single atom, and it exploded, and that's where it all comes from, ha ha ha."

The popular press loved his theory. "Out of a single bursting atom came all the suns and planets of our universe," *Modern Mechanix* marveled. But to physicists, the idea was an abomination: it was simply absurd. The Canadian John Plaskett charged that it was "an example of speculation run mad without a shred of evidence to support it." Lemaître's old mentor, Eddington, called it "repugnant." Although he agreed that the universe was expanding, this was a bridge too far. He preferred to believe in a universe that always existed.

Like Einstein, he had fallen into the "Too Weird to Be True" thinking trap, which could also be called the "Surely Nature Would Never Do *That*" bias. Scientists, like the rest of us, lapse into flawed thinking when they don't double-check their assumptions. Eddington never did change his mind.

As for Einstein, he was also convinced that Lemaître was inspired by Christian dogma. The theory reeked of religion. It was just too suspiciously similar to the biblical moment of creation. Lemaître denied it. Yet, in truth, there is a sense in which Einstein was right. Like a bee to honey, how could a priest and cosmologist not be drawn to how science and scripture each describe the origin of everything? In fact, in 1978, historians uncovered a paper Lemaître wrote in graduate school in which he attempted to prove scientifically that the universe began with light. But Lemaître abandoned that youthful exercise. For the rest of his life, he disavowed the claim that science could ever describe Genesis. "Physics provides a veil hiding the creation," he wrote. In an interview with the *New York Times*, he said, "There is no conflict between science and religion."

"If the Bible does not teach science, among other things, what does it teach?" the reporter asked.

"The way to salvation," Lemaître replied. "Once you realize that the Bible does not purport to be a textbook of science, the old controversy between religion and science vanishes." He saw them as separate, not incompatible ways of understanding the world. Years later, he would ask Pope Pius XII not to use his theory as evidence of scriptural truth.

Einstein would have continued fighting Lemaître tooth and nail, except that he had an Achilles' heel. He knew that the cosmological constant, which he had introduced in his equations to keep his universe from expanding, was a fudge factor. He would later call it his "biggest blunder." In 1933, Lemaître, garbed as always in his clerical collar, lectured in Pasadena, California, where Einstein was in attendance. By now, Einstein, having visited Hubble, seen the data, and conferred with others, saw no way out. Lemaître's clarity prevailed. "This is the most beautiful and satisfactory explanation of creation to which I have ever listened!" Einstein said. Although it is possible that he was speaking ironically, he was soon in agreement with Lemaître. The two even went on a lecture tour together. What Lemaître called initial fireworks, and the astronomer Henry Norris Russell, whom we will meet again, called "*the* Catastrophe to begin all other catastrophes," we know by a catchier name.

Even with Einstein's approval, the theory did not sail smoothly into acceptance. The most famous critic, the maverick British astrophysicist Fred Hoyle, thought it ludicrous. "The hypothesis that all matter of the universe was created in a big bang at a particular time in the remote past [is] irrational," he charged, coining the term *the Big Bang*. Hoyle even called Lemaître "the Big Bang man." But the evidence continued to grow.

It is thanks to Lemaître that we now have an origin story about the birth of everything, including every particle in your body. Nowadays, physicists will tell you that in the beginning there were no atoms, no molecules, no space, or time. That's their way of saying that almost everything we know about the origin of the universe comes from the ten equations in Einstein's general theory of relativity that reveal how space, time, and gravity are intertwined. But it was Lemaître who saw the possibilities in Einstein's work that Einstein himself refused to take seriously. As crazy as it seems, a handful of equations tell us that our entire universe began as a minuscule point that had no volume, but infinite density. That tiny "singularity," as physicists like to call it, contained inconceivable energy. And the Big Bang, the expansion of that infinitesimal point, created time, space, matter, and ultimately us.

That's more than enough to make your head hurt. Intuition is no help. The theory begs a few basic questions: How could the density of a point be infinite? How could there be no time? If space started with the Big Bang, where was everything before that? The answers to all these questions are simple. We don't have a clue.

To try to make sense of it, I asked Avi Loeb, a cosmologist at Harvard University, what came before the Big Bang. Loeb doesn't shy away from tackling big questions. He studies the formation of the universe's first stars and has written papers on when life could have first evolved. (His answer to this is a mere 70 million years after the Big Bang.) But when I asked him how there could be no time before the Big Bang, he grew shy about speculating. "It is the same way that you can't think about what happens after you die, or before you were born."

I pressed him again. "I prefer not to conjecture about things I have no idea about," he replied, as though echoing Lemaître's belief that the Big Bang veils everything that went before.

The problem, he admitted, is that at the very smallest scale, "Einstein's equations break down." No one has been able to reconcile them with another terribly accurate theory, called quantum physics, that predicts the behavior of matter at the smallest level of photons and electrons. We rely on quantum theory in our lasers, atomic clocks, computer chips, and GPS. But its paradoxes—like the impossibility of predicting the exact location of a subatomic particle (known as Heisenberg's uncertainty principle)—make it incompatible with Einstein's general theory of relativity. At least until scientists can reconcile those two theories (in the long-dreamed-of "theory of everything"), the era before the Big Bang will be open to speculation, but remain unfathomable.

At this point, all we can say is that every single attempt to prove Einstein wrong has failed—while countless observations have proved him right. In 1949, the physicists George Gamow, Ralph Alpher, and Robert Herman calculated the extreme heat that would have been necessary for the first elements to form. And they predicted the exact energy of the surviving afterglow that should still be around today. In 1965, a decade and a half later, their calculations were accidentally confirmed by a pair of astronomers who couldn't figure out why the background noise in their radio telescope would not go away. No matter where they aimed their telescope—at nearby stars, distant galaxies, or empty space—they detected low-level electromagnetic waves—even after they eliminated all the suspects they thought could be creating interference, such as the pigeon droppings they cleaned out of their antennae. The discovery of this electromagnetic radiation, called cosmic background radiation, was found in every direction in the sky at the precise frequencies that Gamow and his colleagues had predicted. It was strong proof that the universe began with a Big Bang.

When Georges Lemaître heard the news of this discovery, he was lying in a hospital bed in Brussels, recuperating from a heart attack. He rejoiced at this vindication of his theory. It came just in time. He died eleven months later.

Experimental support for Einstein's equations, and Lemaître's extrapolation of them, continues to roll in, including, in 2016, the detection of gravitational waves, which might make it possible to detect gravitational echoes of the Big Bang.* Incidentally, when you change the channel on an old television, you can actually tune in to the Big Bang. About 1 percent of the flickery snow on the screen is radiation left over from it. The expansion of the universe stretched those electromagnetic waves out long enough that they can be picked up by your TV.

Whether Einstein himself once liked it or not, the equations of relativity tell us that time began with the Big Bang. Physicists have rolled the clock backward (using measurements of the universe's density and the rate at which it's expanding) to find the date: 13.8 billion years ago. That gives us a starting point for our own remarkable journeys. In the instant afterward, space began to expand. A trillionth of a trillionth of a second later, particles of matter and antimatter (particles with the same mass but opposite charges) sprang up in the newly created vacuum of space. Bent on mutual destruction, they immediately wiped each other out.† But obviously our story does not end there. For

---

* Even the fudge factor, the "cosmological constant" that Einstein added to his equations and later considered a blunder, may turn out to be important. It may help predict the properties of dark energy—a mysterious force in the universe that physicists call dark because they understand almost nothing about it.
† As strange as it may sound, antimatter is real. Today we are able to detect it and even make very small amounts in laboratories. NASA scientists dream of making enough antimatter to power their spaceships.

some reason that scientists are still scratching their heads over, there was a slight imbalance in the amount of matter and antimatter that appeared. For every billion particles of antimatter, a billion and one particles of matter were born. We are, to put it simply, leftovers. Those rare one-in-a-billion survivors created the visible universe, including all the atoms within us.

Lemaître helped reveal that long before a calendar could mark the date, cosmic candles were lit on your ultimate birthday. One fine day, 13.8 billion years ago, the most elementary bits of matter inside you sprang into being.

With that discovery, we now knew when our history began. But how much of what followed could be traced through the hazy mists of time? Scientists would have to start by uncovering the nature of the first particles to emerge from the Big Bang. What were the tiny building blocks that would ultimately create our solar system, planet, life, and then us? Strangely enough, the search to find them would begin with another Catholic priest, this one at the top of the Eiffel Tower.

2

"THAT'S FUNNY"

What the Eye Can Never See

The most exciting phrase to hear in science, the one that
heralds the most discoveries, is not "Eureka," but
"That's funny."
—*Isaac Asimov, attributed*

Pity the poor scientist in the age of electric streetcars and horse-drawn carts, who hoped to discover the most elementary particles within us—the ones that we now know sprang out of the Big Bang. How do you find something you can't possibly see with the naked eye or even the most powerful microscope? That was a rub that long troubled physicists. Even before Lemaître and Einstein were grappling with the origin of the universe, others were searching for the identity of the tiniest bits of matter that the universe is made of. Doubt, however, had always clouded the enterprise. Would they ever be able to find them? The ancient Greeks had speculated that everything, even humans, was made of tiny indivisible units they called *atomos*, meaning "tiny and uncuttable." By the turn of the twentieth century, many chemists agreed and called them atoms. But many physicists were dubious.

"Atoms and molecules . . . from their very nature can never be made the objects of sensuous contemplation," the influential physicist Ernst Mach declared. Chemists' experiments suggested that atoms existed in theory, but no experiment had ever directly revealed one. No scientist had ever seen an atom, touched one, or measured one.

Then, two startling discoveries turned doubters into believers. In 1897, at the Cavendish Laboratory in Cambridge, the epicenter of British physics, the forceful J. J. Thomson was investigating a puzzling phenomenon. What was the nature of mysterious cathode rays produced when two electrodes in a glass vacuum tube were zapped with electricity? Curious, he exposed the rays to a magnetic field to see what would happen, and was startled to see it deflect their path. Surprisingly, he had discovered invisible negatively charged particles that we now know are part of the atom. In 1911, the year King George V was crowned king of England and Hiram Bingham explored Machu Picchu, Thomson's brilliant former student Ernest Rutherford made an equally momentous discovery. When positively charged radioactive particles were fired toward a thin gold foil, most passed through as expected, but some bounced back. He was stunned. "It was almost as incredible," he recalled, "as if you fired a 15-inch shell at a piece of tissue paper and it came back at you." The particles must have been repelled by dense positive charges in the foil. He had discovered that atoms also contain positively charged nuclei.

In time, everyone agreed with Rutherford's triumphant conclusion: We now knew the nature of our most elementary particles. Everything in the universe was made of atoms. But the Greeks were not correct that these were the smallest particles of all. At the centers of atoms lay nuclei—objects ten thousand times smaller—that contained dense positively charged particles called protons. Thomson's negatively charged electrons whirled in orbit around them like planets around the Sun.[*]

---

[*] Quantum physicists would soon find that the electrons' orbits were less predictable than that. Now they view their orbital paths as hazy clouds that suggest where the electrons are most likely to be found.

And atomic weights led Rutherford to suspect that nuclei contained one other item: particles without charge that we now call neutrons.

Surely, that was it. No one had reason to think even smaller particles could exist. Mind you, even if they did, there was no way to look for them. Certainly, the most powerful microscopes would be no help. Scientists had as much chance of spotting an atom with a microscope as spotting Pluto with their naked eyes. Trillions of hydrogen atoms would fit on the head of a pin, and a proton was a hundred thousand times smaller still. If even smaller particles existed, it seemed impossible we could ever find them.

A word of warning: We are about to embark on a trip into some of the strangest phenomena around us. Physicists will unearth a rogues' gallery of baffling subatomic particles before they find the most elementary particles of all. But the path will be rocky and winding, for scientists often make their greatest discoveries while searching for something completely different.

The first unexpected clue would arrive out of the clear blue sky.

In the spring of 1910, Father Theodore Wulf, a German physicist and Jesuit priest, emerged from an elevator at the top of the Eiffel Tower toting a breadbox-size device that he hoped would solve a frustrating mystery. Pesky electric charges were following scientists everywhere like faithful dogs. Researchers had succeeded in making their electroscopes—devices that detect electric charge—so sensitive that their instruments were now driving them crazy. Their devices picked up charges even when they were insulated from everything. Scientists put them in thick metal boxes, isolated them in tanks of water; still the troublesome charges refused to go away. So Wulf designed an exceptionally sturdy portable electroscope, and then set out, determined to pin down the source of the charges.

His most obvious suspect was radioactivity. Just a decade earlier, the director of Paris's Natural History Museum, Henri Becquerel, happened to place uranium salts on top of photographic plates in his desk drawer. Days later, he was shocked to find that the salts had created images on the plates. He had discovered that some rocks, like uranium, are radioactive: that is, they cast off charged particles, as well as invisible electromagnetic pulses with wavelengths shorter than visible light. So it seemed obvious to Wulf that the annoyingly persistent charges in the air must be created by radioactive rocks deep in the earth. The radiation they emit must strike molecules in the atmosphere, knocking off electrons and creating charged particles. Carrying his electroscope, Wulf descended deep into caves to prove it, expecting his readings to increase as he drew closer to their source. They didn't. So, with an electroscope strapped to his back, he took the elevator to the top of the world's highest man-made structure, the Eiffel Tower, expecting the charges to go away. They didn't—at least not enough. Too many charges remained. The mystery deepened.

Wulf's experiment inspired a fearless twenty-eight-year-old Austrian physicist named Victor Hess to take up the challenge. Hess figured that the only way to know if the bothersome charges came from the ground was to carry an electroscope much higher still. In 1911, the only way to do that was to take risky high-altitude flights—by balloon.

Enlisting the help of a local aero club, Hess made six flights near Vienna, venturing as high as 6,000 feet. His experiments were frustratingly inconclusive (although one trial during a solar eclipse established that the charges did not come from the Sun). Not one to let a little danger get in the way of a good experiment, he resolved to fly higher. He persuaded the German Balloon Enthusiasts Association in a small Austrian town to volunteer the pride of their fleet, a twelve-story orange-and-black state-of-the-art beauty named the *Böhmen*.

At dawn in a large green meadow, on August 7, 1912, the aero club inflated the massive *Böhmen* from tanks they had hauled on a horse-drawn cart. At 6:12 a.m., Hess squeezed himself into a small wicker basket beside a pilot, a meteorological observer, a small bench, three electroscopes, hand luggage, and, most important, three large oxygen cylinders.

Space was tight, but Hess was well aware that he needed these last items. Our brains consume a quarter of the oxygen that we draw in with every breath. Without enough of it, there's trouble. That fact had been emphatically illustrated over thirty years earlier when three French balloonists attempted to break the altitude record in a craft named the *Zénith*. Wisely, they carried oxygen with them; foolishly, they didn't breathe enough of it. At over 20,000 feet, one recalled, "an inner joy is felt like an irradiation from the surrounding light. One becomes indifferent." He felt "stupefied"; his tongue was paralyzed. Then he blacked out. On the way down, he recovered but found his companions slumped over. A lack of oxygen had made short work of them, leaving him as the sole survivor. The gruesome accounts of their deaths put balloonists off high-altitude flights for another twenty years. Hess planned to climb almost as high—but he planned to live.

At seven a.m., he began his ascent, carried aloft by hydrogen (the same explosive gas that would later doom the German Zeppelin the *Hindenburg*). They rose into a crisp cloudless sky. An hour and a half later, they found themselves drifting across the German border. At 13,000 feet, they were buffeted by 30-mile-per-hour winds. Bundled in an overcoat, Hess gamely continued taking measurements, despite the piercing cold. By 9:15, he felt tired. Sensibly, he decided it was time to breathe some oxygen. An hour later, at the dizzying height of 17,400 feet, three miles up, he was so weak he feared he would faint and decided perhaps it was time to call it quits. He ordered the captain to release some of the balloon's hydrogen. By 13,000 feet, he began to feel himself again.

Back on firm ground in a pasture, Hess was elated. At his highest altitude, he had measured a charge twice as great as that on the ground. There could be only one explanation—the higher he rose, the closer he was to their source. He was sure he had discovered that a stream of electric charges constantly bombarded the Earth from outer space.

Other physicists found that hard to swallow. Wasn't it more likely that Hess's instruments were affected by the freezing temperatures, for instance? The American physicist Robert Millikan was an especially fierce opponent—that is, until 1925, when his own experiments confirmed Hess's measurements. They were soon dubbed Millikan rays, until Hess bitterly objected, at which point another name from Millikan stuck—cosmic rays.

That was unfortunate, because they are not a unique wavelength of electromagnetic radiation like light, radioactive rays, or X-rays, as Millikan supposed. Instead cosmic rays are showers of charged particles that continuously rain down on us.

Physicists still had no idea that these invisible downpours contained hints of even smaller particles. And they would only be able to find them if they could invent new tools to "see" the impossibly small.

Actually, one such tool had been invented by one of J. J. Thomson's researchers years earlier, but it had been dreamed up for an entirely different purpose by a young man infatuated with clouds. Charles Thomson Rees Wilson, the son of a Scottish sheep breeder, was tall, quiet, and soft-spoken. In 1895, the twenty-two-year-old, with a newly minted physics degree from Cambridge, volunteered to work for a few weeks at a primitive meteorological observatory. It was on Ben Nevis, Scotland's highest peak. The stone cabin he slept in was often drenched in fog or racked by rain and thunderstorms. But in the early mornings, he was occasionally rewarded with a glorious sight. On the clouds just below, he saw halo rainbow patterns so majestic, he decided to make artificial clouds in a laboratory to study them.

Once back in Cambridge, Wilson, who was possessed of extraordinary patience, perhaps because of his pronounced stutter, taught himself devilishly tricky glassblowing techniques. After countless breakages, he built an ingenious glass chamber with a piston that could change the pressure inside. Through careful experimentation, Wilson was excited to see that, if he filled his chamber with moist air, and then quickly expanded its volume with the plunger, water vapor condensed on dust particles in the air. He had made artificial clouds.

Then, a discovery in Germany sent his research veering in a completely different direction—one that would transform his obscure tabletop apparatus into a tool that Ernest Rutherford himself praised as "the most original and wonderful instrument in scientific history."

Three hundred miles away, at the University of Würzburg, the physicist Wilhelm Röntgen, like J. J. Thomson, was investigating the rays created by a cathode ray tube. Although he carefully covered the glass tube with black cardboard to prevent any light from escaping, he happened to glance at a nearby screen coated with phosphorescent paint and was startled to see it glow, as if lit up by invisible rays. He was stunned.

Fearing others would think him crazy, he told no one as he began a feverish round-the-clock investigation. When he placed his trusting wife's hand between the glass tube and a photographic plate, a ghostly image of her finger bones and wedding ring appeared. "I have seen my death," she said on seeing the image. Röntgen had discovered that when cathode rays inside the tube struck the end of the glass, something entirely different was emitted. He had stumbled on X-rays—electromagnetic waves much shorter than visible light that are absorbed only by heavy elements, like the calcium in our bones.

Back at the Cavendish Lab in Cambridge, physicists were skeptical of newspaper reports of these see-through rays, until they saw the photographs. "Nearly every professor in Europe is on the warpath," a converted Rutherford wrote of the crowd of physicists now racing to investigate them. Soon, in America, Thomas Edison was X-raying brains and working on an X-ray light bulb. (He would abandon these efforts some years later, after his assistant suffered X-ray burns and died of cancer.)

Enthusiastically, Charles Wilson joined the stampede. On a hunch, he borrowed a crude cathode tube from J. J. Thomson and fired X-rays into a cloud chamber filled with humid air. He was startled to see the X-rays create a dense fog inside. They had knocked electrons off air molecules, creating charged molecules called ions that water vapor condensed on—producing droplets of fog. Wilson was ecstatic. His cloud chamber revealed the tracks of the unseeable: individual molecules so small no one imagined they could ever be detected. When he introduced radioactive particles into his cloud chamber, "little wisps and threads of clouds" almost magically appeared and vanished like vapor trails behind airplanes. It was almost like magic. Wilson painstakingly refined his chamber, and the simple apparatus he invented to study clouds was soon used worldwide as a powerful tool to investigate electrons, ions, and radioactive particles. But the greatest achievement of his cloud chamber—the detection of unsuspected particles smaller than the atom—was still to come.

By 1932, scientists had determined that cosmic rays contained electrons, so a young researcher at the California Institute of Technology named Carl Anderson reluctantly built a cloud chamber to study them. Having just received his PhD, he hoped to move to another university, but his advisor, who happened to be Robert Millikan (the same Millikan who confirmed the existence of Hess's cosmic rays), insisted that Anderson work on this project first. Anderson borrowed components from a Southern Edison junkyard to build a colossal electromagnet that was strong enough to bend the path of any electron flying into the chamber from the sky. Occasionally, his patience was rewarded. He succeeded in photographing the track of an electron curving in the powerful electromagnetic field. But he was puzzled that every once in a while, he would see a similar-size track curve in the opposite direction.

At first, Anderson thought they must be made by electrons moving upward. But Millikan reminded him that cosmic rays come from the sky, not the ground, so the tracks must be from positively charged protons raining down from space. Anderson was unconvinced. Protons, the only known positive particles, were larger, so their tracks should be wider than the tracks of electrons, but these were not. They argued. Anderson modified his experiment. Finally, with new evidence in hand, he bravely announced that he had found a new type of subatomic particle. And this one was a very odd duck. It was identical to an electron, except it had a positive charge.

None of the famous gods of quantum physics—Rutherford, Bohr,

Schrödinger, or Oppenheimer—believed him. Everyone knew that atoms had only three building blocks: negative electrons, positive protons, and just-discovered particles called neutrons, which have no charge. A positive electron just wasn't possible. Except that, just six months earlier, the physicist Paul Dirac had announced that, after several years of wrestling with Einstein's theory of relativity, he was forced to make a strange prediction. Electrons should have twins with identical mass but opposite charge. Even Dirac was skeptical of his own claim, yet Anderson had found one. This new subatomic particle was an *anti*electron— the first antimatter particle ever discovered. It was named the positron.

(If you think antimatter sounds far removed from daily life, you may be interested to know that you're more familiar with positrons than you think. We contain small amounts of naturally occurring radioactive potassium in molecules whose roles include sending nerve signals. About 0.001 percent of those potassium atoms decay every day, emitting positrons. If you weigh 150 pounds, you generate almost four thousand positrons a day. But they don't hang around long. Each swiftly meets an electron and as they destroy each other, they leave a tiny burst of radiation as their calling card.)

Just two years after Anderson stumbled on the positron, he bagged another particle—the muon. Inexplicably, it had the same charge as an electron but was over two hundred times heavier. "Who ordered that?" the physicist Isidor Rabi asked, on hearing the news.[*]

Hess, Wilson, and Anderson all received Nobel Prizes for their discoveries. Physicists had found something that no one thought could exist: new kinds of subatomic particles. Suddenly it was clear that atoms contained more than just electrons, protons, and neutrons. We no longer knew with any certainty what the atoms within us were ultimately made of.

---

[*] Today we know that the majority of cosmic ray particles that strike Earth's surface are muons. About ten of them pass through your body every second. Cosmic rays add about 27 millirem to your yearly dose of radiation, almost as much as three chest X-rays. (Sundermier, "The Particle Physics of You.")

Yet physicists remained almost blind. To find the very smallest constituents of our atoms, they still needed inventive new tools. Fortunately, another crucial one would soon come, courtesy of Marietta Blau, a five-foot-tall, introverted Austrian researcher, whose contribution would for too long be forgotten. Like Wilson, she would pioneer a method of "seeing" tiny objects no microscope could reveal.

Blau's interest in physics was sparked in the 1910s at the preparatory school she attended for girls. Her interest deepened at the University of Vienna, where she earned a doctorate in physics. She was among a number of European women who, inspired by Marie Curie, studied the puzzling new phenomenon of radioactivity. Years earlier, Madame Curie and her husband, Pierre, had discovered an almost magical new element—radium—that was a million times more radioactive than uranium. Scientists were agog that it seemed to provide an "inexhaustible amount of light and heat." During the radium craze that followed, you could buy radium-fortified soap, cigars, toothpaste, and pastries, even radium-infused furniture cleaner and suppositories. (Unaware of the danger, Madame Curie herself died an untimely death from radiation exposure.)* The radium was extracted from uranium ore, and Europe's only uranium mine belonged to the Austro-Hungarian Empire. So perhaps it is not surprising that Vienna founded an Institute for Radium Research, and Blau, a gifted experimentalist, landed a position there.

In 1925, her supervisor, a fellow physicist, presented her with a daunting challenge. Could she use a photographic plate to detect the path of a proton that was ejected when two nuclei collided? This was even more difficult than it sounds. She would be attempting to find a track of a single particle smaller than an atom. Doggedly, systematically, she experimented with thicker photographic emulsions, tested new techniques to develop her plates, and struggled to interpret barely perceptible impressions. After years, she actually succeeded, not just

---

* Even today, her papers are so radioactive that researchers must don protective clothing before examining them.

in capturing the paths of impossibly small particles, but in using them to measure the energy of the particles. It was a remarkable demonstration of the promise of her technique.

Yet in all her years at the Institute, Blau went unpaid. She supported herself by tutoring, by freelancing at medical companies, and with help from her family. When she began to receive international recognition, she finally worked up the nerve to ask for a paid position. That was impossible, she was told—she was a Jew, and a woman.

In the early 1930s, Blau refined her methods and set herself a more ambitious goal: detecting particles in cosmic rays. Yet she also faced increasing troubles. She was always eager to assist anyone in need, so when she met Hertha Wambacher, a young woman who had been unhappy studying law, Blau generously offered to help. Wambacher became her student, assistant, and finally junior colleague. But as time went on, Blau's generosity returned to haunt her. In 1933, the fascist dictator Engelbert Dollfuss seized power in Austria. Calls rang out to expel Jews from academic positions. And Blau's protégé became an early member of the illegal rival Nazi party. To make matters worse, Wambacher began an affair with an even more enthusiastic Nazi, a married physicist named Georg Stetter who would become director of the Institute. Blau and Wambacher continued their work together, but their once warm collaboration became tense.

In 1937, Blau was finally ready to try detecting cosmic rays. Her photographic plates might have great advantages over massive cloud chambers. Her plates were much easier to transport to high altitudes, where cosmic rays were strongest. And she could leave them there for a long time, greatly increasing her chances of capturing a rare particle. Blau and Wambacher simply rode a suspension cable car to a research station that Victor Hess had established on the 7,500-foot peak of Hafelekar Mountain. Expectantly, they placed their specially formulated plates facing the sky.

Four months later, they returned to retrieve them. Back in their laboratory, they peered through a microscope and were elated. They saw long slender lines etched on the plates. They had captured the tracks of invisible particles hurtling down from space. Even more astonishing, in some cases many straight lines emanated from a single point. A cosmic ray had struck and split a nucleus in the photographic emulsion, ejecting up to twelve smaller particles that created a starlike image. Blau's discovery sparked interest from physicists everywhere. Her years of experimentation had paid off. She had pioneered a new technique that would help reveal the smallest particles within us.

Tragically, she would have little opportunity to use it. In 1937, with Austrian anti-Semitism on the rise, Stetter pressured Blau to turn her work over to her junior partner and leave the Institute. Wambacher alternated between treating Blau with rudeness and generosity. Anguished, Blau considered abandoning her research. Then, unexpectedly, she was offered a temporary respite. A former friend and colleague, Ellen Gleditsch, got wind of her troubles and invited her to the University of Oslo for a few months. On March 12, 1938, Blau left on a train, taking her newest plates with her. From her window she saw German troops rolling across the border. It was the Anschluss, the day Hitler annexed Austria. The next day, Hitler himself entered Vienna, to the delight of adoring crowds.

Norway offered only a short-term refuge, but fortunately, Einstein had heard of Blau's plight and, with his help, eight months later she found a teaching position in Mexico City. Fearing war was imminent, she took the first flight out of Oslo. Regrettably, it was on a German airline. When it touched down in the connecting city of Hamburg, Blau was summoned. The Nazi officials seemed to know exactly what they were looking for. They rifled through her bags and seized her photographic plates before they allowed her to leave.

In Vienna, Wambacher continued their research where Blau left off, and began taking credit for their joint work. Blau was devastated. Although she would later have a series of academic appointments in the United States, she would never regain the momentum to return to her research on cosmic rays. In her sixties, her need for a costly cataract operation she could not afford in the United States forced her to return to Vienna. But she no longer felt welcome at the Radium

Institute, where she worked briefly, unpaid again. And she was embittered to find that Stetter, despite his prominent affiliation with the Nazis, had been allowed to return to a prestigious university post in the early 1950s. Blau died in 1970, her achievements unrecognized in her native Austria.

Meanwhile, many others adopted the techniques that Blau pioneered, and reaped the rewards. In 1947, Cecil Powell and Giuseppe Occhialini placed sensitive photographic plates high on a summit in the French Pyrenees and discovered a new subatomic particle—the pion—the first new kind found since Anderson's muon. Pions, like muons and positrons, are strange beasts. They are about 270 times heavier than electrons and can have a positive charge, a negative charge, or none at all.

Powell received the Nobel Prize for their discovery. Blau was not mentioned. She was nominated for the Nobel Prize at least three times, twice by Erwin Schrödinger, himself a Nobel laureate, to no avail.

The beautiful wispy tracks in Wilson's cloud chambers and on Blau's plates opened a Pandora's box. They revealed the existence of new particles smaller than the atom. Yet they hardly clarified the picture. With particles like positrons, muons, and pions, physics seemed even further from finding the most basic building blocks within us. It was as if scientists had looked for the solution at the bottom of a well, and discovered that the well was much deeper than they ever imagined. All they could do was keep lowering the bucket and hope. Yet the picture within the atom would soon grow even murkier.

By the 1940s, physicists knew that cosmic rays were largely atomic nuclei, along with stray protons and electrons, hurtling toward Earth at close to the speed of light. Most are absorbed by the atmosphere. Some high-speed collisions, however, split atoms, creating smaller, more exotic subatomic particles, like pions and muons, that also rain down on the Earth.

So now researchers decided to try a more direct approach. Instead of waiting for cosmic rays to send down a new particle every once in a blue moon, why not smash particles together to see if new kinds would be ejected from the collision, like debris from a highway crash? Particle hunters began building atom smashers—or particle accelerators, as they preferred to call them—to smash electrons against electrons and neutrons against neutrons at staggering speeds.

The first colossal atom smasher, given an optimistically futuristic name, the Cosmotron, was built in Brookhaven, New York, in 1949 by a consortium of eight American universities. It looked something like the innards of a flying saucer that had just been flown in for a tune-up. Physicists fired particles into a two-hundred-foot circular track girdled by 288 six-ton magnets. These steered the particles around the track, while microwave-emitting tubes, located every ten feet, bumped up the particles' speed every time around, as if pushing them along on an accelerating merry-go-round. Researchers operated the 3-billion-volt device from behind a two-foot-thick concrete wall. When they finally got all the kinks out, they could hurl their particles 130,000 miles in just four-fifths of a second, accelerating them close to the speed of light.

These smashups were spectacularly successful. Investigators were elated to find the tracks of much smaller particles. Blau's photographic plates and bubble chambers—close cousins of Wilson's cloud chambers—were arranged around the collision sites. They were so astonishingly sensitive they could detect particles less than a hundred trillionth of an inch in diameter that existed for less than a billionth of a second. Finding the most fundamental particles of all suddenly seemed within reach.

And then it wasn't.

Joy turned to bafflement and consternation as the number of discoveries began to climb. By the late fifties, physicists were scratching their heads over a "zoo" of dozens of odd particles they could make no sense of. Kaons, lambdas, sigmas, xis, hyperons, mesons; the list went on and on. "If I could remember the names of all these particles," grumbled Enrico Fermi, "I'd have become a botanist instead of a physicist." Were particle physicists doomed to become glorified cataloguers? They saw hints that the particles shared some properties, but could find no unifying scheme to connect them. So much for scientists' dream of discovering an elegant simple order beneath the chaotic surface of the universe. The search for the most basic particles of all now appeared to be going nowhere fast.

Things remained a mess until 1961, when Murray Gell-Mann joined the fray.

Gell-Mann was a child prodigy. At age three, he could multiply large numbers in his head. He entered Yale at fifteen. He was one of those know-it-alls who were notoriously bad at finishing papers and often cut classes, yet always aced their exams. By thirty-two, he had earned a doctorate from the Massachusetts Institute of Technology, worked at Princeton's Institute of Advanced Studies, collaborated with the legendary Enrico Fermi at the University of Chicago, and held a professorship at Caltech. He spoke thirteen languages, including Upper Mayan. His collaborator Sheldon Glashow recalled that "on first acquaintance you would soon learn, through his painfully in-your-face erudition, that he knew far more than you about almost everything, from archaeology, birds, and cacti to Yoruban myth and zymology." Photographs usually show Gell-Mann's warm eyes smiling through his heavy dark-rimmed glasses, yet he was also prickly and arrogant. He relished putting down those who didn't agree with him, including another genius down the hall, a collaborator who would turn rival, named Richard Feynman.

Gell-Mann chose to study particle physics, and he had an unusual genius for discerning underlying patterns. After years of work, his eureka moment came when he discovered that he could use an obscure form of algebraic theory to sort all the creatures in the particle zoo into groups based on their properties—their charges, masses, spin, and "strangeness" (a property that seemed to predict discrepancies in the time some particles took to decay into simpler particles). In a stunning demonstration of how nature mirrors mathematics, he saw that the particles fell into sets organized by geometric shapes of eight, called octets. He dubbed his theory the Eightfold Way, a joking nod to Zen Buddhism (this was California in the Sixties, after all). Perhaps it would finally lead particle physicists to enlightenment and end their suffering. Gell-Mann's octets sorted the particles into groups, and these seemed to align with their properties just as spectacularly as the Russian chemist Dmitri Mendeleev's periodic table had organized the elements. Where particles were missing in the geometric pattern, he predicted new ones would be found.

Yet Gell-Mann was apprehensive about publishing his theory—so much so that he withdrew his paper from *Physical Review* twice. He finally submitted a paper on the properties of the particles, with his revolutionary suggestion of the Eightfold Way cautiously tucked away at the end.

As often happens in science, someone else arrived at the same theory at the same time, in this case, another child prodigy, Yuval Ne'eman, a physicist (who later became an Israeli cabinet minister). Collegially, they shared the credit for an obscure discovery that went over at first like a lead balloon. Because their theory suggested that particles should exist that had never been seen in nature, it simply wasn't clear that they were on the right path.

Several years later, however, an experiment at Brookhaven gave Gell-Mann a lot of street cred. He had predicted the properties of a particular particle he called omega-minus, which was missing from one of his geometric sets. Experimenters set out to find it. It took them months to set up the necessary apparatus at the accelerator. Then they began taking thousands of photographs a day, so many that they employed Long Island housewives in round-the-clock shifts to help analyze them. After 90,000 photographs, they had nothing. But in the 97,025th, they discovered a match. Gell-Mann had foreseen the properties of the omega particle. The Eightfold Way was confirmed.

Gell-Mann was pleased with his theory, but not content. He hadn't reduced the ridiculous number of particles that the experimentalists had catalogued. And he had no idea why they fit into the Eightfold Way. There had to be an underlying pattern that explained it. Surely, he thought, the known particles must be made of something simpler, more elementary.

His breakthrough came in March 1963, when he was visiting the physicist Robert Serber at Columbia University. Serber had been mulling over the mathematics of the Eightfold Way, and he suspected that the underlying algebra might reveal a deeper pattern based on threes. Over lunch at the faculty club, he asked Gell-Mann if it was possible that each particle in the octets, was in turn made of three smaller particles.

"That would be a funny quirk," Gell-Mann said.

"A terrible idea," their dining companion, the physicist Tsung-Dao Lee, rejoined. Gell-Mann picked up a napkin and began scribbling away to show why this was a crank suggestion. The problem was that, if each particle was made of three smaller ones, then each would have to have a one-third or two-thirds positive or negative charge. And no one had ever seen a fractional charge. So how could they exist? That hardly seemed possible.

Yet the question nagged at him, and the next day, he found himself returning to it. He began to wonder if there was a strange way in which fractional charges might actually be possible. Suppose fractionally charged particles were imprisoned within a larger particle that had a whole charge. That would explain why we can't detect them. They can never escape the larger particles they are trapped in. It seemed implausible, perhaps even too weird to be true. But then again, he thought, *What the hell, why not?* Teasing the idea out, he toyed with naming these curious particles quacks or quorks, but he settled on quarks, a nonsense name from James Joyce's *Finnegan's Wake*. (He was tickled, as well, to learn that, in German, it can mean "balderdash.")

Gell-Mann thought he'd come up with a clever theory, but not something we could ever observe. He wondered, instead, if his fractionally charged particles might be a useful mathematical fiction. Fearing that his paper might be rejected, he decided not to submit it to the cautious editors of *Physical Review Letters*. Instead, he sent it to *Physics Letters*, whose editor was more open to publishing "crazy" ideas.

Once again, someone else had a similar brainstorm. This time, Gell-Mann's colleague and former student, George Zweig, independently conjured up the idea of fractionally charged particles (although he called them aces, not quarks, and speculated there might be four of them, not three). But Zweig, a younger scientist, was working at CERN in Switzerland, the world's most powerful particle accelerator, and the institution had restrictions on where and in what form papers could be published. Zweig grew so frustrated by the barriers the head of his division put in his way that he gave up. While at CERN, he circulated only two prepublication drafts. The reaction to his papers, he recalled, was "generally not benign." One senior scientist labeled him a charlatan and blocked his appointment to a teaching post at Berkeley. Zweig eventually left physics for neurobiology with a bad taste in his mouth.

Quarks, like aces, were also dismissed at first. Fractional charges in particles that had never been detected—and what's more, never could be—seemed just too far-fetched. That changed in the summer of 1968 when Gell-Mann's rival, Richard Feynman, used quantum physics to show that an experiment at Stanford's accelerator proved their existence. He demonstrated that electrons were bouncing off a proton, as though it contained three hard objects inside. With cloud chambers, Blau's photographic plates, particle accelerators, and Gell-Mann's intuition and mathematical acumen, scientists had finally detected quarks—particles a million times smaller than a grain of sand. Many physicists believed, the *New York Times* wrote, "they [had] begun opening the door to the innermost sanctum of matter."

Over the last fifty years, doubt has turned to conviction. Quarks are widely accepted as the ultimate building blocks of every member of the particle zoo, including all the protons and neutrons in your atoms. Quarks come in six types, and they interact by exchanging particles of force that Gell-Mann called gluons because, well, they glue particles together. Why quarks have fractional charges no one knows. But scientists have discovered a force that prevents the fractionally charged quarks from escaping from protons and neutrons. When quarks move apart, the force that binds them grows stronger and stronger, like a stretched rubber band. Scientists estimate that the pressure confining them is greater than any other known in the universe—a billion billion billion times more than that in the deepest ocean depths. Gell-Mann's theory, extended by others, now reigns unchallenged. For his breakthroughs, Gell-Mann received the Nobel Prize.

Although you could say that Gell-Mann discovered the most fundamental particles that we and everything in our world are made of, to be fair, you contain a lot more of something else—empty space. You may think you're solid, but that's an illusion—99.9999999999999 percent of your atoms are full of nothing. The oceans of empty space in atoms are so vast that if you scaled up a hydrogen atom's nucleus to the size of a tennis ball, its electron would be whirling about a mile away. If you removed all the space between your electrons, protons, and neutrons, you would be no bigger than a large speck of dust. You could fit all of humanity in a sugar cube.

This raises an intriguing question. If our bodies are so empty, why do we feel so solid? The answer is that, when we touch something like a table, the atoms don't actually meet. Instead, the electrons in your fingers and in the table repel each other. So your atoms don't actually *touch* the table, they hover over it, triggering your nerves to create the sensation of touch. Physicists will tell you that something else is also going on. In the world of quantum mechanics, identical particles cannot occupy the same space. So, as atoms approach, their electrons are forced to dance around their nuclei in different patterns, and this dance creates repulsive energy that keeps them from physically touching.

So far, all of our observations in the history of science tell us that, apart from empty space, you, and all known matter, are made of only three fundamental types of particles: electrons, quarks, and gluons. The gluons—massless particles of force—bind quarks together to form protons and neutrons. In one sense, you are a collection of something like 30,000,000,000,000,000,000,000,000,000,000 (thirty octillion) electrons, scores more quarks, and countless gluons, which bind your quarks into particles.

Gell-Mann's discovery allows us to tell the history of our elementary particles from the moment they were summoned into being. Back then, 13.8 billion years ago, there was no universe, no space or time. Then, right after the instant of the Big Bang, quarks, gluons, and electrons sprang out of a tiny point of infinite density. And *bam!* our journey had begun. The particles that would create us whirled, danced, and swarmed in a superhot plasma. Fractions of a millisecond later, gluons began binding quarks together to form protons and neutrons. Within three minutes (according to Einstein's equations and estimates of the total quantity of matter in the universe) the plasma cooled down slightly. That allowed the enormously powerful nuclear forces carried by gluons to come into play. They began gluing protons and neutrons together, creating nuclei of the very first atoms in our universe.

Three-quarters of them were the simplest element of all—hydrogen—with just one proton in its nucleus. That means that every one of the four trillion trillion hydrogen nuclei inside you, about 10 percent of your mass, was already formed three minutes after the Big Bang.

The next simplest elements: helium (with two protons and neutrons) and tiny quantities of lithium and beryllium (with three and four protons and neutrons) were also brewed in that swirling primeval plasma. But that was it.

For about 200 million years, the universe was supremely boring. Those were the true Dark Ages. There was little to see and no light to see it by. Just clouds of four idle elements floating in dark space as the universe expanded.

It is probably a safe bet that those first four elements could never have created life on their own. Your body, after all, contains over sixty others, ranging from iron and selenium to fluorine and molybdenum. And you are here. How is that possible? How were the rest of your atoms created? What were the astoundingly large sources of energy—as large as a trillion quadrillion H-bombs—that brought them into existence?

The first unexpected clues would come from "the best man at Harvard," who happened to be a fiercely determined Englishwoman.

THE BEST MAN AT HARVARD

The Woman Who Changed How We See the Stars

There are three stages in the acceptance
by the world of a new idea.

a. The idea is nonsense.
b. Somebody thought of it before you did.
c. We believed it all the time.
—*Fred Hoyle, paraphrasing Raymond Lyttleton*

In the spring of 1923, Cecilia Payne, a tall twenty-one-year-old student at the University of Cambridge, began to fear for her future. She loved astronomy research so much that she dreamed about it. She was keeping a notebook of research problems she would like to tackle as a scientist. But in her final year of school, she realized she was facing a dead end. In the England of her day, the best a woman of her intellect could look forward to was to become a teacher or principal of a girls' school. "I saw an abyss opening before my feet," she later wrote in her autobiography. "I saw the life of a schoolmistress as 'a fate worse than death.'" But that awful fate never came to pass. Instead, despite the odds against her, she made a crucial discovery that set the stage for a great triumph of twentieth-century science: uncovering how all the elements within us (except hydrogen) were first created.

Payne's interest in science began at age six, when the sight of a meteor made a deep impression on her; but it was cemented at age ten, after she cooked up an experiment at her Catholic school to test the power of prayer. She prayed for high marks on half of her exams and failed to pray for the other half. When she saw no difference in her grades, her belief in the power of reason was confirmed. She later became a Unitarian.

A pious headmistress told her she would be "prostituting her gifts" if she studied science. Her school choirmaster, the then-unknown composer Gustav Holst, who later penned *The Planets*, urged her to become a musician. She had other plans.

Payne won a scholarship to the University of Cambridge and expected to become a botanist. But there, in the heady days of physics after the First World War, she had a transformative experience when she heard the astronomer Arthur Eddington deliver his historic lecture revealing that the Sun's gravitational field bends the path of light—just as Einstein predicted it would. Payne was thunderstruck. "My world had been so shaken that I experienced something very like a nervous breakdown," she wrote later. You could say she fell hard for physics. The day after, she "confronted the College authorities" with her decision to switch from botany to physics. At home, she wrote the lecture out almost word for word and hardly slept for three nights.

The atmosphere at the Cavendish Lab at Cambridge was electric. J. J. Thomson, the discoverer of the electron, and Charles Wilson, the inventor of the cloud chamber, were there, along with the most luminous star in residence: the legendary Ernest Rutherford, who had discovered the nucleus of the atom. Yet for Payne, there was a fly in the ointment. Rutherford did not appreciate women in his classes. Although women no longer needed to have chaperones, they still had to occupy separate benches. Each time she entered the lecture hall, the painfully shy Payne, the only woman, had to sit alone in the front row. Rutherford pointedly began each class with the greeting "*Ladies* and gentlemen." She recalled in her autobiography, "All the boys regularly greeted this witticism with thunderous applause, stamping with their feet in the traditional manner, and at every lecture I wished I could sink into the Earth."

She soon sought out Eddington. Recognizing her drive, he was much more encouraging than Rutherford, and allowed her to take part in research. Payne was also introduced to the radical new theory of quantum physics by one of its discoverers, Niels Bohr. But in her final year, she realized that she was approaching a dead end. Women at Cambridge were not permitted to take advanced degrees. (In fact, they were not even formally granted diplomas or invited to the graduation ceremony.) So against all odds, with some arm-twisting and persistence, Payne managed to wrangle a fellowship for a woman to conduct research at Harvard's observatory. She would work under its director, Harlow Shapley.

The observatory, in Cambridge, Massachusetts, on a modest rise about a mile from campus, was well known for employing women because its previous director, Edward Pickering, had found them, in addition to being diligent and smart, considerably lighter on his budget. In an unprecedented effort to inventory the sky, Pickering had hired over eighty women known as "Pickering's computers"—or more commonly as "Pickering's harem"—to process a collection of photographs that would grow to half a million plates.

Shapley thought that Payne might join their efforts to use the plates to classify and catalogue the stars. But in her first independent research she was eager to tackle a more ambitious problem suggested by her Cambridge professors. She wanted to investigate a gaping blind spot in our understanding of the cosmos: What were stars made of?

Scientists already knew a certain amount. In addition to photographing stars, the Harvard astronomers were also recording their light spectra on photographic plates, and these offered many clues to the elements in stars. That was because a star emits light of every color. However, each element in the periodic table absorbs a unique set of wavelengths. So the atom of any element hovering in the atmosphere of a star will absorb particular wavelengths from the starlight before it reaches us. When astronomers looked at a horizontal band of the star's spectrum, they saw thin black lines wherever those wavelengths were missing, and these revealed the identities of elements that absorbed them. So the glass plates were spectral fingerprints. They resembled cosmic bar codes. And they showed that stars contained many elements found on Earth, such as iron, oxygen, silicon, and hydrogen.

Astronomers had a problem, though. There were anomalies in the patterns of spectral lines that made their interpretation difficult. So, while the plates told scientists which elements a star contained, they didn't reveal how much of each was present.

Nonetheless, astronomers thought they already knew the answer. The stars and planets must be made of the same stuff. At the time, many believed that the planets had been created when a passing star wrenched chunks of hot gases out of the Sun, so they had to have the same composition. In fact, the preeminent expert on stars, Henry Norris Russell, was confident that, like the Earth, the Sun had a massive iron core. He was sure that if you heated the Earth's crust to the temperature of the Sun, their spectra would be almost identical.

That is what Payne wanted to investigate. She wanted to use the plates to determine the proportions of the various elements in stars. And she proposed to do it with a new cutting-edge theory from the brilliant astrophysicist Meghnad Saha in distant Calcutta. In the new theory of quantum mechanics, electrons could only whirl around the nucleus in distinct orbits, with higher energy levels increasingly farther away from the nucleus. Building on this, Saha suggested that in stars of different temperatures, atoms of the very same element may have electrons in different orbital paths (and those in the hottest stars might even lose electrons). These variations would cause otherwise identical atoms to absorb different sets of wavelengths of light—and thus confuse the interpretation of the stars' light spectra.

Payne was eager to take on the challenging work of attempting to apply Saha's equations to Harvard's vast collection of plates. And she was the only one at Harvard with enough knowledge of quantum theory to do it.

In her office, on the third floor of the large brick building that housed the great collection, Payne was up at all hours, analyzing confusing differences in the spectra of thousands of individual stars. Those plates still exist in yellowed sleeves in that same building. There, one of Payne's former graduate students, the astronomer Owen Gingerich, showed one to me. It contained dark bands, each about a quarter of an inch thick, that were laced with many fuzzy light lines of varying intensity that you would need a magnifying glass to see clearly. "As you look at it you are totally bewildered," Gingrich said. "You have to learn to recognize patterns. But after sitting day after day, they become friends." Looking at all those tiny smears, this was hard to believe.

Observatory director Shapley occasionally stopped by Payne's office in the evenings. He would find her chain-smoking as she pored over her plates, struggling to recognize patterns in the blurry lines and to match them against her calculations. "Often I was in a state of exhaustion and despair, working all day and late into the night," she wrote. Months turned into almost a year of "utter bewilderment."

But finally the pieces fell into place. Using Saha's equations, she discovered something completely unexpected. In the first draft of her dissertation, she bravely claimed that despite what virtually everyone believed, the composition of the stars and Earth could hardly be more different. Stars contained little of the most common earthly elements, such as iron, silicon, oxygen, and aluminum. Instead, every star was 98 percent hydrogen and helium. In fact, hydrogen was a million times more plentiful in the Sun than on our planet.

That was strange. It was not what she had been taught at Cambridge, and it surely did not fit with her teachers' understanding of how the Earth formed. "Miss Payne? You're very brave," the physicist Alfred Fowler told her. Shapley proudly sent a draft of her thesis to his old advisor, Henry Norris Russell, a distinguished astronomer at Princeton. His reply was highly complimentary but contained a strong caution. Payne's claim, that stars were almost entirely hydrogen and helium, was "clearly impossible." There were strong reasons to think otherwise, including reasons to think that the Sun contained a lot of iron. Not only were there many more lines of iron than any others in the spectrum of the Sun, but many meteorites were made of iron, and the Earth's core was full of it. To Russell, all of this made it obvious that iron must be abundant in all celestial bodies.

Payne, a mere graduate student, accepted the influential expert's word, or at least felt she had to go along. "His word could make or break a young scientist," she recalled. She wrote a caveat in her thesis that this part of her conclusions was "almost certainly not real." "Throughout her life she lamented that decision," Payne's daughter told the writer Donovan Moore. Yet a few years later, Russell himself, capitalizing on advances in quantum theory and the fact that others were arriving at Payne's conclusion by other means, confirmed her discovery.

Payne's dissertation would long be regarded as the most brilliant doctoral thesis ever written in astronomy. The famed astronomer Edwin Hubble would call her "the best man at Harvard." Yet advancement there would be a long time coming. For years, her lectures would not be listed in Harvard's catalogue. The university's president, Lawrence Lowell, horrified by the thought of a woman on the faculty, swore she would never receive an appointment while he was alive. Indeed, she was appointed a professor only in 1956, years after Lowell died.

Payne's discovery transformed our view of how stars work. Knowing that they were largely made of hydrogen and helium allowed researchers to solve another long-standing puzzle—what stars burn for fuel. Researchers figured out that in stars' pressurized interiors, energy is liberated when hydrogen atoms with single protons fuse to make helium atoms with two protons. That's how our Sun creates heat and light. Thanks to Payne, with this new understanding, scientists were finally in a position to solve the mystery of how the heavier elements were created. The answer would lie in the stars.

It was Fred Hoyle, whom we last met contemptuously deriding the Big Bang, who would be first to discover where our elements came from. Hoyle was of medium height, wore beefy glasses, and had perpetually tousled hair and an impish smile. As a young boy in a Yorkshire village, he was offended by the "stupidity" of his teachers, so he perfected the art of truancy, often for weeks or months at a time. When he was not "ill" he was hanging around canal locks and meandering in woods and fields. At home, he made his way through a chemistry text he found on his parents' bookshelf and impressed his friends with explosions of homemade gunpowder. As a scientist, Hoyle could be difficult and argumentative, gleefully brandishing his contempt for orthodoxy. He inherited this skepticism from his father, who had seen a ghastly mix of machine guns and ineptness in the British high command annihilate his unit in trench warfare in the Battle of the Somme during the First World War. Hoyle went to his grave rejecting the Big Bang. Nevertheless he has been called "one of the most innovative minds the world has ever produced," and "the most creative and original astrophysicist of his generation."

At Cambridge, Hoyle decided to study physics, but only after becoming fluent in mathematics. So when his interest was drawn to astronomy, he had the advantage of having a great facility with statistics and could apply it to the complex reactions of nuclear particles.

In the 1940s, Hoyle began to think about the origin of the elements. The physicist George Gamow had already recognized that the temperatures generated by the Big Bang were staggering. And so Gamow postulated that the nuclei of all the elements were quickly assembled from protons and neutrons during these initial fireworks. It was over "in less time than it takes to cook a duck and roast potatoes," he said. That seemed to make sense; if you are going to build a universe, why not start with all the building blocks at once? Yet Gamow could not make all of his calculations work. After the first few elements, he was stuck.

That was when Hoyle stepped in. He knew that Payne's glass plates revealed that, in addition to hydrogen and helium, stars contained small amounts of oxygen, carbon, iron, and other elements. The fact that the proportions in each star were slightly different suggested to Hoyle that the stars themselves might be factories that churn out new elements. Yet this hardly seemed possible: physicists had calculated that stars were nowhere near hot enough. They didn't have nearly enough energy.

That's because hydrogen, the first element in the periodic table, has a single proton in its nucleus, and each of the 117 elements that follow has one more. Helium has two, beryllium three, and so on. While that may sound easy to produce, it's not, as any would-be modern alchemist knows. Adding a proton to a nucleus requires a ridiculous amount of energy, for a simple reason. By the 1930s scientists had discovered that a nucleus's protons are bound together by a force different from gravity or electromagnetism. It's an incredibly powerful glue they called the "strong nuclear force." But it acts only over extremely short distances. Over longer distances, electromagnetism is in charge. So a stray proton that flies toward a nucleus will be strongly repelled by the electromagnetic force of similarly charged protons inside— unless it has the energy to fly so close that the strong nuclear force can grab it. And astronomers calculated that stars didn't have enough energy to create nuclei heavier than helium, with two protons inside. The temperatures of stars were far too low to add more protons.

Hoyle knew that somewhere there had to be an explanation of how the elements formed. But where?

In 1944, he took a trip that changed everything. As the Second World War raged on, Hoyle was designing countermeasures against radar-guided guns for the British military. A top-secret meeting on radar technology took him to Washington, DC. Once there, Hoyle took advantage of his trip to do some unauthorized astronomical freelancing. In Princeton, he met the great expert on stars, Henry Norris Russell, who advised him to visit the Mount Wilson Observatory, home to the world's most powerful telescope.

It is ironic that in wartime, the most prominent astronomer there, after Edwin Hubble, was technically an enemy alien. Most astronomers had been assigned to war research, but Walter Baade was not, because he had emigrated from Germany in 1931. Instead, he was confined to Los Angeles County. A nighttime curfew for aliens would have prevented him from using the telescope, except that the observatory's director had appealed to the army for an exemption. Baade used this opportunity—with little competition for observing time and a dark sky, thanks to partial wartime blackouts—to great advantage. Among many matters, he told Hoyle of his discovery that a large star doesn't quietly fade away. Instead, it dies in a colossal explosion that he and a colleague had called a supernova. Moreover, Baade told Hoyle, his latest research had convinced him that supernovae create vastly more heat than the hottest stars. Hoyle listened with keen interest.

His thinking about stars deepened when he traveled to Montréal, where he met British scientists who were designing the world's first nuclear energy reactor. At least that was their cover story. In fact, as far as he could tell, they were members of Britain's atom-bomb effort, and were trying to glean wartime intelligence on the Manhattan Project from their Canadian collaborators. Their vague description of their work was enough for Hoyle, who was fully versed in nuclear physics, to grasp the principle of the American plutonium atomic bomb. He realized that a hugely powerful implosion in the bomb produced by a ring of conventional explosives could set off an even more devastating nuclear explosion.

Back in Cambridge, Hoyle began to wonder if supernovae were set off the same way. When a star ran out of fuel to burn, did it implode like a plutonium bomb and set off a much larger explosion?

His calculations showed that, if so, then long before a star exploded, it must gradually grow much hotter than anyone ever dreamed possible. And in that case, perhaps the answer to the origin of the elements was staring us in the face—every night. Even before they blow up, could massive stars called red giants actually be hot enough to turn hydrogen into the other elements? He decided to use his knowledge of nuclear physics and statistics to take a crack at calculating the reactions that might make this happen.

Hoyle assumed that, after the core of a massive star had converted all its hydrogen to helium, the energy released made the star much hotter, so the particles inside it were traveling faster. Now, three helium nuclei, each with two protons, should have enough energy to smash together so hard they could overcome their repulsive forces and fuse into a single nucleus, making carbon, with six protons. And once a star had made carbon, which released more energy, its interior would be hotter and the particles would be moving faster still. In effect, the star was pulling itself up by its own bootstraps. At first his calculations appeared to show how successive reactions that increased the temperatures and pressure would enable it to make all the other elements up to iron, with twenty-six protons. He seemed to have made a stupendous breakthrough.

But then he smacked into a brick—or rather carbon—wall. The energies of the reactions that created a carbon atom did not add up. And since a star could not make heavier elements without first making carbon, his entire theory was cast into doubt. Now Hoyle was stuck.

Some years later, Hoyle was invited to spend a few months at Caltech, in Pasadena, California. There, as he prepared to give a lecture, he reexamined his dilemma. And he began to think that the energies to make carbon could work out if two helium atoms collided, and then a third struck them almost instantaneously. But he still had a problem: this reaction would create an especially energetic form of carbon that couldn't exist.

That's because an atom can have a different amount of energy, depending on how its protons and neutrons are arranged. And he calculated that his reactions would produce carbon atoms with more energy than any actual carbon atoms ever detected by experiment. He even calculated the precise amount: 7.65 million electron volts.

Still, Hoyle sensed that he was onto something. He knew that carbon is the sixth most common element in the universe. It makes up almost 23 percent of our bodies' mass. He couldn't see how it could be assembled anywhere but in stars. Perhaps, he thought, carbon exists at that energy level and we simply haven't found it.

Excited, Hoyle made a bold prediction. He claimed that carbon was produced in large stars when three helium atoms, each with two protons, collided to create a carbon atom with six protons. But it was unstable. So it lost a bit of oomph, emitted a little energy, and turned into the common less energetic form we find around us. Since we exist and we are made of carbon, he would later claim, his prediction had to be true. The universe was full of carbon and there was no other way it could possibly be made.

Luckily, Hoyle was now at Caltech, where some of the world's greatest nuclear experimentalists were working with particle accelerators. One day he barged into the office of the group's leader, the burly Willy Fowler, and asked him to test his prediction. Fowler was put off by this interruption from an odd duck with a British accent. "Here was this funny little man who thought that we should stop all this important work that we were doing . . . we gave him the brush-off. Get away from us young fellow, you bother us."

But Hoyle would not stop pestering Fowler's group. Finally they realized that, though Hoyle's prediction seemed extremely unlikely, it would have momentous import if it proved true. And so they agreed to put his theory to the test. They would yoke a spectrometer that could measure an atom's energy to a small particle accelerator in which they would try to create the carbon.

Hoyle was tense as he waited for the results. "I felt the hot wind on my neck as I crept each day into the laboratory," he recalled. He was like a prisoner in a dock, waiting for the foreman to announce if he was innocent or guilty.

After several months of tests, the physicists were flabbergasted to find a form of carbon at the exact energy level Hoyle had predicted. Fowler was stunned. No one had ever done anything like this before: used theory from astrophysics and stars to predict something so microscopic—the precise nuclear structure of an atom. When Hoyle heard the verdict, he wrote, the scent of California's orange trees that he loved smelled even sweeter. He had begun to unravel the mystery of how the elements were created.

The stage for creating new elements was set after the Big Bang, when great clouds of hydrogen wafted into space, which had already vastly expanded. About 200 million years later, the relentless vise of gravity began squeezing atoms in the largest clouds so tightly together that the wrenching heat and pressure converted their cores into nuclear reactors. Hydrogen fused into helium. Torrents of energy were liberated that ignited the clouds into blazing stars. They radiated heat, lighting our pitch-black universe for the very first time.

That much had already been understood. Hoyle's insight was that, after the core of a very large star consumed all its hydrogen, it entered a furious new stage. Gravity compressed the helium in the core so tightly that this too began to burn. In that superhot maelstrom, helium nuclei collided so violently that they overwhelmed their repulsive forces, creating the carbon within us. Once carbon was created, a star metamorphosed like a cat with nine lives. The heat released by that fusion pushed out the star's layers, transforming it into a massive "red giant." Although our Sun is too small to create elements heavier than helium, it will experience a similar fate about 6 billion years from now. Its surface will balloon out so far that it will incinerate our planet, burning Earth to a nice crisp.

Hoyle realized that, once massive ancient stars turned into red giants, the universe became a much more interesting place. Due to the stars' ever higher temperatures, new elements were created and destroyed in onionlike rings around their cores—carbon transmuted into oxygen, oxygen to neon, neon to magnesium, magnesium to silicon, all the way to iron with twenty-six protons in its nucleus.

That's where most of our atoms came from. Over 88 percent were cooked up in a fiery inferno, a massive red giant. They include the oxygen fueling our movements, the calcium in our bones, the sodium and potassium sending our nerve signals, and all the elements but hydrogen in our DNA. The most versatile of all, carbon, is your second most abundant by mass. (Oxygen is first.) If you removed all the water from your body, carbon would make up less than 1 percent of your bones but about 67 percent of your tissues. Carbon chains are the backbones of your proteins, sugars, and fats. When you blacken a steak on a grill, you're looking at carbon that was created in the stars. Convection currents carried all the elements up to iron from the interiors of the red giants to their surfaces. Stellar winds—mainly streams of electrons and protons—wafted them in giant clouds into space.

Yet, when Hoyle tried calculating how the elements heavier than iron formed, he hit another wall. This one stopped him cold. As the elements up to iron were created, the reactions liberated loads of energy that could be used to make even heavier elements. But after iron there was no more free lunch. Hoyle's calculations showed that a star could not forge elements heavier than iron unless it somehow found an almost inconceivable amount of additional energy.

Hoyle was mystified. There are sixty-six elements heavier than iron on Earth. Our bodies contain trace amounts of over forty of them, including copper and selenium in our enzymes and the fluorine in our tooth enamel. Where did the energy to make these come from? An event somewhere in the universe must have liberated much more energy. But no one could tell Hoyle where.

In a windowless room at Caltech, next door to a particle accelerator, Hoyle persevered. He teamed up with his old foil-turned-collaborator, William Fowler, and the astronomers Margaret and Geoffrey Burbidge. Wielding slide rules, primitive calculators, and their collective knowledge of nuclear physics, they investigated the old age of massive stars, originally much larger than our Sun, that turned into red giants. They discovered that, after the core of one of these red giants was fully converted to iron, it went cold turkey; the reactions in the core abruptly stopped. That was bad news. The massive star was no longer producing enough energy to support its outer layers. Instead, in a breathtaking single second, its core shrank to a ball less than thirty miles in diameter and three hundred thousand times denser than Earth. (A marble made of it would weigh more than a billion tons.) This sudden contraction sent cataclysmic shock waves through the star's layers and triggered the King Kong of all explosions—one of the most powerful in all the universe—the supernova. This was the explosive demise of a star, first discovered by Walter Baade.

For days the massive dying star burned as bright as a hundred billion suns, and it continued to burn, perhaps for months. You might think those temperatures were hot enough to make all the remaining elements, but Hoyle's team couldn't be sure. They grew more confident after they came across newly released data from the first hydrogen bomb test, just four years earlier. The explosion had made short work of Elugelab, a small island of coral and sand in the Marshall Islands. It kicked up a mushroom cloud a hundred miles wide. In the debris, scientists found a very heavy radioactive element with a half-life of sixty days, which the blast had created. Hoyle's team found evidence that light from supernovae dimmed at the same rate—apparent confirmation that a supernova was hot enough to produce it. This turned out to be a false lead, as many do. Nonetheless, it seemed to Hoyle's team they were on the right track, and it spurred them on.

One day at Caltech, Hoyle had some time to kill before lunch, so he idly decided to try using the team's equations to predict the abundance in the universe of the elements heavier than iron. With each calculation he grew more excited. They matched estimates that researchers had made by studying the Earth and meteorites and by measuring the composition of the stars with Cecilia Payne's technique. In the few cases where Hoyle's numbers were off, the measurements, not the calculations, turned out to be wrong.

"Synthesis of the Elements in Stars," the team's 107-page masterpiece, is considered one of the greatest scientific papers ever published. It laid out eight different nuclear pathways that created the elements. It outlined in exquisite detail how fusion within the depths of red giants created all of your elements up to iron. It showed how, when the sudden collapse of a monstrous red giant triggered a supernova, particles fried at dumbfounding temperatures—billions of degrees. This was thousands of times hotter than the paltry 27 million degrees in the center of our Sun. Neutrons, protons, and nuclei slammed together at close to the speed of light, and when they reassembled, they had created every element in the universe, including the heaviest elements in our bodies.*

Hoyle was rewarded with accolades, awards, even a knighthood for his groundbreaking work. But the greatest badge of honor a physicist can receive always eluded him. In October 1983, Willy Fowler, his collaborator, was thrilled to hear that he had won the Nobel Prize, and stunned and disappointed to learn that Hoyle would not share it. The reason was never explicitly stated, but was generally understood. By then, Hoyle had become the bad boy of physics. "It is better to be interesting and wrong than boring and right," he once said, and followed that maxim with a vengeance. He never lost his contempt for authority or hesitated to insult prominent colleagues who disagreed with him. Moreover, although he continued producing brilliant science, he also publicly championed some provocative, even outrageous, theories. His claim that life on Earth was seeded from outer space—panspermia—no longer seems so crazy. But others clearly were, such as his argument that the famed *Archaeopteryx* fossil at London's Natural History Museum was fake, and that space was full of clouds of viruses and bacteria that periodically sparked diseases such as the 1918 flu, cycles of whooping cough, and even

---

* A supernova employs a trick to create these heavy elements. The explosion sends neutrons flying around at much faster speeds. As neutrons lack charges, they're not repelled by the charged protons in a nucleus, so a neutron traveling fast enough can sneak into one. There, an extra neutron can be a Trojan horse because it actually contains a positively charged proton, a negatively charged electron, and a bit of additional energy. If this neutron decays, its electron will be ejected from the nucleus, leaving a proton that has turned the nucleus into a heavier element.

Legionnaires' disease. He died in 2001, still unwilling to accept the Big Bang.

With Payne's and Hoyle's discoveries, we can begin to trace the journeys of all our atoms from the beginning of time. Every hydrogen atom in your body was created when protons were produced from quarks and gluons, less than a second after the Big Bang. Two hundred million years or so later, the rest of your elements up to iron started to form in the depths of fiery hot red giants. In their ever increasing heat, nuclei collided violently, and protons and neutrons smashed into them at close to the speed of light, transmuting elements into heavier ones. Hydrogen from the Big Bang and elements cooked up in red giants (principally oxygen, carbon, nitrogen, calcium, and phosphorus) make up 99 percent of you. Your remaining 1 percent are heavier elements, like zinc and manganese, that were born when red giants exploded in supernovae. They were baked at billions of degrees in some of the most powerful explosions in the universe.[*]

Now, as your atoms sailed away from the stars that forged them, they found themselves trapped in swirling masses of dust and gas that were recycled into new red giants. Or they were swept up in larger stars that later exploded in other supernovae. Over billions of years, your atoms may have been reworked in a great many stars.

Finally, about 5 billion years ago, they were streaming in a huge cloud toward the vicinity of empty space that now contains our solar system. At last they were on their remarkable journeys to becoming you.

But it would not be a smooth ride. In fact, their troubles had just begun.

---

[*] In 2017, we discovered that a very small proportion of the heaviest atoms within us, such as gold, were also created by the extremely violent collisions of neutron stars—exotic dense stars made largely of neutrons, which form in the aftermath of supernovae.

4

CATASTROPHES
TO BE
THANKFUL FOR

How to Make a World from Gravity and Dust

My own suspicion is that the universe is not only queerer
than we suppose, but queerer than we can suppose.
—*J.B.S. Haldane*

Over 4.8 billion years ago, the atoms that would create us sailed in great clouds of gas and dust, toward . . . well, nothing. There was no solar system, no planets, no Earth. In fact, for a long time, scientists could not explain how our solid planet, not to mention one so hospitable to life, appeared at all. How was our now-rocky planet conjured, like magic, out of an ethereal cloud of gas and dust? How and when did Earth become so welcoming to life? And what travails were our molecules forced to brave until life could evolve? Scientists would learn that our atoms could finally create life only after they endured wrenching collisions, meltdowns, and bombardments—catastrophes that beggar any destruction ever witnessed by humankind.

Explaining how our planets were created seemed so difficult that, by the 1950s, most astronomers had given up. Their theories appeared to lead nowhere. Two centuries before, the German philosopher Immanuel Kant and the French scholar Pierre-Simon Laplace had begun, promisingly enough, by correctly theorizing that gravity reeled in a massive spinning cloud of gas and dust so tightly that fierce temperatures and pressures ignited it into a star—our Sun. But how did the planets form? They posited that a disk of stray dust and gases still remained spinning around the Sun, and this broke into smaller clouds that created the planets. However, no one could convincingly explain how the disk broke up or how the planets formed from these lesser clouds.

In 1917, the Englishman James Jeans took an inventive new tack that, as we saw, Cecilia Payne's contemporaries endorsed. Jeans surmised that the gravitational pull from a passing star was so strong that it wrenched massive chunks of gas from the Sun's surface—and these became the planets. Others thought that our planets were debris left behind by the collisions of stars. But how nine distant planets formed from such a collision was anybody's guess. It seemed as likely as if you had put wet laundry in a dryer and then opened it to find your clothes not just dry, but neatly folded. Only a few astronomers continued to take the question seriously. It was a matter fit only for "innocent entertainment," or "outrageous speculation," observed the astronomer George Wetherill. It simply wasn't clear that we could ever see so far back in time.

Nevertheless, in the Soviet Union in the late 1950s, at the height of the Cold War, a young physicist decided to tackle the problem head on—with mathematics. His name was Viktor Safronov. Safronov was slight in stature and struggled with malaria, a legacy of his military training in Azerbaijan during World War II. He was modest, humble, and uncommonly smart. At Moscow University, he distinguished himself with advanced degrees in physics and mathematics. Recognizing his talent, the mathematician, geophysicist, and polar explorer Otto Schmidt recruited him to the Soviet Academy of Sciences.

Schmidt himself, like Kant and Laplace before him, was sure that our planets had been created from a disk of gas and dust that orbited the Sun. He wanted someone with the technical skill to help him puzzle out how, and the soft-spoken Safronov was a brilliant mathematician.

In an office at the Academy of Sciences, Safronov started at the beginning. He took upon himself the daunting task of trying to explain how trillions upon trillions of gas and dust particles could build a solar system. He would try to do it with mathematics—primarily statistics and equations of fluid dynamics, which describe the flow of gases and liquids. All this without computers. In fact, his lack of a computer may have even helped, by forcing him to sharpen his already formidable intuition.

Safronov began by assuming that our solar system first took shape when the vast primordial cloud of dust and gas, which in the previous chapter we left floating in space, was transformed by the relentless pull of gravity into a star. Almost all of it (99 percent, we know now) became our Sun. But lingering remnants were too far away to be dragged into the Sun, yet not distant enough to fully escape its clutches. Instead, gravity and the centripetal force of rotation flattened this cloud into a disk of dust and gases orbiting around the Sun.

Safronov, who dazzled colleagues with his gift for making quick mathematical estimations, set out to compute what happened when tiny particles within the disk smacked into one another and then struck their neighbors. With pencil and paper and a slide rule, perhaps in the quiet of a library where Soviet scientists often retreated from the hubbub of large common offices, he doggedly attempted to estimate the effects of trillions upon trillions of collisions. That was an incredibly daunting endeavor, with or without a computer. By comparison, one would think that calculating the path of a hurricane from the initial water droplets forming in clouds would be child's play.

Safronov realized that the swarm of cosmic dust and gas orbiting the Sun would be traveling around at roughly the same speed and direction. Sometimes, when the particles bumped into their neighbors, they would stick together like snowflakes. More collisions begat bigger and bigger clumps, until they were as large as boulders, ocean liners, mountain ranges, and eventually mini-planets. Building on his insight, Safronov single-handedly outlined most of the major problems scientists would need to solve in order to explain the origin of our planets. And with mathematical bravado, he conquered many of them.

For years, he had the field of planetary formation, which he had created, virtually to himself. Most Soviet colleagues were skeptical and uninterested; his research appeared so speculative, so far removed from any evidence. Then, in 1969, Safronov published a slim paperback, a retrospective of his decade of lonely work. He presented a copy to a visiting American graduate student, who passed it on to NASA with a recommendation that they have it published. Three years later, an English version appeared in the West.

It lit a spark in George Wetherill at the Carnegie Institution of Washington. Wetherill had grown into a scientist after his fundamentalist mother tolerated his early fascination with natural history. He studied physics and was working in geochemistry, but he was also a strong mathematician. Wetherill was intrigued by Safronov's theory, and more important, he realized that he could test it. Unlike scientists in the Soviet Union, Wetherill had access to a primitive computer, so he could adapt Safronov's equations into a simulation. And he could incorporate a groundbreaking program he had access to that used a new statistical tool, the Monte Carlo method, which was first developed to make the atom bomb. It would add elements of chance, so if he ran trials over and over again with different starting assumptions, he could learn the most likely outcomes.

Wetherill was astonished by the results. His simulation showed that if trillions upon trillions of tiny collisions created hundreds of Moon-size bodies orbiting the Sun, they would proceed to form a much smaller number of planets that were similar in size and location to our own.

Moreover, his simulations with the planetary scientist Glen Stewart revealed an astonishing picture of how it happened. Once a space rock grew to about a hundred miles in diameter, its gravity became so powerful that it began to reel in all the objects around it. That triggered a runaway effect. As it grew quickly, like a rolling snowball gone wild, its larger size caused it to collide with other objects more often. Soon it swiftly vacuumed up all the smaller objects in its feeding zone: its orbit around the Sun. Once it got rolling, planetary formation, it turned out, was not a slow, gradual process; it happened very quickly.

The picture was not a pretty one. In fact, to astronomers it was not just disconcerting, it was terrifying. Our planets did not assemble peacefully. In the inner solar system, a large number of bodies, perhaps even a hundred between the size of the Moon and Mars, once jostled about as if they were in a roller derby, elbowing one another and perturbing one another's orbits. Some crashed into the Sun. Others were hurled out toward Jupiter, by far the largest planet. If they did not collide with it, Jupiter's powerful gravity disrupted their orbits enough to sling them right out of the solar system. Meanwhile, our own hapless molecules, trapped within massive rocks and miniplanets, suffered countless violent collisions as the Earth formed.

Safronov's theory, as elaborated by Wetherill, solved another confusing puzzle: How did the rocky inner planets and the outer planets, which are gas giants, all evolve from the same cloud? The answer, they realized, was that, just after the Sun formed, the nearby temperatures were so hot that the lighter elements remained vaporized; only the heavier ones could condense into solid particles. These created the rocky planets: Mercury, Venus, Earth, and Mars. Most of the lighter molecules in their regions were pulled into the Sun, or drifted out of the disk and were lost to outer space. But farther away, the temperatures were cold enough for water, methane, and carbon dioxide to freeze, roughly doubling the solids there. This mix of rock and ices helped bulk up the cores of planets like Jupiter and Saturn. They grew so huge and the pull of their gravity became so strong that light gases like hydrogen and helium collected in their atmospheres, giving birth to the gas giants.

Safronov and Wetherill's theory even accounted for why, although all of the planets orbit the Sun in the same plane and direction, Venus spins backward, Uranus spins almost on its side, and the Earth spins on a tilt that gives us seasons. Our hemispheres now enjoy summer when they incline toward the Sun and winter when they tilt away. As the planets were forming, massive impacts with other giant bodies knocked their orientations out of sync, even as they continued orbiting the Sun.

One of the greatest triumphs of all was that Safronov's theory helped explain the mystery of our Moon. For years, scientists were at a loss to explain why the Moon does not have some of the light elements that Earth does. Then, in the 1970s, two astronomers at the Planetary Science Institute, William Hartmann and Donald Davis, and two others at Harvard, Alastair Cameron and Bill Ward, recognized that when the Earth was almost fully formed—about 90 percent of its current size—it must have been violently assaulted. A Mars-size body slammed into it at tens of thousands of miles an hour—vaporizing and fragmenting the impactor and knocking off a huge portion of the Earth's mantle. That colossal smashup ejected a massive cloud of rock vapor and debris into the Earth's atmosphere, while the impactor's heavy iron core merged with the core of the Earth. Much of the debris that was spewed into the atmosphere, including some of our atoms, fell back down. But that Earth-shattering collision, known in technical parlance as the Big Whack, the Big Thwack, or the Big Splat, also hurled enormous amounts of vaporized rock into space.

Some of the very lightest elements, like hydrogen, sailed away and were lost, but the heavier ones, trapped in Earth's orbit, created the Moon. That's the reason the Moon has fewer light elements than Earth. Though the theory was ignored at first, refined versions of it have prevailed as the best explanation for the origin of the Moon. The Big Whack increased Earth's mass by about 10 percent, effectively marking the birthday, 4.5 billion years ago, of our full-grown planet.

It was as if blindfolds had dropped from astronomers' eyes. Before Safronov, they were in the dark. Now they had a model of planet formation, and a way to test it. Their time machines—or rather, computer simulations—could be refined and matched against astronomical observations to reconstruct the evolution of the solar system and Earth.

Ironically, by the 1970s, Safronov had largely been left behind because, like most Soviet scientists, he still had no access to computers. Yet he often visited the West, where he was revered. "His contributions are of overwhelming proportion, and I am not worthy to touch the hem of his garment," Wetherill wrote with sincerity. It is hard to overstate how dramatically Safronov transformed our understanding of our planet's history.

Four and a half billion years ago, our atoms finally had a home, the Earth. You might think that our fledgling planet would quickly solidify, oceans would form, and life would soon strut across the stage. Alas, it was not that simple. Scientists would discover that, after our planet had been violently assembled, it experienced another series of awe-inspiring catastrophes before life could thrive. The first one was a doozy: a complete meltdown that, ironically, would make our own existence possible.

Although he didn't know it, as early as the 1600s, William Gilbert, a natural philosopher and physician to Queen Elizabeth, found a hint of this disaster when he discovered that the Earth had a huge iron core. In the annals of firsts, Gilbert, who was a major influence on Galileo, is often hailed as the "first scientist." He was among the earliest to claim that if you want to understand the natural world, you can't just theorize. You have to experiment. His investigations dispelled sailors' age-old beliefs that eating garlic near a compass would prevent it from detecting a magnetic field, and that their compasses were being moved by far-off magnetic mountains. Gilbert found that, if he moved a compass needle around a large magnetic rock with two magnetic poles, it continued pointing north-south, just like a ship compass traversing the globe. He concluded that the Earth, like his rock, was enveloped by a magnetic field that emanated from an iron core at its center.

Where did that core and field come from? By the 1970s, scientists understood that, after the Earth had fully formed, gravity compressed the atoms in its interior so tightly together that it grew hotter than the gates of hell. Thanks to Safronov, they also knew that impacts raised the Earth's temperature, particularly the Big Whack, which melted much of the mantle. And they calculated that radioactive elements like uranium, which was more prevalent then, made the planet even toastier. Putting this all together, it became clear that Earth could only have obtained its iron core in one way: it grew red-hot and completely melted down.

Scientists aptly call this period the Hadean Era. Our planet's surface was a roiling ocean of magma, and there was a sorting of molecules by weight. Many light gas molecules, like nitrogen and carbon dioxide, rose into the atmosphere. Inside the molten globe, elements of life—such as phosphorus, sodium, and potassium—bonded with silicon and rose to become part of the Earth's crust. Meanwhile, like sediment in a pond, heavy metallic iron, with a bit of nickel, sank to the center of the Earth. Even today, much residual heat from that colossal meltdown remains in the Earth's core. The temperatures there remain scorching—almost 10,000 degrees Fahrenheit—about as hot as the surface of the Sun. Geologists call this meltdown the Iron Catastrophe. Yet we should be thankful for that planetary disaster; without it we wouldn't be here.

Because the outer region of our planet's largely iron core is molten, it spins within the rotating Earth. Circulating electric currents within it generate a vast invisible magnetic field around our planet. This enormous force field, extending far beyond the atmosphere, shields us from energetic cosmic rays that would otherwise tear our DNA to bits. It also protects life from another peril—solar winds from the Sun (largely electrons and protons)—that would chip away at our atmosphere and send it sailing into space. That's what happened to Mars. It was too small to maintain its magnetic field, so when the molecules in its atmosphere were struck by particles in solar winds, they went off into space.* Without the young Earth's meltdown, our planet would not have a magnetic shield.

Paradoxically, for us, it was a lucky break.

Just how quickly after the meltdown did our planet turn from a danger zone into an Eden hospitable to life? In the 1960s, geologists scoured the far reaches of the world, from the high Himalayas to the most remote deserts, looking for clues. They returned with little for their troubles. The oldest rocks they could find were only about 3.5 billion years old, although the Earth itself was created 4.5 billion years ago.† Unfortunately for geologists seeking the Earth's oldest rocks, the movement of tectonic plates had shoved the edges of colliding plates into the Earth's hot mantle, leaving little trace of the first billion years of our planet's history.

Which is why geologists began to turn their gaze so longingly at the Moon. It was too small and cold to have had continental drift. So there, tantalizingly, ancient rocks still lay on its surface. Researchers were sure they would tell stories about the earliest days of both the Earth and the Moon, if only we could reach them.

On July 20, 1969, a warm humid day at NASA's Johnson Space Center in Houston, Texas, a crack team of geologists, astronomers, and biologists were biting their nails. Just two months earlier, the Apollo 10 astronauts had orbited the Moon. Now, two Apollo 11 astronauts, Neil Armstrong and Buzz Aldrin, were about to attempt the first Moon landing.

The scientific team had been recruited to work in the first facil-

---

*   Here on Earth you can see the impact of the solar wind in the colored light shows called aurorae borealis and aurorae australis at the North and South Poles.
†   They knew the age of the Earth because 4.5 billion years is the date that radioactive elements give for the origin of meteorites—debris left over from the formation of the planets. The planets formed quickly after meteorites, perhaps in only about 100 million years.

ity designed to analyze rocks from outside Earth. Their anticipation was building. "As far as I'm concerned this will be the most exciting thing that's happened in the history of science," Dr. Elbert King, an enthusiastic young geologist, told a writer from the *New Yorker*. King had helped persuade NASA of the need for the $8 million, 83,000-square-foot facility, and he had been given an optimistic title, Lunar Sample Curator, although that day his shelves still lay bare.

To many politicians, the success of a Moon landing would demonstrate to the world the undeniable superiority of the United States' economic and political system over its Communist rivals. But the scientists in Houston were on tenterhooks for a very different reason. The astronauts' principal scientific objective was to bring rocks from the Moon directly back to their laboratory at the Space Center. That made the Apollo project, at $25 billion ($160 billion in today's currency), by far the most expensive geology field trip ever bankrolled.

In 1969, we knew so little about the Moon that scientists were debating whether the massive craters that pitted its surface had been blasted out by volcanoes or by enormous asteroids or comets. Our ignorance was so great that an eminent planetary scientist issued a stark warning: a lunar lander would be swallowed by a deep layer of Moon dust. That would make the safe return of a spacecraft carrying astronauts extremely unlikely. Fortunately, an earlier, unmanned mission had ruled this danger out, but there was no denying that the astronauts would face great unknowns.

At 7:59 p.m. Houston time, television broadcasters reported that the spiderlike lunar lander, *Eagle*, had separated from the command module, which would remain in the Moon's orbit. The lander began its descent.

Inside, Neil Armstrong and Buzz Aldrin stood slightly hunched at their controls because giving them seats would have added too much weight. To navigate, they were placing their trust in a primitive guidance computer developed at MIT. They had faith in it, just not complete faith. As they approached the silvery Moon, Aldrin read out the computer's calculation of their altitude, and Armstrong peered out the window to confirm that they were on course.

For a time, everything checked out. Then suddenly, just 6,000 feet above the crater-flecked surface, a yellow alarm flashed on Aldrin's console, and buzzers sounded in their headphones. The scientists watching television back home did not know it, but the bottom had fallen out of their chances of ever studying Moon rocks. Mission flight controllers were stunned. A digital code told them that the guidance computer was in trouble. Aldrin watched helplessly as his computer shut down, then quickly rebooted. Several minutes later, a similar alarm appeared. Mission controllers had no idea why. And they had little time to decide if they should make the risky decision to abort.

Then, a quick-thinking flight engineer consulted a handwritten list of alarm codes. He saw that, although the same alarms had appeared in a simulation, the computer had quickly righted itself. Mission controllers made the split-second decision to ignore the alarms and hope for the best.

They made a good call. The computer settled down.

But Armstrong, understandably, had been distracted. Now, with just minutes of fuel remaining, his body set off its own alarms as he peered out his small windows. The guidance computer had overshot their landing site by about four miles, and was about to set them down in a crater full of boulders the size of Volkswagen Beetles.

NASA had chosen the astronauts from a pool of test pilots for good reason. Armstrong instantly activated his manual controller. Turning sharply, he accelerated from 8 miles an hour to 55, and made a desperate dash for a distant clearing. Mission controllers watched helplessly as Armstrong's pulse doubled. With just twenty-five seconds of fuel left in his descent tanks, he brought the lander down in a cloud of dust. "We're breathing again," Mission Control told the astronauts.

The scientists in Houston cheered. The first part of the mission had succeeded. They might actually get their hands on Moon rocks after all—thanks to Armstrong's nerves of steel, a fantastic feat of engineering—and NASA's extremely healthy dose of good luck.

Giddily, the astronauts experimented with walking and hopping in the Moon's low gravity. Back in Houston, the researchers watched impatiently, wondering what else could go wrong. The astronauts planted a flag, chatted with President Nixon, and installed several scientific instruments. Then, to the scientists' immense relief, Armstrong and Aldrin finally got down to their most important job—filling two suitcase-size aluminum boxes with rocks to haul home.

Three days later, the command module *Columbia*, trailed by navy ships and helicopters, splashed down in the Pacific Ocean southwest of Hawaii. From that moment on, the team at the Lunar Receiving Laboratory anxiously monitored their rocks' transport like bank officers tracking a large shipment of gold. NASA's chief administrator joked that the rocks were more valuable than the rarest jewels. He priced them at $24 billion—the cost of the entire space program. That made them worth a cool $400 million a pound.

The astronauts filled out a US customs form, as a joke. Meanwhile, the two aluminum boxes, holding between them sixty pounds of rock and soil, were flown on separate C-114 jets to Ellington Air Force Base in Houston. From there, they were to be transported just four miles to the Lunar Receiving Lab. But lunar sample curator Elbert King was taking no chances. This was 1969. Antiwar and civil rights protests were raging. The Reverend Ralph Abernathy had brought a mule cart to Cape Canaveral to protest the vast sums that the government was spending on Apollo instead of on poverty programs. In preparation for the Moon rocks' arrival, police closed the road to the Space Center and prepared to escort the van. Yet King still feared that "a radical group of hippies" might want to cause trouble. So for extra security, he went home and placed six bullets in his long-barreled Smith & Wesson .357 Magnum. Then he followed the van in his Plymouth Valiant, with his loaded revolver in a bath towel beneath his seat. Nothing was going to keep those boxes from reaching his laboratory.

In the end, it was an uneventful ride. The rocks arrived peacefully at Building 37, the Lunar Receiving Laboratory. The new facility was crammed with many small labs, biological isolation systems, and a quarantine area for the astronauts, because scientists could not rule out an apocalyptic scenario: contamination by alien life. In Michael Crichton's famous science fiction thriller *The Andromeda Strain*, published just two months earlier, a microorganism hitched a ride to Earth on a space capsule and almost destroyed mankind. So the rocks, like the astronauts, were to be quarantined for three weeks, while scientists searched for signs of deadly lunar germs. Warily, they tested the effects of exposing algae, plants, houseflies, great wax moths, German cockroaches, oysters, fathead minnows, shrimp, white mice, and Japanese quail to lunar dust and soil.

But King and his colleagues did not have to wait out the quarantine. Great pains had been taken to design sealed steel vacuum chambers. Black rubber gloves protruding through glass windows would allow scientists to handle the rocks inside. Team member Bill Schopf recalled that, if there was a large spill in the chamber, contingency plans called for the scientists to rush outside to a grassy area where a helicopter would pick them up and transport them to an air base. From there an awaiting plane would spirit them to the Bikini Islands in the Pacific to be quarantined.

Shortly after the boxes arrived, a team of scientists showered in a cleanroom. They changed into surgical gowns and caps. Then, with gas masks by their sides, they approached the chamber, as a colleague on closed-circuit television narrated a blow-by-blow description of their progress for their colleagues. Placing his hands inside awkward stiff rubber gloves, a technician scrubbed the exterior of one of the rock boxes with a powerful germicide. Then he opened it and stood back, allowing the scientists to peer through the window.* At last, the moment of truth had arrived. What would the rocks reveal about the

---

* Bill Schopf recollected that they were in for a bit of a surprise. When the technician tried opening the case, he found that it was already unsealed. The latches had been too stiff for the astronauts to close on the Moon, so by the time they splashed down, they looked like coal miners; their faces were streaked with Moon dust.

Moon and the early Earth? The researchers eagerly craned forward, like explorers about to glimpse a new continent for the very first time.

They were sorely disappointed. The Moon rocks were bone dry and coated in dark dust. "I was the second guy in the world to look at them," Schopf, a geologist, told me. "I mean, big deal. They looked like chunks of coal." Once the scientists removed the dust, the rocks brought something else to mind: ordinary earthly basalt, the common gray lumpy rock formed by cooling lava.

Their first appearance was undeniably humdrum, yet the rocks soon began to offer up startling revelations. For starters, they settled the age-old controversy over the origin of the Moon's craters. What the astronauts had brought back were not the remains of volcanic craters, as some scientists expected. Instead the rocks came from enormous circular scars left by asteroids or comets. But the ages of the rocks were the real surprise. They would bring to light a catastrophic chapter in the history of Earth and of our own atoms.

The discovery would be made after NASA couriered samples of the Moon rocks to the lab of Gerald Wasserburg, a high school dropout–turned–Caltech geochemist, who was something of an instrument fiend. In the early 1960s, the National Science Foundation had turned down his request for a grant to build the first digital mass spectrometer. So Wasserburg raised the funds himself. His machine would measure with new precision the quantities of a radioactive element and the elements resulting from its decay, allowing him to calculate a rock's age more accurately. He connected his device through tunnels and two-thirds of a mile of wire to a mainframe IBM computer. His spectrometer was soon obviously the Rolls-Royce of dating technology, thirty times more accurate than anyone else's. While working on the Moon rocks, he christened it the Lunatic I. He operated it in a sterile cleanroom, identified by a brass plaque in the hallway as the Lunatic Asylum.

Wasserburg hoped that, because the Moon did not have plate tectonics, he would find rocks that were as old as the Moon and Earth: 4.5 billion years. But the first ones the astronauts brought back were only 3.6 to 3.9 billion years old. He hoped that older rocks would come from the next trips.

Three years later, in 1972, his team finally had rocks from five Apollo missions. Now Wasserburg and his fellow "inmates" reviewed the ages of the rocks brought back from every crater the astronauts ever visited by foot or buggy.

They could hardly believe their results. In fact, Wasserburg was so surprised that he took his team out to a Pasadena bar to celebrate.

Although none of the rocks were as old as the Moon and the Earth, they had found something else. If you look at a high-resolution photograph of the Moon, you will see that its surface is completely covered by massive craters. Strangely, every one of these that the astronauts visited was 4.1 to 3.8 billion years old. None were older. And none were younger.

Shockingly, this suggested that 4.1 billion years ago, 400 million years after the Moon and Earth formed, the Moon was suddenly bombarded by huge comets and asteroids that dug out craters and reworked its entire surface. Then, just as suddenly, about 3.8 billion years ago, the intense bombardment stopped. "It must in any event have been quite a show from the Earth, assuming you had a really good bunker to watch from," Wasserburg and his colleagues wrote. They called this devastating period the Lunar Cataclysm, although it is now known by a less inspired name, the Late Heavy Bombardment. The Moon rocks seemed to offer evidence that hundreds of millions of years after the Moon formed, our neighborhood suddenly looked less like a quiet suburb, and more like a busy shooting gallery. It was inundated by a barrage of massive asteroids or comets—and the Earth would have offered a much larger target than the Moon. Our planet must have suffered the same cataclysm.

Naturally, astronomers wondered where this huge volley of asteroids or comets came from. In 2005, they believed they'd found their source. Although we like to think of the planets as having stable orbits, early in our solar system's history, the gas giants Jupiter and Saturn were still settling into their paths around the Sun. Simulations by the astronomer Alessandro Morbidelli and his colleagues suggested that these may have disturbed space rocks in the Asteroid Belt, an orbital path between Mars and Jupiter. Today, millions of rocky asteroids still live there. The largest, Ceres, is 585 miles across. The astronomers proposed that the Asteroid Belt once contained many more, and during the Late Heavy Bombardment, Jupiter's movement scattered them like billiard balls in all directions. It sent twenty times the mass of all the space rocks now in the Asteroid Belt smashing into the Earth.

For a decade, this scenario was commonly accepted. Today many are not so sure. Nowadays, Morbidelli thinks this emptying of the Asteroid Belt occurred earlier, even before the Moon formed. And many are skeptical that the Late Heavy Bombardment existed at all. It has become a point of controversy. But don't let that confuse you. Almost all agree that the Earth and Moon were pounded by a hail of giant asteroids and comets until 3.8 billion years ago. Some just don't believe that, after the Earth formed, a long lull in impacts was followed by a sudden uptick of them. They don't think there was a lull at all. In other words, skeptics agree on the "Heavy," just not the "Late." Instead, they suspect that the traces of the older craters were simply wiped out by newer impacts.

Fortunately, we have no reason to fear another round of searing bombardments on that scale. By 3 billion years ago, most of the debris left over from the formation of the planets had settled into stable orbits. Some large rocky bodies are still in the Asteroid Belt. And another collection of cosmic bodies, this one largely of orbiting comets, lies on the outer fringes of the solar system, past Neptune, in a region called the Kuiper Belt. Morbidelli believes that the orbits of Jupiter and Saturn are now stable; there will be no more large flurries of asteroids or comets hurtling in our direction. (Although others have estimated that, about 3.5 billion years from now, there is a small chance that Mercury could wander slightly, upsetting the orbit of the neighboring planets enough that Mercury, Venus, or Mars could collide with the Earth.)

What is evident, many scientists believe, is that, for many hundreds of millions of years after our planet formed, any molecules beginning to organize themselves into life would have had few places to hide. A 250-mile-wide asteroid, cruising at 38,000 miles an hour, would have excavated a crater over four-fifths the area of the United States. It would have flung mountains of rock into space. The asteroid's abrupt stop would convert much of its energy of motion into heat. Rock vapor, heated to a sizzling 7,000 degrees Fahrenheit, would have sterilized the atmosphere as a red-hot wave of magma rippled out from the crater in every direction. In comparison, the impact of the 9-mile-wide asteroid or comet that extinguished the dinosaurs would have seemed like the splash of a tiny pebble. The punishing impacts that largely ended 3.8 billion years ago may have destroyed any early attempt to create life. Or at least greatly slowed it down.

In the early 1960s, when Viktor Safronov was struggling with pencil-and-paper calculations behind the Iron Curtain, he had little inkling that his theory would reveal so much of our planet's turbulent history—and the history of our own molecules. The firm ground we stand on was once an ethereal cloud of dust and gas, our molecules among them, dimly illuminated by the light of distant stars. With Safronov's tools, scientists discovered that 4.5 billion years ago, collisions between dust grains created larger particles, pebbles, huge rocks, and then a monstrous one that rapidly swallowed everything within its orbit and became the Earth.

That was just the beginning of our atoms' travails. Once Earth formed, it melted and our atoms had to swim for it. Molecules sorted themselves as heavy iron sank to the center. But if any of our atoms were unhappy that they had not yet reached our planet's surface, they would soon be thankful. About 50 million years later, a Mars-size planet plowed into the Earth, vaporizing much of its mantle. Any of your atoms near the surface were flung high into space before floating back down. After that, Earth may, or may not, have had time to collect itself, but 400 million years later—about 4.1 billion years ago—a barrage of pulverizing asteroids was bombarding the Earth. They sent out massive tsunamis of magma and choked the planet's atmosphere with hot rock vapor. If any life evolved during that time, those searing impacts likely destroyed it. Or at least forced it to remain in hiding.

It may not have been until 3.8 billion years ago, by then a good 700 million years after the Earth formed, that our solar system finally calmed down. The planets had settled into stable orbits. The shooting gallery of rogue asteroids and comets had tapered off. Earth's surface had cooled. A thin, rocky crust had formed.

At long last, conditions were favorable for life.

PART II

LET THERE BE LIFE!

In which we learn the surprising ways in which the building
blocks of our bodies—water and organic
molecules—found their way to Earth, and how, in
perhaps the most astonishing hat trick in the history
of the universe, life was born.

DIRTY SNOWBALLS AND SPACE ROCKS

The Biggest Flood of All Time

> If there is magic on the planet, it is contained
> in the water.
> —*Loren Eiseley*

We take the water within us for granted, while sipping and gulping it all the time. Yet billions of years ago when our planet was just forming, our neighborhood was hot enough to vaporize any water, allowing it to drift away even as our planet was being assembled from colliding rocky particles. Still, all earthly life is dependent on water. Water is to life as a canvas is to a painting. It's so vital that NASA's motto in its search for life elsewhere is "Follow the water." We certainly would not be here without it, which is why scientists are so eager to explain how the water running through our veins made its way here in the first place. Over the decades, they've found that simple question surprisingly difficult to answer.

In fact, water is one of the most unusual molecules in the cosmos. We live in our solar system's only water world—that is, on the only planet whose surface is dominated by water. As Arthur C. Clarke put it, "How inappropriate to call this planet 'Earth' when it is quite clearly ocean." Deep blue water covers 70 percent of our planet to an average depth of two and a half miles. Without it, Earth would lose much of its magic. No other molecule covers nearly so much of the Earth's surface or occurs naturally on Earth in three states: as a solid, liquid, and gas. In our solar system, the only other place with clouds, rivers, and lakes is Titan, one of Saturn's moons. But it rains methane on Titan, not water. Does life based on an entirely different chemistry live there? Possibly; we can't know until we visit. Here on Earth, life needs water.

In terms of circulation, water has no remote rival on our planet; it freezes and melts, runs, flows, trickles, drops, and climbs to dizzying heights. That allows life to exist everywhere: in caves miles beneath the Earth's surface, in rock deep within the ocean crust, and in rivers, streams, hot springs, puddles, fog, dew, and clouds that contain many millions of water droplets in every cubic foot. In 2010, a University of Georgia chemist arranged for his students to hitch a ride on a NASA research aircraft. A dizzying 30,000 feet above Earth, they found more than a hundred species of bacteria and fungi at home in storm clouds.

We too are completely dependent on water. You may think of yourself as living on dry land, but you're aqueous to the core. When you left the womb, about 75 percent of you, by mass, was water—roughly the same as a banana. About 60 percent of an average man is water, 55 percent of an average woman. In old age, you'll lose about 10 percent more. That sounds like a lot of water, yet when you lose just 2 percent or so, your brain wastes no time in telling you that you're thirsty. Though you can survive about a month or more without food, you'll only last about a week without water.

Water does us countless favors. It aids reactions that store and liberate energy in food. It helps our DNA and proteins assemble and maintain their inordinately complex shapes. Water helps our cells break down molecules. And it carries out the trash every time you empty your bladder. Which is why you need to replace about eleven cups of water a day.

Above all, water is the matrix, the ocean within you where your molecules meet and mix. Ironically, that's only possible because water is such a weakling; the bonds between its molecules are easily broken. A water molecule is made of two tolerant hydrogen atoms stuck in an unequal relationship with an oxygen atom. And the heftier oxygen doesn't share well. It pulls the electron it shares with each hydrogen a little closer to itself, creating a slight negative charge on the oxygen and a slight positive one on the hydrogen. Those slim differences in the charges of oxygen and hydrogen atoms on neighboring molecules hold liquid water together in feeble alliances called hydrogen bonds.

That fortuitous weakness means that the bonds between the water molecules within us are constantly breaking and reforming every 1.5 *trillionth* of a second. This allows other molecules to race through with ease, which allows you to think. Electrically charged ions can send signals at up to 350 feet per second in your soggy brain. When you're dehydrated, you don't think clearly—possibly because your water-deprived neurons shrink, or your water-deprived veins don't bring them enough oxygen. If you ever feel foggy in the head, perhaps you just need a drink of water.*

All of which, naturally, gave scientists even more reason to wonder

---

* Any ice fisherman will appreciate why water is helpful to life in yet another way. When water cools into a solid, it doesn't sink. It floats. Ice does that because, as water freezes, its odd hydrogen bonds force its molecules to stretch out into less dense crystalline patterns. If ice sank, like most solids, then our rivers, ponds, and oceans would begin freezing at their lowest depths, and if temperatures grew cold enough, all life below would be encased in ice.

how the water within us found its way to Earth. In their search for answers, they were repeatedly bushwhacked by new evidence that threw their accepted theories into doubt.

One of the first clues to water's baffling appearance on our planet came in the 1950s from an astronomer who, among other claims to fame, penned a book with the rocket scientist Wernher von Braun on how to send an expedition of fifty scientists and technicians to the Moon. Fred Whipple didn't think of becoming a scientist in his youth. That was plan C, born of boredom and failure. Failure came first.

The son of a farmer, Whipple received much of his education in the early 1900s, in a one-room schoolhouse in Red Oak, Iowa. His family lived far from a hospital, and suffered a series of medical problems that traumatized him. When Whipple was four, his two-year-old brother died of scarlet fever. A year later, Whipple himself contracted polio. And his parents, frightened by the struggles they saw others having with appendicitis, decided that the entire family would receive preventative appendectomies. "For a number of years," he recalled, "I had the gruesome sight in the medicine closet of three bottles of formaldehyde: Daddy's appendix, Mommy's appendix, and Baby's appendix." Despite these scares, he excelled in school. In 1922, recognizing his unusual promise, and fearing an impending drop in corn prices, his family decamped for California, where he would have better academic opportunities. It was there that Whipple fell in love with tennis. At the University of California, Los Angeles, he decided to major in mathematics. It was undemanding enough, he figured, that it would leave him plenty of time to hone his tennis game before turning pro.

That was not to be. Whipple's run-in with polio left his left leg an inch and an eighth shorter than the other. Forced to accept reality, he abandoned his dreams of tennis glory. Now as he looked ahead, he feared that a career in mathematics would be as deadly boring as any fate he could imagine, so he turned to astronomy.

In 1930, in graduate school at Berkeley, Whipple and a fellow student were the first to calculate the orbit of Pluto, just weeks after it was discovered. He was pleased to find that he enjoyed the "orbit computing business." After receiving his PhD, he went to Harvard, where he directed a sky-observing program at a new observatory located twenty-six miles from the bright lights of Cambridge. One of his duties in the small brick building adjoining the telescope was to check the accuracy of the camera by peering through a magnifying glass at every new photographic plate. Whipple decided to use the opportunity he had been handed to hunt for comets. Over ten years, he examined seventy thousand images. It might have been tedious for some, but apparently not for Whipple. For his pains, he won the glory of discovering six new comets. (The record holder is Jean-Louis Pons, an eighteenth-century observatory doorkeeper–turned–astronomer, who bagged over thirty of them.) But finding new celestial bodies would not be Whipple's chief claim to fame. He would make the contribution for which he will forever be known after he began puzzling over the curious behavior of comets.

In antiquity, a comet's fiery blaze in the night sky was often considered a portent of doom. Sky watchers saw a bright ball trailed by an eerily long shimmering tail (hence its name, from the Greek word *kométēs*, meaning "head of long hair"). But in Whipple's time, scientists no longer thought that these celestial objects traveling in long elliptical orbits around the Sun were quite as impressive. They were merely "flying sand banks"—clouds of sand or gravel particles along with a bit of gas loosely bound by gravity. That long-prevailing theory had been refined by a distinguished astronomer at Cambridge, Raymond Lyttleton, a collaborator of Fred Hoyle.

What troubled Whipple was that, as comets approached the Sun, they developed long tails that trailed behind them for hundreds of thousands or millions of miles. These were made of gases from the comet's surfaces. Yet he calculated that while Comet Encke had already orbited the Sun over a thousand times, it showed no sign of running out of the gas that created its long tail. And there was another thing. Comets were unreliable. Although they traveled in stable orbits around the Sun, they never returned when they were expected, unlike the punctual planets and asteroids. Some comets arrived slightly early, like Comet Encke that came back every 3.3 years, but always a half hour to an hour before it should. Others, like Halley's comet, reappeared every 76 years, but each time ambled by a few days later than expected. To Whipple, something just didn't add up.

One day in 1949 at Harvard, he sat down to calculate the trajectories of meteors—fragments of comets that enter our atmosphere. As he analyzed the forces they experienced, he realized that their leading edges would heat up. It occurred to him that if they contained ice, a bit of it would turn into a stream of gas. "My God, this is what's happening to comets!" he thought. It was his eureka moment. Suddenly, comets' strange behavior made complete sense.

Whipple realized that as a comet orbits the Sun, it also spins like all orbiting bodies, just as a baseball does on its way to home plate. If a comet were a massive ball of ice, not a cloud of sand, then as it approached the hot Sun, some ice on its leading edge would turn into gas and stream to the side like a tiny jet thruster. It was this gas stream that explained why comets were so unreliable. Depending on the direction a comet was rotating, the gas would make its orbit slightly larger or smaller, and speed up or delay its journey.

In 1950, Whipple kicked up a ruckus by boldly claiming that comets were not the loose collections of sand and gas that Lyttleton and others believed. The clouds of dust and gas surrounding them hid the fact that, beneath these veils, they are massive balls of ice. In other words, they are dirty snowballs in flight. That's why a comet never runs out of gas; its icy nucleus is miles across—so large that it loses just a small amount of gas as it approaches the Sun. Their cores, Whipple suspected, were mainly frozen gases such as methane, ammonia, carbon dioxide, and ice, all of which were common far from the Sun when the solar system was forming.

Were overgrown ice balls really journeying hundreds of millions of miles and tens of thousands of years, or longer, around the Sun? Not surprisingly, more than a few scientists were dubious. Whipple's suggestion was fiercely rejected by Cambridge's Lyttleton, who had a detailed mathematical theory of how sand and gravel created comets. Lyttleton acidly commented that Whipple's theory only survived because it was impossible to prove it wrong; the icy nuclei, which Whipple claimed were in the heads of comets, were too small to be detected from Earth.

It was not until three decades later, when Whipple was seventy-nine, that researchers could finally test his theory. In 1986, after an absence of three-quarters of a century, Halley's comet, which had been observed by the ancient Chinese and likely by Babylonians and Greeks before them, was about to reappear.

Anticipating its historic return, researchers wasted no opportunity to study it. When it was 89 million miles away, about the distance of Mars, a NASA aircraft flying at 41,000 feet tracked it with a spectrometer. The Soviets sent twin spacecraft to fly within 5,000 miles of Halley. The Japanese launched two spacecraft of their own. And the European Space Agency (ESA) debuted their very first space probe—*Giotto*—the only spacecraft that would probe its interior.

On March 13, 1986, in Darmstadt, Germany, teams of scientists from around the world assembled in the ESA's dimly lit mission control room and in numerous adjoining labs that they had wired with receiving instruments. After a decade of planning, the investigators nervously awaited the encounter. Whipple was among them. "Well Fred," one scientist said, "tomorrow is your moment of truth."

Many of *Giotto*'s engineers were dubious that the fragile spacecraft would survive its "kamikaze mission" long enough to reveal the comet's secrets. That's because, in order to reach the nucleus at the heart of the comet, *Giotto* would have to pass through billowing clouds of dust and gas while the spacecraft and comet raced toward each other at over 40 miles a second.

For protection from the flying debris, *Giotto* was outfitted with Whipple shields on its front and rear. Each was made of slightly separated thin layers of aluminum and Kevlar. Whipple himself had invented this lightweight bumper for spacecraft back in 1947, when space travel was still a Buck Rogers dream. But even with Whipple shields, the researchers knew that *Giotto*'s sensitive instruments would have to withstand a withering sandblasting. And unlucky hits by small grains of dust could make short work of them.

When *Giotto* was 62,000 miles away from the comet's center, the pelting by dust grains commenced. By 10,000 miles, the blows were constant. The spacecraft continued closing in. Then, at 493 miles from the nucleus, a particle with a mass less than four hundredths of an ounce smashed into *Giotto*. It set the half-ton machine wobbling. Even more alarming, engineers lost radio contact. Many feared the worst. They reestablished communication less than thirty seconds later, but dust grains had disabled the camera, destroyed part of the shielding, and knocked out some sensors.

Despite all the damage, the scientists, Whipple among them, were overjoyed. Photographs taken just seconds before the collision revealed a striking ghostly outline. The core itself was a potato-shaped nucleus roughly nine miles long—about the size of Manhattan. And data from the mass spectrometer showed that 80 percent of the gas that the comet cast off was water. It turned out, however, that overall, the nucleus of the comet had much more rock and dust than Whipple expected. Only about a third of it was water ice, which meant it was not so much a dirty snowball as a snowy dirt ball. Nonetheless, it was clear that comets held a great deal of water.

To planetary scientists wondering how the ancient Earth was hydrated, humongous snowy dirt balls now became the prime suspects. It seemed that a hail of comets fell on our planet billions of years ago, delivering an ocean of water.

But where did these comets come from?

Scientists had long believed that comets were born in the cold regions beyond Mars. By the 1990s, it seemed evident that most of these comets must have been vacuumed up by the growing planets. Nonetheless, the Dutch astronomer Jan Oort had suggested that trillions of comets may have survived at the edge of the solar system. Too distant for the planets to reel in, they encircled the solar system in a huge spherical shell, known today as the Oort Cloud. So there were plenty of icy comets out there that could have filled the Earth's oceans. The problem was they were thousands of times the distance of the Earth to the Sun—too far away to have traveled here.

A few researchers wondered if some comets managed to survive somewhere closer in our solar system, beyond the orbit of Saturn. That is a thousand times closer than the Oort Cloud. Yet this was mere speculation. No one was foolish enough to think that they could spot a comet tens of miles across or smaller that was so far away.

No one, that is, except a young MIT professor, Dave Jewitt, and his graduate student Jane Luu. Jewitt, who has a high-domed head, a ready smile, and a droll British wit, was the son of a London factory worker and a telephone operator. When he was a boy, a chance sighting of a meteor in the night sky kindled his fascination with astronomy.

In 1985, he had the brainstorm to attach new digital light detectors, called CCDs, to a telescope to search for small objects like comets in the distant realms of our solar system. Just because we couldn't see them didn't mean they weren't there, he reasoned. But he was seeking funding to search for objects that few others suspected might exist. His grant proposals were denied time and time again. Over thirty years later, he quickly grew heated when he recalled the dismissive reviews of his applications. "The usual thing you get is, 'You haven't demonstrated that the proposed measurement is possible,'" he said. "I mean, Jesus Christ, that's the most stupid comment in the world. The whole point is to try to do something that hasn't been done. And maybe it's not possible, but the idea is to try." Some of his reviewers may have fallen for the "If Our Current Tools Haven't Detected It, It Doesn't Exist" bias—the assumption that because we haven't found anything yet, there must be nothing there.

Refusing to give up, Jewitt and Luu surreptitiously borrowed telescope-observing time from their other projects to search for dubious tiny objects billions of miles away.

For years they found nothing. One and two years passed into four,

five, and six. Then, one summer night in 1992, they were working at
the Mauna Kea observatory on the Big Island of Hawaii. They were
close to concluding that their five-year search had been pointless,
when they detected a very faint dot. And they realized it was just
barely moving. "It can't possibly be real," Jewitt thought. But it was.
Jewitt and Luu had discovered an orbital path lying beyond Neptune
that would turn out to have millions of comets. It would be named the
Kuiper Belt, after a Dutch astronomer who discussed such a possibility in the 1950s (but, ironically, did not believe it could exist).

With the detection of hordes of comets in the Kuiper Belt, the
origin of the water inside us seemed settled. Sometime after Earth
formed, comets from the Kuiper Belt, and perhaps a small number
from the more distant Oort Cloud, carried the water here that now
covers our planet. These reservoirs of flying icebergs were plenty large
to have filled the Earth's oceans. This theory was soon accepted and
taught by most everyone. Mystery solved.

Or was it? The trouble began again in 1995. At a stargazing party
near Phoenix, Arizona, Thomas Bopp, a parts manager at a concrete
supply company, took a turn at a friend's telescope. He noticed a fuzzy
glow in the corner of his eyepiece. That same evening, from his drive-
way in Cloudcroft, New Mexico, the astronomer Alan Hale spotted
the same object. Their newly discovered comet, one of the brightest
ever seen, would henceforth be known as Hale-Bopp.

The next year, Dave Jewitt was among a team of scientists who returned to the Mauna Kea observatory, this time to train a powerful
radio telescope on Hale-Bopp. In uncomfortably thin air at 14,000
feet, they took spectrographic measurements in thirteen- to sixteen-
hour shifts at night. They wanted to compare the proportion of a
rare form of water in the comet with the percentage in the Earth's
oceans.

You may or may not be aware that water comes in several forms.
Most of it is made from hydrogen atoms with only one proton in their

nuclei. But another form of water exists that is 10 percent heavier. Its hydrogen atoms are isotopes, called deuterium, whose nuclei contain a neutron in addition to a proton. This heftier water is rare; in our oceans, it's only one of every 6,400 molecules. So, as the Mauna Kea team prepared to take measurements of Hale-Bopp, they were confident they would find the same proportion of heavy water. After all, Earth's water had come from comets.

But that wasn't what they found. Hale-Bopp had twice as much heavy water as our oceans. And that was a problem. Earlier astronomers had detected a similarly high proportion in Halley's comet, but had shrugged it off as an anomaly. Then scientists measured the same levels in a comet called Hyakutake. Now, three readings were in agreement, and it was hard to ignore the evidence that the composition of the water in comets and our oceans did not match.

"How did astronomers react to the Hale-Bopp measurements?" I asked Jewitt.

"They freaked people out," he said, referring to the implications that were dawning on them. "It was kind of an awakening of consciousness. It was New Agey." He laughed, and added, "I wish I hadn't said that." His meaning was clear; researchers had been caught flat-footed. Suddenly it was apparent that you couldn't just melt comets to get oceans. Although Whipple was right that comets are full of water, our oceans came from somewhere else in the solar system. But from where?

That's when Jewitt's mind, like many others, turned from icy comets to the massive rocks floating in space called asteroids.

You probably think you can't wring water from a rock. But you can; at least from some of them. If you heat rocky meteorites—fragments of asteroids that fell to Earth—wisps of water vapor will rise from water molecules trapped in their crystalline structure. Scientists had known for years that asteroids hid water, and that these rocks varied widely in the amount they secreted away. Most of the asteroids that formed close to the Sun contained almost no water. But those created in the regions of deep freeze beyond Mars can hold up to 13 percent.

If the asteroids that struck Earth were large enough, Jewitt and others realized, that could add up to a *lot* of water. Not only that, but astronomers also knew that a large collection of asteroids lay in orbit between Mars and Jupiter, in a region called the Asteroid Belt. And the proportion of heavy water in asteroids, unlike comets, matches the ratio in our oceans and our bodies. All of this suggested that our planet's water could have come from space rocks.

That may sound like case closed. But the Asteroid Belt was 300 million miles away. Imagine standing there and trying to aim a pool ball at Earth. You would have to be insanely good to make that shot. What are the chances that enough asteroids were launched at precisely the right angles to cover the Earth with water? And how could we ever possibly know?

In 1998, at France's Nice Observatory, less than a mile from the sun-drenched sands of the Riviera, Alessandro Morbidelli thought he saw a way to put the asteroid theory to a test. Instead of making astronomical observations with a telescope, he'd rerun the evolution of the Earth, at his desk. Morbidelli was a newcomer to the planet evolution game. He was inspired by George Wetherill and John Chambers, who had just created a new, much more sophisticated computer program to simulate the late stages of planetary formation. Morbidelli asked Chambers if he could work with him and use the program to track the sources of all the chaotic bodies that collided to form the Earth—including asteroids and comets.

Like Sherlock Holmes's nemesis, Moriarty, who was brainy enough to calculate the complex orbits of asteroids, Morbidelli also had the right mathematical stuff.* As a college student in Italy, Morbidelli, gangly and confident, intended to follow his teenage passion for sky

---

* "Is he not the celebrated author of *The Dynamics of an Asteroid*, a book which ascends to such rarefied heights of pure mathematics that it is said that there was no man in the scientific press capable of criticizing it?" Holmes asked Watson in *The Valley of Fear*. Moriarty may have been modeled on the famous Canadian American astronomer Simon Newcomb, who published numerous papers on the subtle dynamics of asteroid orbits and was "more feared than he was liked."

watching, but his professors warned him off. "I was strongly advised not to do that," he said. "I was told that with a degree in pure astronomy it would be complicated to find a job. It would be too risky." Disappointed, he studied physics instead. After graduating, he approached professors at his university to do a master's thesis in astronomy. No one would take him. Feeling let down again, he did his PhD in mathematical physics and chaos theory, and kept an eye out for a way to cross over. In his postdoc, he worked on mathematical problems that happened to be relevant to celestial mechanics. That work was noticed by astronomers in Nice, who wanted to add more sophisticated mathematics, such as chaos and chance, to their models of how asteroids reached Earth from the Asteroid Belt. They offered him a job, and he never looked back.

Morbidelli, Chambers, and their colleagues tweaked their computer simulation to estimate the odds that enough asteroids or comets struck the Earth to gift our globe with oceans. The answer, when it came to asteroids, was—very likely. Their simulations appeared to reveal that in the late stages of Earth's formation the Asteroid Belt, between Mars and Jupiter contained many more inhabitants than today. As massive Jupiter settled into a stable orbit, it scattered a huge number of asteroids like pool balls in all directions, including some on deadeye 300-million-mile trajectories right toward us.

What Morbidelli found equally compelling was their discovery that a comet has a much smaller chance of hitting the Earth than an asteroid. "That part of the paper is often overlooked," he said, "but it's possibly more important, because we showed that the probability that an object that is beyond Jupiter will hit the Earth is less than one in a *million*." That was the aha moment. A comet from the more distant Kuiper Belt has to travel from so much farther away that you are a hundred times more likely to be hit by an asteroid than by a comet of the same size.

Morbidelli and Chambers showed that it was highly improbable that icy comets delivered most of the water in our oceans. Comets were out; asteroids were in. That seemed conclusive enough. The water in our bodies had been delivered to Earth by massive rocks from space.˙ The End.

Yet once again, the debate was far from over. At the University of Arizona, a British-born geologist, Mike Drake, an expert on the chemical composition of the Earth, remained skeptical. Drake, whom fellow grad students once teased for keeping an immaculately clean desk, thought he might be able to tidy up a messy theory. The Asteroid Belt is 300 million miles away. Did most of the water within you really travel from so far? Maybe we didn't need asteroids to carry it to us. Perhaps there might be a simpler explanation.

In his mind, Drake retraced the journey of water to Earth from the earliest days of the solar system. He thought about what our neighborhood looked like when collisions in a vast disk of dust and gas began to form our planet. What, he asked, were the most abundant gases in that disk? Hydrogen, followed by helium and carbon dioxide. And the fourth most common? Water. So as Drake saw it, the cloud that formed Earth was basically a lot of dust surrounded by water vapor. Admittedly, the Earth's neighborhood was extremely hot then. Even so, did all that water really just float away?

One day, as he was having a cold drink, as one often does in sunny Arizona, he noticed that a thin layer of water droplets had condensed on the outside of his glass. A question immediately bubbled up in his mind. Did droplets of water condense in just the same way on our planet's smallest building blocks? Did water become stuck to the tiny

---

˙ You may be wondering if asteroids also delivered water to our neighbors. They did. Venus probably received as much water as Earth and may even have been habitable. But it was closer to the Sun and about 30 degrees Fahrenheit hotter, so convection currents of water vapor rose high above Venus's atmosphere and the ultraviolet rays there gradually broke Venus's water down.

Mars also had oceans, but because it was smaller, its feebler gravity did not hold its atmosphere as strongly. Nor did its weaker magnetic field shield it from ultraviolet rays. Thus its surface water was vulnerable to gradual destruction, too. Even so, some of its water still survives as ice on its surface, and possibly below.

grains of dust that were colliding to form the Earth? If so, maybe we didn't need asteroids to bring us water after all. Perhaps most of it was trapped within Earth's rocks from day one. Drake even calculated that olivine, the most common rock in the Earth's mantle, can hold a great deal of water.

But he still had to explain how water broke out of its widely dispersed hiding places in the Earth's interior. That was a more straightforward problem. Volcanic gases are more than 60 percent water. As rock in the upper mantle heated and turned molten, the light water molecules hidden within it would have pooled and vented into the atmosphere. Drake, who collaborated at times with the geologist Kevin Righter, calculated that when the Earth formed, dust grains could have held on to many times the amount of water in our oceans. Even today, the Earth's mantle probably holds more water than exists on its surface, perhaps several times as much. It was clear to Drake that most of the water circulating in the oceans had been hiding here all along. It was just trapped—in rock created from colliding particles of dust.

Not all rushed to embrace Drake's theory. It remained hard for Morbidelli, Jewitt, and many other astronomers to swallow. It was certainly possible that some water might have reached the Earth this way, but did it really account for most of it? For one thing, Morbidelli wondered, even if water vapor condensed on tiny grains of dust, would it have remained stuck to them during the countless violent hot collisions that they endured as our planet formed?

I asked Jewitt which theory most scientists subscribe to today. "Everyone's theories are commonly accepted, by themselves," he said dryly. He went on to say that a majority probably agree that our water came from many far-flung places. As Drake suggested, a small amount likely condensed on dust grains and was trapped within the Earth from the very start. A modest amount came from comets in the Kuiper Belt. Less may have arrived from the distant Oort Cloud, just outside our solar system. But the lion's share of the water within you probably hurtled in on asteroids. In Morbidelli's latest simulation, these came not from the Asteroid Belt, but from around Jupiter, and were sent in our direction by a change in Jupiter's orbit or by its growth.

Once all that water reached Earth, how did it make its way into our oceans? That's a story as dramatic as the biblical flood. If you looked down at Earth in its earliest days, you would have seen it glowing deep red. A hot magma ocean, roiled by tides, covered its entire surface. If you returned sometime later, a thin outermost layer would have cooled into a crust of dark rock. As it thickened, water vapor from magma below continued pouring out of numerous volcanoes and fissures, creating thick dark clouds that hung menacingly over the Earth. Meanwhile in these early days, asteroids and probably a lesser number of comets continued battering the Earth, bringing more water, which was absorbed by magma or flung into the atmosphere.

Finally, as the Earth and the atmosphere cooled, the threatening clouds grew so heavy that they unleashed the biggest flood of all time, one that would have astonished even Noah. The water now residing within our bodies was once swept up in this monstrous deluge. Rain poured down for thousands or tens of thousands of years, fueled by water vapor that continued to vent from magma below and replenish the water in the atmosphere. In these days before plate tectonics, our planet had no high mountains or deep basins. When the rain ceased, an ocean of water over a mile deep encircled the entire globe.

Reconstructing Earth's early history is not for the fainthearted. "Nothing that far back is without controversy," said the geologist John Valley. Scientists could now plausibly explain how water arrived on our planet. Yet figuring out how early a stable body of water appeared that was suitable for life turned out to be as difficult as finding the source of the water itself.

It seemed easy at first. Until 2001, most scientists believed that they had the story down. As we saw, after the Earth and Moon formed

4.5 billion years ago, an asteroid and comet bombardment continued to pulverize the Earth. Any water unlucky enough to be on the Earth's surface then was likely vaporized or sterilized by massive fiery impacts, or absorbed into hot magma. Only after the Late Heavy Bombardment ended, 3.8 billion years ago, could a stable ocean or pond emerge.

However, the discovery of chemical signatures in ancient crystals no larger than a grain of sand upended scientists' tidy narrative of our planet's history. It forced them to wonder if an ocean fit for life might have existed much earlier—perhaps remarkably soon after the Earth formed.

The crystals in question first surfaced when the geologist Simon Wilde mapped a remote region of Australia for a geological survey. From Perth, Wilde drove almost 400 miles north through desert to a town called Cue, then bumped along an unpaved road for another 125 miles until he reached a scrubby red and brown range known as the Jack Hills. There, as he was mapping a small rise, he spotted an exposed, green-tinged outcrop, about six feet to a side, of a rare quartz pebble conglomerate. He was electrified. A conglomerate is a mix of rocks bound together by sand, mud, and other chemical glues. And the particular kind that Wilde came across is famous for making men rich. If you ever discover one, it may be worth your while to investigate. A similar formation in South Africa, called the Witwatersrand, has produced 30 to 40 percent of all the gold ever mined.

In the Jack Hills, Wilde painstakingly separated and strained the smallest specks of his rocky conglomerate. He found nothing to make him a wealthy man, but although it would take some time to know it, he had stumbled on a scientific gold mine. Among the assorted rocks were a few rare crystals of an incredibly durable mineral called zircon. And these turned out to be much older than any other rocks on Earth. They were 4.1 billion years old, a date later revised to 4.3 billion years.

That was big news. It would get even better.

A decade later, in 1999, the geologist John Valley, at the University of Wisconsin, and his graduate student William Peck decided to ask Wilde for permission to study the ancient crystals. They wondered if these tiny bits of rock might preserve chemical clues about conditions on the early Earth when the rock crystallized from hot magma. Valley and Peck flew to Edinburgh, carrying about a hundred of Wilde's crystals, among them five of his very oldest, each no larger than a period on this page. A Scottish colleague had offered them the use of an expensive new ion probe, a device that looked something like a particle accelerator shrunk to the size of a car. By shooting a beam of ions at the crystals and analyzing the molecules that it chipped off, it would reveal the amount and type of oxygen inside. The terribly sensitive instrument ran more stably at night—when no one else was in the building and the elevator wasn't running—so they worked in fourteen-hour graveyard shifts. At three in the morning on the tenth day, Valley began analyzing the most ancient crystals. He immediately saw that their oxygen isotopes were higher than they should be if the crystals had formed at very high temperatures. When Peck arrived with his morning tea, Valley joined him to drink his "about to go to bed" bottle of British ale, and they puzzled over the results. "We spent the next four days trying to figure out what I had done wrong," Valley said.

Having ruled out every conceivable source of error, they were left with the only possible conclusion: their readings must be accurate. The shockingly high levels of a rare isotope of heavy oxygen could only have been present (for reasons too complicated to delve into here) if the zircon had formed from sedimentary rock that had once, at a surprisingly early date, been on the Earth's surface in the presence of liquid water.*

Others soon confirmed their finding. Less than a dozen crystals, each about the size of a grain of sand—ancient relics, which, as far as

---

* The isotopes revealed that rock on the Earth's surface had been broken down into clay by water that also changed the clay's oxygen isotope ratio. After the clay was buried in sedimentary rock, high pressures underground created zircon crystals from it.

we know, survive only in small outcrops in the Australian outback—rewrote our understanding of Earth's history. They tell us that 4.4 to 4.2 billion years ago—only 100 to 300 million years after a massive impact cauterized Earth and created the Moon—our planet already had water on its surface, and very possibly an ocean.

Did massive asteroids repeatedly vaporize Earth's first ocean until the Late Heavy Bombardment finally ended 400 million years later? Or did portions of that ocean remain intact enough to provide a safe harbor for life? Reasonable scientists disagree, but many researchers now think that the Late Heavy Bombardment may have been milder than first thought. That means that the ocean in which our cellular ancestors first evolved could have appeared very quickly after the Earth itself formed.

Obviously, many questions remain; there is much we still don't know. What we *can* say with certainty is that some of the water flowing through your veins condensed on the dust whose collisions first created the Earth. Other water molecules within you braved long journeys on comets from the Kuiper Belt, a region between Neptune and Pluto. A smaller amount of your water journeyed tens of thousands of years to reach here from the even more distant Oort Cloud at the outer edge of our solar system. But most water within you probably arrived on massive rocky asteroids from the vicinity of Jupiter. And sometime 4.4 to 3.8 billion years ago—anywhere from 100 to 700 million years after the Earth was born—all the water that arrived from these disparate places pooled in a vast ocean.

In that ancient primordial landscape, occasional islands poked their heads above crashing waves. Volcanoes tirelessly spewed gases, lava, and ash into the air. Streaks of bright lightning rent the skies. From afar, Earth was no longer a molten red magma ocean or fully covered by black and gray volcanic rock. Now sparkling water flowed, crashed, rippled, and ebbed on our blue planet. Our molecular ancestors finally had a watery home in which they could meet. The stage was set for life to evolve . . . if, that is, one other vital ingredient could now be found.

THE MOST FAMOUS EXPERIMENT

The Search for the Origin of the Molecules of Life

I have lived much of my life among molecules.
They are good company.
—*George Wald*

In 1918, the citizens of Moscow, the new capital of Communist Russia, struggled to maintain a semblance of normal life. It wasn't easy. A brutal civil war between the White and Red Russian armies was raging. The West had imposed a trade war. The capital was aswirl with revolutionary ideas, new ways of thinking about equality, justice, and history. Those of means who had not fled were demoted to ordinary citizens and forced to share their wealth and homes with the less privileged. Despite all the revolutionary fervor, Alexander Oparin, a young biochemist steeped in radical scientific ideas, received disappointing news. The censorship board would not permit him to publish a manuscript that speculated on how life arose from mere chemicals. Though the Bolsheviks had overthrown the tsar a year ago, their revolutionary ideology had not yet filtered down to the censors, perhaps because they were not yet ready to directly antagonize the Russian Orthodox Church.

Nonetheless, Oparin's radical ideas would not be suppressed long. They would spark a quest to find the origin of our ancient chemical ancestors—the organic molecules that are the building blocks of life. It would be the first step, he hoped, of an effort to tie "the world of the living" to "the world of the dead."

Oparin grew up in Uglich, a country town of traditional log houses, dirt roads, and horse-drawn carriages. A budding plant collector, he delighted in the fantastic variety of trees, grasses, flowers, and insects he found in the forests of spruce, birch, and pine. In 1914 he enrolled in Moscow University to study botany. There he was captivated by the teachings of the charismatic biologist Kliment Timiryazev, who, though expelled from the university because of his opposition to the tsar, continued lecturing to students in his apartment. As a twenty-six-year-old, Timiryazev had been so inspired by Darwin that he'd made a pilgrimage to England. Lodging at a pub near Darwin's house, he staked out the retiring scientist's home for a week until Darwin finally agreed to meet him. Timiryazev became one of his great popularizers—preaching that Darwinian evolution and Marxism fit hand in glove and heralded a single "scientific" worldview. Darwinism revolutionized our understanding of our biological past, just as Marxism rewrote our understanding of human affairs. Evolution and communism both appeared to him to be inevitable consequences of history.

In 1917, the year the Bolsheviks seized power, Oparin took up graduate studies in plant physiology. He adopted a Lenin-like goatee and mustache, and began working with the distinguished scientist and revolutionary Alexei Bakh, whose scathing, widely read pamphlet, *Tsar Hunger*, had popularized revolutionary socialism. Under Bakh, Oparin studied photosynthesis in algae.

The more he learned, the more he was convinced of another revolutionary idea: that chemical evolution could explain the origin of life.

Even half a century after Darwin published *The Origin of Species*, few others agreed. In England, many prominent scientists had long been men of the cloth who regarded their mission as revealing the majesty of God's creation. It was heretical to suggest that life could arise from inanimate chemicals. But in the new Russia, Oparin's speculation along these new lines was positively encouraged (although not yet by the censorship board).

Still, in trying to retrace our chemical origins, Oparin faced a glaring problem: the molecules in your body and all life are entirely different from the inorganic ones found in the rocks around us. If you analyzed your composition, you'd find that about 60 percent of you is water. Another 1 percent is ions—charged molecules made of elements like sodium, potassium, and magnesium. Everything else within you, from your fingernails and skeleton to your muscles and brain, is fashioned from organic molecules—molecules built around chains or rings of carbon.

If carbon has a personality type, it's an extroverted connector. In fact, if we ever discover life elsewhere in the universe, many scientists believe that it too will be built around carbon. Carbon's versatility stems from the fact that it has four electrons in its outer shell. That, and its small size, means that through neat tricks of geometry it can bond easily in four directions, creating long, stable rings and chains. These are the backbones of your organic self. Your sugars, fatty acids, amino acids, and nucleic acids are all built around carbon. When these link together, they make carbohydrates, fats, proteins, and DNA— your larger organic building blocks. Your heart, for instance, a large muscle, is about 70 percent protein (not counting the water)—in other words, 70 percent amino acids.

As far as scientists knew, however, these organic molecules could be made only by living things. You won't find them in Earth's rocks, no matter how long you search—except in sedimentary rocks, like coal, that were created from organic matter. That posed an obstacle to explaining the origin of life, to put it mildly. You couldn't very well understand its appearance if you didn't know where its building blocks came from. Scientists were baffled. The gulf between the inorganic molecules in dead rocks and the complex organic ones in life was as problematic to scientists then as explaining how the molecules in our brains create consciousness is today. Many believed that organic molecules could be created only by a "vital spark"—an inexplicable force found only in living organisms.

When I was a student, I always thought vitalism was ridiculous. How could any scientist put stock in it? But it's easier to understand if you walk in scientists' shoes. As far back as Aristotle, many great thinkers believed in a form of vitalism. If you had no theory of how simple molecules became organic ones, no powerful electron microscopes to visualize cells or the structures within them, and no idea how heredity was transmitted, then the leap from dead chemicals to living creatures might appear magical. Consider this: If you break a stone in half, nothing further will happen to either piece. If you slice a planarian flatworm in half, both sections will regenerate into identical wholes. How do you explain that? "In living Nature, the elements seem to obey different laws than they do in the dead," wrote the eighteenth-century Swedish chemist Jöns Berzelius. Inanimate matter appeared to lack a life energy. The brilliant nineteenth-century physicist Lord Kelvin (also known for opining that heavier-than-air flying machines could never be possible) wrote, "Dead matter cannot become living without coming under the influence of matter previously alive. This seems to me as sure a teaching of science as the law of gravitation." In the twentieth century, Niels Bohr, a founder of quantum physics, speculated that we might need to discover new types of physical phenomena to understand life. Even Darwin himself, who had shown how new species arose, was at a loss to explain how the first life sprang from a pool of chemicals. "It is mere rubbish thinking at present of the origin of life," he wrote to the botanist Joseph Hooker. "One might as well think of the origin of matter."

Many nineteenth-century scientists were so frustrated that they punted. Lord Kelvin's solution was to propose that the universe and life had always existed. The famed scientist and philosopher Hermann von Helmholtz was of the same opinion. Life was immemorial, they believed—as old as matter itself. It must have existed elsewhere in the universe long before it turned up on Earth. How it found its way here remained a mystery, although they speculated that it might have hitched a ride on meteors or comets. "Who knows," Helmholtz argued, "whether these bodies, which everywhere swarm through space, do not scatter germs of life wherever there is a new world?" But the theory of panspermia (meaning "seeds everywhere"), which Kelvin, Helmholtz, and others proposed, merely kicked the can down the road. It did nothing to help unravel the mystery of life's origins.

In 1922, some years after Oparin's rejection by the censorship board, he was working in a Moscow laboratory with his Bolshevik hero, Alexei Bakh. He also received a teaching appointment. He would long be remembered for the imposing, incongruous figure he cut at the university. He had been sent abroad to study briefly, and in contrast to his students' worn, shabby clothes, he wore a sharp European suit, always with a bow tie, which bestowed a note of elegance and authority. Living conditions were tough in the new worker's paradise. The economy was in tatters, and many in Moscow were starving. Oparin began applying his biochemical knowledge to improve the production of bread and tea.

Even in this time of great need, though, he couldn't shake his fascination with deeper scientific questions. He too recognized that Darwin's masterpiece, *On the Origin of Species*, was "missing its very first chapter," but Oparin thought something could be done about it. He decided to return to first principles. Was it really possible that organic molecules could be made only by living organisms? If so, then the very first cell—the first membrane-enclosed collection of molecules able to produce energy and replicate—must have been so fantastically sophisticated that it could also manufacture the very materials of which it was made. Clearly, this was much too great an evolutionary jump to contemplate. To Oparin, it made much more sense to presume that the first cell arose from organic molecules that already existed around it. But where did they come from?

He already knew one fact that makes the origin of life appear deceptively simple. Nineteenth-century chemists had already established that, despite the large number of elements in the periodic table, almost all of our mass comes from just six of them: carbon, hydrogen, oxygen, nitrogen, sulfur, and phosphorus.

Your fats and carbohydrates are chains of molecules made exclusively of carbon, hydrogen, and oxygen. Your proteins are built from carbon, hydrogen, oxygen, nitrogen, and sulfur. And your DNA is made only of carbon, hydrogen, oxygen, nitrogen, and phosphorus. Those six elements compose roughly 99 percent of everything within you. A 150-pound person contains by mass 94 pounds of oxygen, 35 pounds of carbon, 15 pounds of hydrogen, 4 pounds of nitrogen, almost 2 pounds of phosphorus, and a half pound of sulfur.

Those six elements also happen to be among the most plentiful in the universe. Hydrogen is the most abundant of all; oxygen is third; carbon, sixth; nitrogen, thirteenth; sulfur, sixteenth; and phosphorus, nineteenth. In one sense, that makes understanding the origin of life a game of chemical Scrabble. You simply have to explain how those few elements combined to make organic molecules.

Of course, that turns out to be devilishly difficult. Atoms are picky about whom they bond with. And the number of potential combinations of these six elements is mind-boggling. Carbon is so promiscuous, so talented at contorting and bonding, that there are more than ten million organic molecules known on Earth.

In 1924, in a Red Russia now eager to convince the populace that God did not exist, Moscow Worker published Oparin's seventy-one-page manuscript as a pamphlet, with "Proletarians of the World Unite!" splashed across its front cover. Twelve years later, Oparin published a book that expanded his argument and incorporated more recent science.

Oparin's first groundbreaking insight was that, to understand how life first arose, he needed a clear picture of the Earth billions of years ago. Curiously, almost no one thinking about life had considered that before. After reviewing the latest findings in astronomy and geology, he realized that when Earth first formed, it looked nothing like it does today.

Most important was what it lacked. Many scientists assumed that oxygen had always been present, but Oparin understood that the oxygen in our atmosphere was produced by photosynthesis. Our atmosphere had no oxygen before life arose. You and I couldn't have survived there for a second.

He claimed that the Earth's early atmosphere was more like Jupiter's, which astronomers had just discovered was full of ammonia and methane. Remarkably, from basic ingredients—simple hydrocarbons, such as methane (CH_4), along with ammonia (NH_4), hydrogen (H_2), and water (H_2O)—Oparin outlined on paper a detailed series of chemical reactions that might create more complex organic molecules, proteins, and life. Life, he argued, could be understood as the culmination of chemical evolution. Modestly, he titled the book *The Origin of Life*, a fitting name for a prequel to Darwin's *On the Origin of Species*.

What did that first life look like? Some of Oparin's contemporaries claimed that it was photosynthesizing algae. To Oparin, that was patently impossible. As a plant biochemist, he had a healthy appreciation for the complexity of photosynthesis. There was no way that the first organisms to evolve were already so sophisticated; that was too much of an evolutionary leap. Instead, for the first life-form, he nominated clusters of organic molecules in the ocean that slowly evolved into bacteria.

In England, the flamboyant freethinking evolutionary biologist, biochemist, mathematician, and prolific author J.B.S. Haldane developed a similar theory independently, which appeared in a journal, the *Rationalist Annual*. Many scientists dismissed it at first as "wild speculation," and Haldane largely moved on to other weighty matters. But Oparin continued to work on the origin of life for the rest of his career.

He rose to great prominence in Soviet science and was awarded a chestful of medals, including the Hero of Socialist Labor, the Order of the Red Banner of Labor, and the highest civilian honor of all, the Order of Lenin. In later years, when he visited the West, he would be showered with accolades there too.

However, his reputation would be tarnished when another side of his rise to the height of Soviet science came to light. In the 1940s, he allied himself with the power-hungry biologist Trofim Lysenko, whose hopelessly flawed Marxist theory of heredity fell into favor with Stalin. Lysenko claimed that the traits of plants, like those of people, are shaped by their environment, not by "genes," whose existence he denied. He ruthlessly persecuted anyone who subscribed to Mendel's rival theory of genetics. Many who refused to fall in line lost their jobs, were shipped off to Siberia, or were eliminated. Regardless, Oparin was a supporter of Lysenko, and a friend. They even had neighboring vacation dachas.

Years later, the writer Loren Graham confronted Oparin about his support for Lysenko. "If you had been here in those years," Oparin replied, "would you have had the courage to speak out and be imprisoned in Siberia?"

However opportunistic Oparin was as he maneuvered to keep his position, and head, in Stalin's murderous Russia, his contribution to science was not just groundbreaking—it set off a scientific explosion.

Oparin had laid out a theoretical framework for investigating the origin of life, but no one had tried to test it. The techniques available for detecting organic chemicals were limited. Then in 1951, a nervy, ambitious American graduate student named Stanley Miller arrived at the University of Chicago.

The University of Chicago was a scientific powerhouse where many eminent scientists had come to work on the atomic bomb and stayed. As luck would have it, during his very first semester, Miller attended a lecture by the renowned chemist Harold Urey. Urey had won a Nobel

Prize for his discovery of deuterium (the same hydrogen isotope that was being manufactured as fuel for the hydrogen bomb and that Dave Jewitt and his colleagues would later detect in comets). Urey had directed the program for separating uranium isotopes for the Manhattan Project, but haunted by the horrors of the atomic bomb, he opposed the further use of nuclear weapons. "All the scientists I know are frightened—frightened for their lives—and frightened for your life," he wrote in *Collier's* magazine. He acquired a voluminous FBI file for his "leftist" views, and he switched to a field of life more suitable for a pacifist—the chemistry of the planets, the Moon, and the Earth.

As it happened, Urey's views on the composition of the atmosphere of the early Earth were quite similar to Oparin's. In the lecture that Miller attended, Urey remarked offhandedly that someday someone should try testing Oparin's theory. Miller took note, but he put it aside. As a new PhD student, he was looking around for a problem for his doctoral thesis, but his undergraduate research had convinced him that he should avoid experiments at all costs. They were messy, time-consuming, and less important than theoretical work, he thought. Instead, he eagerly accepted a plum opportunity to work out how the elements were manufactured in stars—under the controversial physicist Edward Teller, "the father of the hydrogen bomb." Just six months later, however, Teller decamped for the Lawrence Livermore Laboratory in California to develop new nuclear weapons, leaving Miller in the lurch. In retrospect, it was a lucky break for Miller. The physicist Fred Hoyle and his colleagues had a head start, and would have scooped Miller a few years later, when they spectacularly worked out the details of how the elements are produced in stars.

With Teller gone, Miller was back to square one. Casting about for a new topic, he recalled Urey's lecture and asked if he could work with him to test Oparin's theory. Miller proposed that they simulate the first atmosphere on Earth to see if it would actually cook up organic molecules as Oparin claimed it would.

Urey was dubious. "The first thing he tried to do was talk me out of it," Miller recalled. It seemed too risky. Miller was more than a year into his studies and needed to complete his PhD as swiftly as possible, yet this might tie him up in messy inconclusive experiments for many years. How could a process that many assumed took billions of years to unfold yield its secrets in just twelve months? But Miller persisted. The payoff seemed too good not to give it a try. He had no interest in a safer, more boring topic. Finally, Urey relented—but only on one condition. He gave Miller just six months to a year. After that, Urey told him, he would have to pick a new topic and start all over.

In his "dungeon" basement laboratory, Miller set about re-creating the first atmosphere on Earth. For funding, Urey diverted small bits of money from other projects, a practice common then, as now, for scientists committed to experiments others consider foolhardy. Urey and Oparin envisioned an ancient world in which massive dark clouds, fueled by volcanic eruptions and pierced by bright lightning, rolled over a vast ocean. Miller simulated this with a tangle of glass that would have looked at home in the laboratory of a mad scientist. He connected his "ocean," a large round flask partially filled with water, to his "atmosphere," another flask filled with hydrogen, methane, and ammonia gases (H_2, CH_4, and NH_3). A small flame beneath the ocean would produce water vapor (H_2O) that would rise into the atmosphere. And a condenser in the atmosphere would turn the vapor back into "rain" that would return, via glass tubing, to the ocean. Urey gave a tour of Miller's lab to a young undergraduate named Carl Sagan, who left impressed and excited. The idea that complex organic molecules would simply assemble themselves seemed so unlikely, but Miller was going to give it a try.

Late one evening in the fall of 1952, Miller was ready. He warned his lab mates, who—wisely—promptly cleared out. He lit a low gas flame beneath his ocean to begin creating water vapor. Then, because safety was not high on his agenda, he prepared the coup de grace, which he hoped wouldn't blow up in his face—literally. He would simulate lightning in his atmosphere by routing a 60,000-volt electric current between two electrodes, creating an undulating spark that would have made Dr. Frankenstein feel right at home. It should work fine: if, that is, he successfully evacuated all the explosive oxygen from his glassware before adding the gases. And if there were no leaks. If there were, the volatile mixture of gases and air would detonate like a bomb.

Miller turned the spark on. There was no explosion. Relieved, he watched the flasks for a few hours, then left for the night.

The gods smiled on him. Within two days, the water in his "ocean" turned yellow, and he saw black scum on the walls of the flask by the electrodes. Too excited to let the experiment continue, he stopped it to analyze the water. He was astounded to find that he had made glycine (NH_2-CH_2-$COOH$), the simplest amino acid in our bodies. Miller was ecstatic. His ancient atmosphere had spontaneously assembled a molecule that is a building block of our proteins, acts in our brains as a neurotransmitter, and composes a third of the collagen fibers that hold our bones, skin, muscles, and tissues together.

Excited, he ran the experiment again. This time he cranked up the heat under his "ocean" to simulate an atmosphere that volcanoes filled with more water vapor. And he resolved to let it run for a week.

Miller watched in suspense as, day by day, the water in his ocean turned pink, deep red, and then yellow-brown, while a black oily substance dripped from the electrodes. A fellow graduate student, Gerald Wasserburg (who, as we saw, later dated the Moon rocks), was unimpressed. "It looks like fly shit," he said, slyly implying that Miller had not cleaned his flasks properly. But flies had nothing to do with it.

Just three and a half months after beginning his project, Miller completed his analysis. "He was three feet off the floor," his brother recalled. It was as if he had closed the door on a pile of lumber and nails in a garage and returned to find a new table and chairs inside. He had created a number of organic molecules, including two of the twenty amino acids that our cells link to make proteins, and perhaps several others. Years later, with more sensitive instruments, he would detect at least eight more. With energy from sparks, organic compounds had assembled themselves. And these amino acids were just the kind that Oparin predicted would first arise on Earth.

Urey was bowled over. He encouraged Miller to swiftly write up his results, and in a gesture recounted in virtually every retelling of this story, he gave Miller sole authorship of the paper. That noble offer was a reflection of Urey's generous nature—as well as the fact that he could afford to be magnanimous. He already had his Swedish prize.

As a Nobel laureate, Urey had no difficulty phoning the editor of *Science* to ask him to rush Miller's paper into print. Still, the process was slow. One scientist who was asked by the editor to review the paper found Miller's results so improbable he didn't even bother sending comments back. Miller grew increasingly alarmed that publication was taking so long. Fearing he might be scooped, he withdrew it from *Science* and sent it to a less prominent journal. He agreed to resubmit it to *Science* only after the editor personally wrote to assure him that publication was imminent. His results were so unexpected, Miller said later, that "if I'd submitted it to *Science* on my own, it would still be on the bottom of the pile."

That spring, in a packed large lecture hall, the twenty-three-year-old nervously presented his findings to the University of Chicago's most illustrious scientists, several Nobel Prize winners among them. Carl Sagan, one of the undergraduates in the audience, was shocked by their first reaction. "They didn't take it seriously," he wrote later. "They kept suggesting that he had been sloppy, leaving amino acids all over his laboratory." Even Oparin did not believe it when a colleague came rushing in with the news. Miller's results just seemed so unlikely.

As others have recognized, 1953 was an annus mirabilis for biology. Jonas Salk announced that he had created a vaccine for polio; a neurosurgeon identified the hippocampus as the brain's critical region for memory formation; sperm was successfully frozen and revived; and Miller's paper, "A Production of Amino Acids Under Possible Primitive Earth Conditions," was published just a few weeks after Watson and Crick revealed the structure of DNA.

Miller's paper ignited the public's imagination. It immediately became the most famous experiment in all of biology. The experiment,
which Miller was fond of saying even a high school student could
do, suggested that the very first organic compounds on Earth—our
molecular ancestors—were effortlessly brewed in a prebiotic soup. "If
God did not do it this way, then he missed a good bet," Urey told his
university colleagues. It seemed so easy.

Just four years later, Oparin invited Miller to Moscow to address
the First International Conference on the Origin of Life. Oparin had
once written, "The road ahead is hard and long but, without doubt,
it leads to the ultimate knowledge of the nature of life. The artificial
building or synthesis of living things is a very remote, but not an un-
attainable goal, along this road." In 1936, this seemed like wishful
thinking. No longer. Scores of scientists were now assembling glass
tubes and flasks on their workbenches, with the intoxicating prospect in the air that they might uncover the secret of how life began.
Among them, the young Carl Sagan, who was now convinced that
science was on the verge of discovering not just how life emerged on
Earth but also how it arose throughout the universe.

Alas, in the decade that followed Miller's breakthrough, scientists'
sky-high expectations began fluttering back to Earth. To their disappointment, brewing the elements of life turned out to be much harder
than they had hoped. They could manage to brew only about half of
the twenty amino acids that link up to make proteins in all living
organisms. Even more troubling, they had great difficulty cooking up
and linking the molecules called nucleotides that are the basic units
of DNA and RNA. No one could see how to make them with the
ingredients present on the early Earth.

Miller would continue trying to synthesize organic molecules from
atmospheric gases, but in the 1960s, the rug was pulled out from under him. New evidence suggested that the Earth's first atmosphere
was *not* full of hydrogen, methane, and ammonia, as Miller, Urey, and
Oparin believed. They had reasoned that these gases formed because
hydrogen, the most common element in the universe, was present in

spades. But later researchers realized that light hydrogen would have drifted off, while other gases would have been blasted away by bombarding asteroids and UV rays. Earth's first atmosphere, therefore, came from volcanic gases: primarily nitrogen, carbon dioxide, and water vapor. Unfortunately for Miller, with a bit of heat, these generate smog, not the building blocks of life. So as intriguing as Miller's experiment was, some soon claimed it revealed nothing about life's actual origins.

Miller, the "father of prebiotic chemistry," held fast to his theory, but he grew somewhat embittered as he saw others abandoning it. Many of those who had plunged into the new field were now feeling adrift. They needed some other way to understand how life's building blocks appeared on our planet. They needed new ideas to reinvigorate their quest, new hope.

It would come unexpectedly, from an outsider.

If organic molecules were not created in our atmosphere, then where in the world were they made? By the mid-1960s, there was one location that most scientists agreed was the last place it made sense to look for them—the vast regions of outer space. The simple reason is that space is so harsh. It's shot through with energetic ultraviolet radiation, X-rays, gamma rays, and various injurious particles from the Sun and stars, all of which are more than happy to make short work of fragile organic molecules.

We are able to survive here on Earth only because we are sheltered by two enormous protective canopies. The first one, the magnetosphere—a vast magnetic field generated by the Earth's iron core—wraps around the planet and deflects dangerous subatomic particles, which we call cosmic rays. The second, a layer of ozone high in our atmosphere (made of the oxygen molecule O_3), absorbs damaging ultraviolet radiation. Our cells have also evolved clever mechanisms to reverse many of the worst blows from these ultraviolet rays. In each of our skin cells, hundreds of thousands of enzymes swarm over our chromosomes like worker ants, repairing DNA chains wherever they're broken. DNA damage in our skin triggers chemical messages, signaling that it's time to produce the chemical melanin, which can harmlessly absorb UV rays. If your skin tans in the summer, that's a sign that your DNA has suffered damage and your body is trying to prevent more (which is why doctors and parents harp so often on the need for sunscreen).

Despite the known harshness of space, by the 1950s, scientists had already searched for a few simple molecules there. That was because two Dutch astrophysicists, Jan Oort (of Oort Cloud fame) and his colleague, Hendrik van de Hulst, recognized an astounding fact. Though you might not expect it, every type of molecule emits a unique radio wavelength. When molecules collide, that sets their atoms vibrating and rotating. And because atoms are so tiny, and the forces binding them so springy, they wiggle back and forth *billions* of times a second, like a Slinky on steroids—creating tiny electromagnetic waves. (That's an incredible degree of springiness. To appreciate the enormousness of a billion, consider this: a million seconds adds up to eleven days, but a billion seconds adds up to thirty-two years.) It stood to reason, therefore, that a large cluster of molecules might emit a signal strong enough to be picked up by a radio telescope. In fact, by 1967, clouds of a few simple molecules made of just two atoms had been detected in deep space. But scientists knew that larger molecules could never survive those harsh conditions.

The physicist Charles Townes was less sure of what he knew. Townes, the son of a lawyer, had graduated college at age nineteen, earned a doctorate in physics, and held positions at Bell Labs and then Columbia University. Early one bright spring morning when he was thirty-five, he was sitting on a park bench when he had a "revelation," a moment he later likened to a religious experience. He realized how it might be possible to build an apparatus that amplifies the faint waves emitted by gas molecules. That led him to invent a device that inspired the laser, won him a Nobel Prize, and incidentally spurred him to think about how one could detect signals emitted by gas molecules in outer space.

As early as 1957, Townes had published a paper that described how a radio telescope might be able to detect the presence of some complex molecules in space—if, for some odd reason, they existed there. He even predicted their precise frequencies. As the years passed, he wondered why no one bothered to look for them.

He didn't know that a few young researchers had wanted to but had been talked out of it. One graduate student at Harvard, for instance, had been persuaded by a Nobel Prize winner that, if any larger molecules did manage to survive in space, they would be so few that they would be undetectable.

Townes was fifty years old in 1965 when he decided to switch fields. To stay intellectually sharp, he began reading astrophysics, took a few courses at Harvard (including one with Carl Sagan). Then he relocated to the sunnier climes of Berkeley, which had top-notch telescopes. There, the Nobel laureate, flush with research funds, was looking for new projects when he met a young electrical engineer named Jack Welch.

I asked Welch, who remained at Berkeley for most of his career, how they found each other. "When he came, he asked around, 'Is there anything interesting going on in radio astronomy?' And somebody said, 'Well, Jack Welch thinks he is going to look for molecules. I know it's kind of crazy.' So anyway, he came around to see me."

Welch was in the process of erecting a twenty-foot radio telescope at the university's Hat Creek Observatory, about three hundred miles to the northeast. He planned to use it to study the Earth's atmosphere, but he had other hopes, as well. A few years earlier, he'd come across Townes's paper. It inspired Welch to give a small presentation on how a radio telescope might detect any large molecules that were drifting between the stars. "When I gave this little talk to the astronomers," Welch said, "one of them later said to me, 'You know, your talk was kind of embarrassing. There was no way you could find a molecule of more than two atoms out in space.'" Welch laughed. "This was a really smart guy, but sometimes you are too smart for your own good."

Welch told Townes about the reception his talk received. Townes chuckled. Soon after, he told Welch that when he was a young professor conducting research at Columbia, the heavy hitters in his department descended on him to warn that he was wasting his time, and the university's money. "You know it's not going to work," they told him. "We know it's not going to work. You're wasting money. Just stop!" Townes stuck to his guns. "They walked out kind of angrily. They couldn't stop me," he said. That research won him the Nobel Prize. He told Welch, "Don't listen to people who think they know what they're doing." Townes's unusual philosophy also sprang from his experience working on projects with teams of top-notch physicists and engineers. He had seen how experts could be blinded by their knowledge. Although they had a good grasp of what they knew, such as quantum physics or how amplifiers work, sometimes they lost sight of how much they didn't know. Some engineering outcomes appeared unthinkable because the experts were too sure that they already knew what was possible.

At Berkeley, Townes immediately asked Welch, "Are you interested in looking for molecules?" and offered to put up the cash to build a spectrometer for Welch's telescope. "I got the feeling that most of the Berkeley astronomers thought my idea was a little wild," Townes recalled with understatement.

He recruited a PhD student and a postdoc for their speculative project and decided to search for the four-atom molecule ammonia (NH_3) a precursor of organic molecules. They set to work building a device to amplify the precise radio frequency that Townes calculated ammonia would send out if it was broadcasting its existence from thousands of light-years away.

At Hat Creek, one night in the fall of 1968, they were finally ready to point their telescope at the sky. "The question was, where do you look?" Welch said. "No one had any idea." They decided to direct the telescope at the center of our galaxy.

No signal appeared. But they kept at it. A few days later, they redirected the telescope toward Sagittarius B2, a dust cloud a bit farther away, and there it was: a huge amount of ammonia floating in space, likely formed by collisions of hydrogen and nitrogen in the cloud.

Why had it been so easy to discover? And why had so many distinguished scientists been so spectacularly wrong? It simply hadn't occurred to the experts that molecular clouds could be so immense that the molecules in their interiors would be shielded from destructive UV rays. A few stray molecules would have a hard time surviving in space, but not huge numbers of them on dust grains in a cloud that was millions of miles across. The scientists had failed to recognize how little we knew. They fell prey to the "As an Expert, I've Lost Sight of How Much Is Still Unknown" bias.

The next year, Townes's team retuned their amplifier to search for water. This time, they didn't even travel to Hat Creek. They simply called the telescope operator to tell him how to hunt for it. "Boom, there it was," Welch recalled. As soon as the operator began his search, he found it.

Astronomers around the world raced to their telescopes. "By then, the whole radio astronomy community was on fire," Welch said. They would discover over two hundred organic molecules.

Many of these are surprisingly familiar. Their ranks include nail polish remover, or rather acetone (C_3H_6O), which you produce when you break down fat. There is cooking and heating gas, or methane (CH_4); the principal ingredient in vinegar—acetic acid ($C_2H_4O_2$); and formic acid (CH_2O_2), which sets off pain receptors in your skin if you touch a stinging nettle or are bitten by a black carpenter ant. When hydrogen chloride (HCl), another molecule found in space, meets water, they form hydrochloric acid, which our stomachs use to digest food. Clouds of formaldehyde (CH_2O) also float in space. We use it to preserve the dead, but you may not know that our bodies also produce about 1.5 ounces of it a day. We break it down to produce the chemical formate, which we use to make DNA and some amino acids. Pregnant women, in particular, need folic acid (a form of vitamin B_9) to make formaldehyde, which they use to make DNA's building blocks. (Exposure to other sources of formaldehyde, however, can damage DNA, so the compound is a double-edged sword.)

The most notorious of all organic molecules in space is hydrogen cyanide (HCN). Plants produce it naturally in the pits of fruits like cherries and peaches (about ten peach pits contain enough to do you in). The yellow-spotted millipede secretes HCN to keep predators from dining on it. And farmers began using it as an insecticide in the 1880s. If you breathe hydrogen cyanide, it will interfere with an enzyme that transports oxygen. Your red blood will turn purple, and you'll die from oxygen starvation. The Nazis used a form of hydrogen cyanide called Zyklon B during the Second World War to murder over a million people in gas chambers. Yet HCN is made of the life-friendly elements hydrogen, carbon, and nitrogen. And when hydrogen cyanide combines with hydrogen sulfide (another molecule found in space), they can produce some amino acids, precursors to fats, and a building block of RNA. Under the right conditions, hydrogen cyanide also produces adenine, a building block of DNA.

For scientists hip deep in a sticky quagmire as they tried to explain the origin of life, the possibility that space harbored a rich reservoir of organic molecules brought up an obvious question that raised new hope. If the first molecules of life weren't created in the atmosphere and ocean, as Stanley Miller had tried to prove, could they have been visitors from outer space?

As if to answer that question, at 10:45 on the morning of September 28, 1969, just a few months after Townes's team discovered ammonia in space, a blazing orange fireball streaked across the sky above the village of Murchison, Australia. "We heard this *ba-boom, ba-boom, ba-boom*," one woman said. Others recalled hearing a hissing sound like "truck tires on a wet pavement."

A space rock weighing more than 250 pounds had exploded over their heads, scattering meteorite fragments over five square miles. One fist-size chunk punched through the metal roof of a shed and landed on a pile of hay. It smelled like methylated spirits, a cleaning solvent. Villagers raced through pastures and fields to collect hundreds of fragments, which they sold to rock shops for ten dollars an ounce, a third the price of gold. Some pieces found their way to museums and universities. And a few ended up in the hands of the geochemist Keith Kvenvolden at California's NASA Ames Research Center.

"We were just so excited to have them," Kvenvolden told me. That's because, for over a century, many scientists had claimed to have found organic molecules in meteorites. Yet skepticism abounded. It was notoriously difficult to rule out contamination. "One investigator pointed out," Kvenvolden recalled, "that the distribution of amino acids that were being found in some samples were very reminiscent of those you would find in fingerprints. And the conclusion was that what we thought was extraterrestrial life was actually a little pitter-patter of human fingers." In the early 1960s, scientists even battled over evidence that meteorites contained life itself. One microbiologist actually cultivated living bacteria from a space rock that he suspected were of alien origin. But, he admitted, he couldn't rule contamination out. Other investigators discovered tiny "microfossils" of alien life in meteorites, some of which turned out to be New York City ragweed.

In 1969, when fragments of the Murchison meteorite arrived at NASA Ames, Kvenvolden was in the midst of searching for organic compounds in the first rocks just brought back from the Moon. They turned out to be a disappointment. The Moon rocks harbored no hints of organics, other than minute traces of methane gas. However, as their new laboratory had been carefully designed to avoid any contamination of the lunar samples, Kvenvolden and his colleagues now had an unimpeachably clean state-of-the-art facility to analyze meteorites. And they had just been handed pristine fragments—not musty rocks that had weighed down museum shelves for years. They had high hopes they would find something new.

Dressed in white cleanroom garb, they selected the largest chunks with the fewest cracks. The rocks were smooth, and black as night from the heat of the explosion. Encouragingly, their interiors were also black, a telltale sign of carbon.

The group's chemical analysis revealed that about 2.5 percent of the meteorite was organic. Nothing suggested that it contained living organisms or that the molecules had been created by life. But remarkably, the wide assortment of molecules included amino acids. Many of these were forms not found on Earth, a reassuring sign that they hadn't contaminated their samples. "It was one of those things as a scientist that you dream about," Kvenvolden said. "It was probably the most exciting time in my life. When you make a discovery, and you realize that you and your team are the only ones that know of this result. It is an experience that is almost incomparable." Just as astonishing, they detected seven of the twenty amino acids that our bodies use to build proteins and enzymes; later they found two more.

Who knew that we had so much in common with meteorites? That rocks whizzing around in far-off deep space have many molecules that we can't live without? They include the amino acid valine, which helps modulate our brain's serotonin levels and supply our muscles with glucose; aspartic acid, an excitatory neurotransmitter that also plays a role in making testosterone and other hormones; and glutamic acid, the most common excitatory neurotransmitter in our brain, which is found in over 80 percent of our synapses. It helps us learn and create memories. We can also be grateful to glutamic acid for umami—our fifth taste, after salty, sweet, sour, and bitter. We savor it in foods like soy sauce and cheese, not to mention the controversial but tasty additive to foods monosodium glutamate (MSG). A bit of meteorite on your breakfast toast, anyone?

Astonishingly, many of the amino acids that Kvenvolden detected were the very same ones that Stanley Miller brewed in his laboratory experiments. So, the reactions that Miller, Urey, and Oparin believed had once occurred on the ancient Earth were certainly going on in space—most likely when rocks containing ice were heated by collisions or radioactive decay. Since then, others studying fragments of the Murchison meteorite have found two nucleotides—building blocks of DNA—that Miller and others were unable to coax out of their benchtop experiments. More sensitive instruments have detected tens of thousands of other kinds of organic molecules. Scientists estimate that the meteorite could contain very faint traces of millions more. Many meteorites, made principally of rock, metal, or both, don't contain organics, but the Murchison belonged to a particular type (called a carbonaceous chondrite) that can be rich in them.

Once it was clear that space rocks contained organic compounds, researchers also began searching for them in icy comets. The satellites they sent out to meet them revealed that a comet harbored even higher concentrations, perhaps up to 20 percent of its mass.

That was thrilling news, and led researchers to wonder—even if Kelvin, Helmholtz, and Hoyle were wrong, and life itself didn't arrive from space—did the first organic molecules, just like water, come from far away?

This seemed possible, even likely to many, but only if they could resolve a vexing problem. The clouds of organic molecules that astronomers had been overjoyed to find drifting between the stars were trillions of miles away, far too distant to have traveled here. Small amounts of organics might have survived here and there in little rocks like the Murchison meteorite. But these scattered traces would hardly be enough to create life. And you would think that the large quantities of fragile organics hitching a ride on massive comets or asteroids, traveling at 38,000 miles an hour, would have met their deaths in the hot molten magma and superheated gas that instantly materialized when their journeys came to a sudden screeching halt. So a nettlesome question remained—how could enough organic molecules from space have possibly survived a trip to Earth?

In 1992, the astrophysicists Chris Chyba and Carl Sagan pulled a rabbit out of a hat with a potential answer. Although they're too small for us to see, meteorites and comets cast off tiny particles of dust that constantly rain down on every part of Earth. Scientists have collected falling space dust at 65,000 feet in the air—in specially designed trays mounted beneath the wings of retired U-2 spy planes. These invisible specks—called interplanetary dust particles—are too small to plummet swiftly through the Earth's atmosphere. Instead, they parachute gently down without burning up. And they carry with them tiny bits of organics. Something on the order of forty thousand tons of cosmic dust salts the Earth every year. Over a period of several hundred million years, that would have added up to an awful lot. The amount of interplanetary dust that Chyba and Sagan estimated fell on the young Earth would have amounted to ten to a thousand times the mass of all organic material found in life today.

Others have proposed a different way that organics from space could have found their way here. It's a "glass half empty, glass half full" kind of theory. While a violent impact probably demolished any organics in a large asteroid or comet, their molecular fragments might have recombined to form new ones in the searing heat. Cosmic bodies bring forth organic molecules, and cosmic bodies taketh away. Some laboratory experiments seem to back up this theory of organic give-and-take. If not interplanetary dust, then large impacts might have seeded our planet with the molecules of life.

Has the problem been solved? You can sense a palpable excitement when some scientists discuss the possibility, even the probability, that life was first created from organic molecules that fell from the sky. If so, then other planets were also seeded by organics, making the existence of life elsewhere more likely. Nonetheless, their enthusiasm obscures a stubborn problem. Even if potential building blocks of life arrived here intact, that doesn't prove that they were the very same ones that actually created life. The suggestion that our organic chemical ancestors crash-landed and hatched life here remains a tantalizing but unproven conjecture.

So where does that leave us? What we can say with certainty is that organic molecules are ubiquitous; the cosmos is sprinkled with them. If you press some researchers, they'll tell you that life's first building blocks, our most distant organic forebears, may have come from multiple sources. It's possible that some may have arrived from space on asteroids, comets, and space dust. As we'll see, others may have been home-brewed right here on Earth. Indeed, researchers have proposed that they could have been created in volcanic plumes or hot geysers; in deep-sea vents; in the cracks between continental plates where new ocean floor is made; or even in asteroids' impact craters, which could have been warm incubators for thousands of years. Many of the organic molecules within us are just not very difficult to make.

What researchers do agree on is that, once organic molecules in the right mix were abundant in water, they encountered new types of molecules, whose attractive forces snapped their atoms into new configurations, and new compounds swiftly appeared. Then structures emerged that began to resemble life.

How that seemingly miraculous event occurred is one of the greatest mysteries in all of the universe. That befuddling question provokes not just intense controversy, but some of the fiercest sparring with sharp elbows in all of science.

THE GREATEST MYSTERY

The Puzzling Origin of the First Cells

Life is a cosmic imperative.
—*Christian de Duve*

If you were to hold a family reunion and invite everyone who is even remotely related to you, and if you did not discriminate on the basis of species, you would have to provide seating for . . . well, about a quintillion guests, the vast majority of whom would be bacteria. We can thank Darwin for that insight—the realization that all of life is linked by a single vast family tree. Every organism that ever existed on Earth descended from another. We can trace that continuity along thin chains of DNA to an ancient lineage. The routes that your molecules took to find you were blazed by pioneering generations of earlier creatures. But what was the origin of that very first cell, the most ancient grandmother of us all, who gave rise to life's stunning complexity? How did molecules on Earth conspire to create the most basic unit of life: a self-sustaining, replicating cell? There are few fields of scientific inquiry in which opinions are so varied and debates so contentious.

As we saw, the search for the origin of life received a stunning boost in 1953 when Stanley Miller discovered that he could easily cook up some amino acids from simple gases and sparks. Alas, the ease of his success turned out to be misleading. Miller's techniques could not make all of the amino acids found in life. Nor, as we saw, did the early Earth's atmosphere contain all the ingredients his recipe required, and making DNA's basic units—fragile nucleotides—appeared more difficult still.

In fact, a decade after James Watson and Francis Crick discovered the structure of DNA, some researchers had become so depressed by the complexity of the chemistry needed to explain the origin of life, they just gave up. It simply wasn't obvious which molecules one would start with, or how they could have been assembled. Like a crime scene thousands of years after a crime, most of the evidence had been wiped clean. It hardly seemed that it could ever be solved.

It was in those dispiriting days that Alec Bangham, a researcher in England, found one small part of the puzzle. Sturdy, broad-faced, with an unruly shock of brown hair, Bangham had a contagious scientific enthusiasm. He never tested well in grade school. His parents often saw "could do better" on his report cards. He failed his qualifying exams twice before gaining admission to medical school. But he possessed an exuberant curiosity that after some years compelled him to "renege," as he put it, on his duties as a pathologist. He turned instead to research at the Institute of Animal Physiology near Cambridge. There Bangham, who was trained as a blood specialist, began thinking about puzzles like why red blood cells don't stick together as other cells do. How do they remain independent? That question, in turn, compelled him to investigate the properties of cell membranes.

In 1961, the institute acquired an electron microscope, which were just becoming widely available, and Bangham took it on a test run. Looking around his office for something to examine, he settled on a lipid, a fat called lecithin, found in cell membranes. If you add lecithin to water, it has the captivating property of forming globules like the mesmerizing bubbles in lava lamps. Bangham decided to examine these globules with his powerful new microscope. Peering at the green glowing display in his darkened laboratory, he was amazed to see that the globules were composed of tiny microspheres with thin walls. They looked suspiciously like cell membranes.

Bangham was electrified. No one had any idea how membranes evolved—how the very first cells learned to build floppy globes around themselves. Now it appeared obvious. Membranes build themselves. They arise because one end of a lipid is attracted to water and the other end is repelled by it. So if you place lipids in water, their water-hating ends swiftly pivot like bar magnets to face each other in defensive postures, while their water-loving ends face out. The water-hating ends protect themselves even more by lining up next to pairs of others. Their attractions and repulsions instantly create dense arrays—globes of lipids two molecules thick with the water-hating end of each hiding in the middle. That's what the membranes around our cells look like, Bangham realized. They are globes of fat, with walls just two molecules wide, arrayed back-to-back in formation to make their water-hating and water-loving ends happy.

"Membranes came first," Bangham was fond of saying. Making them was so easy, they must have been the first part of a cell to evolve. In a single stroke, he had made it simpler to imagine the origin of life. It was clear that when the right ingredients were present, some structures could swiftly assemble themselves.

Yet while building a membrane for the first cell appeared so straightforward, making everything inside the cell did not. A cell, made of organic molecules, is constantly taking in new materials in order to generate energy, build structures, and reproduce. The molecule that carries the instructions for doing all of this is DNA. (We'll see how James Watson, Francis Crick, and Rosalind Franklin discovered its structure in a later chapter.) DNA tells the cell which proteins to make—and proteins in turn do all the rest of the work in the cell.

A confounding chicken-and-egg problem now stopped scientists dead in their tracks and tied them up in knots. Which came first, DNA or proteins? The problem was this: since DNA carries the instructions for replication, which obviously is essential for life, you would think that it must have evolved before anything else. Yet DNA is built by proteins, and that creates a dizzying circle. DNA contains the instructions to make proteins, and proteins make DNA. You can't have one without the other. So how could either arise? It makes one wonder if life is possible at all.

In the mid-1960s, three of biology's heavy hitters came to the rescue. Carl Woese, Leslie Orgel, and Francis Crick hit upon the same solution independently. They proposed that the very first cell was not created around DNA, but around its kid brother RNA, which, crucially, could also replicate.

Until then, RNA had seemed less important, second fiddle, because DNA is so much longer and contains so much more information. RNA appeared to be a mere intermediary. That's because DNA contains all of your hereditary information in a staggeringly long sequence of three billion nucleotides. In contrast, an RNA molecule is a copy of just one gene. It's a tiny portion of DNA on the order of only a thousand nucleotides long. Moreover, once an RNA molecule is made in the nucleus, it travels to a chemical factory that translates its code into a sequence of amino acids to make a protein. But the cell doesn't even bother to keep its dull, faithful RNA molecules around. Once their proteins are no longer needed, the RNAs are destroyed. Their lives are "Greek tragedies," one scientist said, because their deaths are ordained at the moment of their birth.

Now Woese, Orgel, and Crick began to look at RNA with new respect. While DNA is a long double helix composed of two helixes joined in the center, RNA is just a single helix. This would have made it much easier to assemble. And they were intrigued by the recent discovery that RNA could contort itself into gnarly origami-like shapes—just like the crucial proteins, called enzymes, that bring molecules together and vastly accelerate chemical reactions. Enzymes make the reactions in our cells happen at insanely fast speeds—on the order of a hundred times a second instead of once every million to billion years. We could hardly exist without them. So the trio cut a Gordian knot by speculating that back in the day, when the first cells formed, our little RNA sidekicks were superheroes with dual powers. RNA carried the instructions for replication, the work now done by DNA, and it sped up reactions, the work now done by enzymes. It was a jack of all trades.

Many of those puzzling over the origin of life now breathed a sigh of relief. They could finally stop talking about chickens and eggs. This vastly simplified any scenario of how life evolved. The very first cell did not need either DNA or proteins. It was created around RNA. Only later did double-stranded DNA (which is almost a million times more stable) and enzymes (which are much more efficient) appear. In life's earliest days, RNA was no disposable assistant. It was the molecule that sparked life, the great-grandmother of us all.

Once again, though, any buzz of excitement soon wore off. No one had ever caught RNA acting like an enzyme by instigating and accelerating chemical reactions. Presumably it lost that role when more efficient enzymes made of proteins came along, making any speculation about RNA's once-mighty powers remain just that—speculation.

A decade later, in the late 1970s, Thomas Cech, a thirty-one-year-old assistant professor at the University of Colorado, was paying no attention to the origin of life. In these exciting times, genetic engineering was just coming of age and scientists were using it to rapidly uncover the intricate mechanisms that translate the genetic code. Cech had just received his first university appointment, and when he wasn't skiing crisp snowy trails in the Rocky Mountains, he was trying to understand the details of how RNA molecules are copied from strands of DNA. For convenience, he was working with the genes of the pond-dwelling protozoan *Tetrahymena thermophila* (a curious bug with seven sexes and as many genes as humans). Conveniently, these single-cell creatures bred fast, and it was easy for him to obtain certain types of RNA because they made lots of it.

Cech had discovered that, in order for a *Tetrahymena* to produce one particular strand of RNA, the cell first had to excise a short unneeded section of nucleotides in the middle of the RNA molecule. He decided to learn how the cell did this, so he began hunting for the enzyme that snipped out the extra sequence, completely unaware that it would lead to a breakthrough in the search for the origin of life. Frustratingly, every time he tried to isolate the full original strand of RNA, the extra piece in the middle was already missing. Cech decided that the enzyme that removes the unneeded section must be bound very tightly to RNA, so he and his coworkers searched for it again and again: with the same results. They could not find it, or figure out their mistake.

Baffled, over the course of a year, he returned to the problem with increased ferocity, as if his sanity depended on it. His new plan was to cripple the hidden enzyme before it could do its work. The team boiled the RNA. They added detergent. They introduced an enzyme that destroyed others. Yet they could not isolate the RNA molecule with its middle section still intact. "We were getting more and more desperate." Finally, as they cast about, Cech recalled, "we were driven almost by desperation to the opposite hypothesis." They wondered if the RNA molecule was doing all the work itself. Was it acrobatically contorting its shape to snip out its extra middle section, and then repatching itself? To test their strange theory, they made an artificial copy of the RNA that they knew had never been exposed to enzymes. "It immediately worked," Cech said. "It did the very same reaction that the RNA from *Tetrahymena* was doing. It was a 'thank God moment,' because we were out of alternatives. That was our one explanation." The RNA was acting like an enzyme. It was initiating and accelerating a chemical reaction.

Soon after publishing his results, he received an invitation to speak to the Origin of Life Club at UCLA. "I didn't even know what origin of life research was at the time," he recalled. "I had never thought much about it," he told me. "I was so disconnected from the origins-of-life community that I really didn't get it. The whole evening, I was talking one language about the chemical mechanism of the reaction, and they were interested in things that happened 3.8 or 3.9 *billion* years ago." He was surprised to learn that he had confirmed what the elder biologists Woese, Crick, and Orgel had suspected would be found long ago. "Unknown to us," Cech discovered, "there was a whole group of people out there who were just waiting for it. They sort of knew that this would come at some point if they would only live long enough." A year later, the Yale biochemist Sidney Altman announced that another RNA molecule was also acting like an enzyme. He and Cech would jointly win the Nobel Prize.

Since then, researchers have found more than a dozen types of RNA molecules that act like enzymes in cells. They're likely vestiges of RNA's role in running cellular life before our ancestors had proteins. In our bodies, the vitamins B_1 and riboflavin contain short units of RNA. Moreover, scientists unexpectedly discovered that long segments of RNA lie in the center of ribosomes, the cellular factories that make proteins. They may testify to the lives of our vanished molecular ancestors. We came from an "RNA world," as the molecular biologist Walter Gilbert famously put it, a bygone world where the first cells were run by RNA.

At this point in our story, perhaps the clouds should part and a shaft of brilliant sunlight suddenly appear. At last, scientists could finally get down to the business of explaining the origin of life. They simply needed to explain how the first RNA, or proto-RNA, evolved and how it was captured by a membrane to form the first cell. And they needed to work out how, as RNA replicated, small copying errors gradually created new kinds of molecules, including ultimately proteins, DNA, and our cellular machinery.

The gods of chemistry did not let scientists off so easily. Explaining how the first RNA molecule formed, not to mention how a cell evolved around it, remained remarkably difficult. A quarter century after Stanley Miller's initial groundbreaking experiment, many origin-of-life researchers felt like they were treading water.

Then came the discovery of a dreamlike alien world, and all kinds of new possibilities.

In February 1977, a scientific expedition passed through the Panama Canal and made its way to a destination in the Pacific Ocean 250 miles northeast of the Galápagos Islands. When they reached the designated spot, Jack Corliss, the chief scientist, saw nothing but sky and sparkling water in any direction. Corliss, a burly young Oregon State University geochemist, had three research vessels at his disposal. The first was the capacious 279-foot R/V *Knorr*, which was outfitted with multiple scientific laboratories, a kitchen, a dining room, a library, and a machine shop. The second was a large catamaran named *Lulu*, which was a platform for launching the third vessel—the famed *Alvin*. This was a twenty-three-foot-long submersible, from the Woods Hole Oceanographic Institute, that could tolerate bone-crushing pressures on the deep-sea floor. The geological expedition's crew included the undersea explorer Bob Ballard, who would later find the *Titanic*, and over twenty geologists and geophysicists, none of whom expected to make one of the most momentous breakthroughs in recent biology.

The National Science Foundation had funded their journey here, at considerable expense, so they could explore the ocean depths to shore up the still-controversial theory of continental drift. If the theory was correct, you would expect that, as vast tectonic plates spread apart beneath the ocean, water would seep into the newly formed cracks. Geologists surmised that the water would sink down until it reached hot magma below, at which point it would become superheated, and erupt back onto the ocean floor. The discovery of hot springs beneath the ocean would support the theory of plate tectonics, and it would also help explain how the Earth vented heat as it cooled down. Yet no one exploring the ocean depths had ever seen these supposed hot springs. Did they actually exist?

Corliss had sailed to this watery address because, a year earlier, geologists on a Scripps Institute of Oceanography vessel had towed a device along the ocean floor and detected a zone of warm water. Their camera revealed a desolate landscape—except for one spot where, curiously, it captured a cluster of massive empty clamshells. Could it be the site of a hot water vent? There was also a beer can, suggesting that the shells might simply be debris tossed overboard after a shipboard feast. They named the spot Clambake and marked it with transponders.

A year later, Corliss's team prepared to investigate. At dawn on February 17, Corliss, the geologist Tjeerd van Andel, and the pilot Jack Donnelly emptied their bladders, and then climbed down the narrow conning tower into the *Alvin*. They crouched by the small portholes and prepared for a descent of 1.7 miles in the titanium submersible, built to withstand pressures of nine thousand pounds per square inch. Through the thick glass, they saw the choppy water suddenly still around them. The light dimmed as the color of the water faded from blue green to dark blue, to darker blue, then pitch black. For an hour and a half, they saw nothing, just the occasional flitting flash of a ghostly bioluminescent creature.

At long last, they reached the bottom.

In the first moments, their searchlights illuminated only black flows of pillow lava, formed when cold seawater met molten rock. But then, as they approached the Clambake location, they saw something no one else had ever seen. Though the water nearby was 36 degrees Fahrenheit—almost freezing—here they saw cloudy blue water, shimmering with minerals, rising from the ocean floor. They would learn that the temperature in some places was a balmy 63 degrees, comfortable enough to enjoy without a wetsuit, if it weren't for the crushing pressure. They had for the first time found a hydrothermal vent.

Looking through a porthole, Corliss saw a sight that forever impressed itself in his memory. Using his acoustic phone, he called his graduate student Debbie Stakes on the *Lulu* above.

"Debra, isn't the deep ocean supposed to be like a desert?"

Stakes took a moment to consult her fellow geologists. "Yes," she replied.

"Well, there's all these animals down here," he said.

He was gazing at clams as wide as dinner plates, giant mussels, albino lobsters, and orange and white crabs.

It made no sense. He was over eight thousand feet down, cut off from all sunlight and food from above. Frantically, Corliss and van Andel scrambled to collect data and capture a few specimens with *Alvin*'s robotic arm.

The next dives revealed more vents, and creatures even more outlandish: spaghetti-like worms, large pink fish, and seven-foot tubeworms with red plumes swaying languidly like flowers. Back on the R/V *Knorr*, the scientists examined their finds with wonder. Kathy Crane, the expedition's navigator, radioed biologists at Woods Hole to ask for help in identifying the strange creatures. They couldn't.

The startled geologists had little to preserve them with, except a small jar of formaldehyde a graduate student had brought and some Russian vodka they'd purchased in Panama. They would have to store the creatures in Tupperware and plastic wrap. Sometime later, one of the expedition leaders received a message from Woods Hole: "RE-TURN TO PORT INSTANTLY. . . . BIOLOGISTS COMING." Needless to say, Corliss did no such thing. He had no intention of being scooped.

Gradually the atmosphere on board grew more exuberant, as the explorers realized that they'd discovered an oasis of unknown life-forms. Unlike animals and plants on the Earth's surface, these creatures were not dependent on sunlight and photosynthesis. Instead, they were sustained by minerals and heat from the Earth's depths. When researchers opened a collection jar in a laboratory on the R/V *Knorr*, they got a literal whiff of how it worked. The smell of rotten eggs sent them running to the portholes, as air-conditioning shared the noxious aroma with the rest of the crew. It was the smell of hydrogen sulfide. They soon realized they had found a unique ecosystem almost as foreign as life on Mars. "We all started jumping up and down," John Edmond recalled. "We were dancing off the walls. It was chaos. It was so completely new and unexpected, that everyone was fighting to dive."

Some years later, a microbiology graduate student at Harvard, Colleen Cavanaugh, collaborating with the Smithsonian's curator of worms, Meredith Jones, and the Woods Hole microbiologist Holger Jannasch, showed that bacteria on the ocean floor created energy and sugar with a process similar to photosynthesis. Instead of stealing energy from the Sun, they liberated it by breaking the bonds in hydrogen sulfide. With this energy, they combined carbon dioxide and water to create sugar, just as in photosynthesis. In short, the bacteria

made energy by eating hydrogen sulfide created when sea water interacted with hot magma below the vents. All the other creatures in the
eerie food chain down below depended on these lowly chemical eaters
for survival.

The unforgettable sight had a profound effect on Corliss. He changed
his field from geochemistry to biology. Together with graduate student Susan Hoffman and microbiologist John Baross, he advanced an
astonishing new theory: that our most ancient forebears—the very
first life—evolved at hydrothermal vents.

It was enough to turn our thinking about the origin of life on its
head. It had always been understood that life arose on the Earth's
surface. In Stanley Miller's scenario, lightning and ultraviolet rays
sparked the formation of organic molecules in the atmosphere. These
fell into the ocean or pools of water, creating a prebiotic soup from
which life emerged. But Corliss and his colleagues argued that life
evolved not on the Earth's surface, but at crushing pressures in the
darkest ocean depths.

Their new theory had obvious merits. For one thing, during the
era of the Late Heavy Bombardment, hundreds of millions of years
after the birth of our planet, massive asteroids and comets were pulverizing Earth's surface. The ocean depths had perhaps provided a
bomb shelter, a sanctuary from the devastating impacts above. Another discovery dramatically bolstered their case. More than a decade
earlier, biologists were astounded to discover microorganisms thriving
in ridiculously high temperatures, like Yellowstone's hot springs that
reach a blistering 163 degrees Fahrenheit. Intriguingly as well, the
most ancient genes that researchers have since traced back in time
belonged to an organism called LUCA (for Last Universal Common
Ancestor) that lived at high temperatures near deep-sea vents.

Nevertheless, when Corliss and his collaborators submitted their
paper, it was roundly rejected by the prestigious journals *Nature* and
Science. It finally found a home about a year later in an obscure periodical, *Oceanologica Acta*. Yet it lit a match. Their suggestion that life
arose at hydrothermal vents didn't simply smolder, it caught fire and

blazed through the scientific community. It offered an exhilarating new way to think about life's origins just when progress had stalled.

In 1979, the discovery of a new kind of deep-sea vent, a "black smoker," fanned the flames of excitement. These new vents were much hotter: about 650 degrees Fahrenheit. And they were huge. The top of one vent, named Godzilla, was forty feet across and towered fifteen stories above the ocean floor. It wasn't hard to imagine that these craggy hulking chimneylike towers with their commingling hot water, minerals, and dissolved gases could be bioreactors churning out organic molecules. Certainly, the profusion of life around them was stupefying. The population densities rivaled the richness of coral reefs.

But to Stanley Miller, the founder of the field, the notion that life could have evolved in the extreme temperatures of deep-sea vents seemed ludicrous. He tried to "head off" dissenters. Surely, he pointed out, if any fragile organic molecules form at vents, the intense heat there would swiftly have torn them apart. Molecules like RNA, amino acids, and sugars degrade at high temperatures. "The vents would have been important in the destruction, rather than the synthesis of organic compounds in the primitive oceans," Miller and his colleague Jeffrey Bada wrote. So began the battle between the "Millerites" and the "ventists."

Of course, you may recall that Miller's theory also had a potentially fatal flaw. Scientists had discovered that the Earth's early atmosphere didn't have the hydrogen, ammonia, and methane that Miller believed made the first amino acids, although Miller insisted that there must have been some places where these ingredients were available. That was why others began wondering instead if interplanetary dust or comet and meteorite impacts first brought organic molecules to Earth.

For some scientists, deep-sea vents now seemed an even more likely birthplace for life. The Earth's heat sent superheated water, rich with minerals and gases, bubbling up to the ocean floor. And a variety of chemicals could mingle at a range of temperatures around the vents, creating fertile conditions for organic compounds to evolve. Still, scientists remained hard-pressed to explain the particulars of how organic compounds, and life, arose.

It was at this point in the late 1980s that a patent lawyer and origin-of-life hobbyist named Günter Wächtershäuser stepped in and turned the alluring suggestion that life arose beneath the ocean into a detailed theory that was hard to ignore. Wächtershäuser was working full-time at his law practice in Munich. For fun, he decided to delve into the philosophy of science and evolution. "Ideas fly around, and they search for a brain, and they usually find the busiest brain," he explained. Describing a patent attorney as logical is like calling a monarch regal. It comes with the territory. Colleagues have called Wächtershäuser argumentative and combative, qualities that serve any lawyer well. He enjoyed the exercise of poking holes in patent applications. He also happened to have a PhD in organic chemistry, which he had abandoned for law more than two decades before. When he looked at current theories about the origin of life, he was thoroughly unimpressed.

Wächtershäuser was heavily influenced by the views of a friend he met at a scientific summer school in Austria. This was the renowned philosopher of science Karl Popper. Popper famously argued that a proper scientific theory must be falsifiable. That is, it should make predictions that, at least in principle, could be disproved by evidence. By that yardstick, Wächtershäuser didn't think that any prevailing theories on the origin of life were up to snuff. He was unmoved by experiments that mixed up batches of chemicals and added energy to see what turned up. It seemed that the supposedly essential ingredients of the "prebiotic soup" just kept changing, depending on which molecules scientists had been able brew that week.

At home, and in his legal office on the Tal, a street leading to Munich's medieval city gates, the careful attorney set out to produce a falsifiable theory of how life arose. In between lawsuits on patents for antibiotics and other disputes, he tried to work out the most likely reactions that could have created the molecules of life. He decided to look very carefully at which compounds were available just beneath the ocean floor.

Wächtershäuser concluded that this would have been the perfect cradle for life because it had all the right stuff. First, the hot water oozing from the depths of the Earth contained the precursors of organic molecules: gases like hydrogen sulfide, ammonia, carbon dioxide, and hydrogen cyanide. They were under pressure, which would help encourage reactions. And when he looked at the enzymes that accelerate reactions in bacteria and our own cells, he discovered something else. At their reactive cores lie metals like iron, nickel, zinc, and molybdenum. All would have been plentiful around the ocean floor—especially iron sulfide (FeS). Intriguingly, clusters of iron and sulfur atoms lie at the heart of many of our most critical enzymes and at the centers of mitochondria, the power plants in our cells that create energy. In fact, genetic defects that interfere with our ability to make clusters of iron sulfur cause heart disease and muscle weakness. Wächtershäuser wondered if it's just a fluke that the minerals found just beneath the ocean floor also happen to be vital to us, and all life.

Just as important, the surfaces of iron sulfide minerals produced there, known as fool's gold, were positively charged. That made them chemically sticky, so any organic molecules born there would have clung to these surfaces. They would have hung around and partied with others instead of drifting off.

To Wächtershäuser, the ocean floor seemed to be the birthplace of life. Under the right conditions, increasingly complex molecules appeared there that created a basic metabolism: a way of generating energy and processing chemicals to sustain life. He even predicted how amino acids, proteins, and RNA may have formed. Finally, he argued, as our molecular ancestors beneath the ocean grew more advanced, they were captured by membranes. Eventually, the intrepid cells that evolved there left their homes. In a nutshell, he proposed, we evolved on mineral surfaces like fool's gold beneath the ocean floor.

Wächtershäuser was reluctant to publish, however. He feared that, as an amateur, he would face ridicule. "I came as an outsider," he said wryly, "who was maybe even considered a lawyer, which is not a very positive term." But with encouragement from Popper and others, he worked up the nerve to write a paper. He didn't shy away from laying out the failures he saw in the theories of others. In fact, he delivered a forceful broadside from the get-go. "The prebiotic broth theory has received devastating criticism for being logically paradoxical, incompatible with thermodynamics, chemically and geochemically implausible, discontinuous with biology and chemistry, and experimentally refuted," he wrote in an opening salvo.

Not surprisingly, Stanley Miller and Jeffrey Bada were neither pleased nor won over by this assault. "Some Like It Hot, but Not the First Biomolecules," Bada titled a rebuttal in *Science*. They questioned why anyone would consider the theory in the first place. "The vent hypothesis is a real loser," Miller complained to a journalist. "I don't understand why we even have to discuss it." Wächtershäuser's model was "not relevant to the question of the origin of life as we know it," Bada said. It was "paper chemistry," an example of some researchers' penchant for writing down hypothetical chemical reactions and claiming that they had some relevance to the earliest life.

But Wächtershäuser, who had the measured calm of a lawyer about to eviscerate a witness, knew how to stand his ground. "As far as I'm concerned," he told a journalist, "the soup theory is more of a myth than a theory, because it doesn't explain anything." When we spoke many years later, he recalled these battles with surprising equanimity. "I was what you call a counterpuncher," he said. "Science is the field of controversy. If, in a scientific topic, there is no controversy, you have no science. So I wouldn't say that I've been treated badly. I know people have attacked me, but," he said, his tone rising with amusement, "consider what I am doing to *them*!"

Despite Miller's and Bada's efforts, there was no way for them to stop what they saw as "runaway enthusiasm for the vent hypothesis." Wächtershäuser's theory breathed new life into the search for our molecular ancestry. And it would spark another theory that appeared to reveal the very key to life's origins.

The geologist Mike Russell admired Wächtershäuser's scheme, but thought he could do him one better. When we spoke by Skype in 2018, he was at his office at the Jet Propulsion Laboratory, where he was helping NASA think about how to search for life on other planets. As he talked, he often held his head and struck poses reminiscent of a Shakespearean actor. They seemed not so much meant for his audience (me) as to allow him to fully experience the intensity of his own thoughts. Ideas and enthusiasm poured out of him. He spoke with the passion of a man who knows that he's found the secret of life.

Russell spent years as a field geologist surveying and prospecting for ores. In the 1980s, academia drew him back to pursue an alluring, then-controversial idea that most of the metals we mine, from copper and uranium to gold, are found at the sites of former deep-sea vents and hot springs.

To investigate, Russell, then at the University of Glasgow, was studying intriguing formations in a lead mine that he suspected were created by ancient hot springs. Curiously, he saw that the rocks were full of tiny holes. One evening at home, his eleven-year-old son, Andy, was playing with a chemical science kit in an aquarium, and he was excited to see that when the minerals precipitated out of the solution, they formed hollow rocky tubules. As Russell looked at them, he realized with a start that they resembled the holes in the rocks in his lead mine, and in a flash he grasped how the ancient structures had been created.

The next day, Russell and a colleague re-created the growth of the structures in his laboratory. They predicted that some deep-sea vents should make similar formations that were different from those at black smokers. The rock in these new vents, which they called "alkaline vents," would be full of tiny hollow chambers. Less than a decade later, alkaline vents were discovered. Russell was thrilled, not least because they appeared to make the origin of life much easier to envision.

For one thing, they were much cooler than black smokers. They were about 60 degrees Fahrenheit, instead of a sizzling 300 degrees or more, so it was easier to imagine fragile organic molecules forming there. The new vents seemed to solve yet another vexing issue—the dreaded concentration problem. That is a sticky question confronting all origin-of-life researchers. If the molecules of life first appeared in a large body of water, or even a pond, what kept them from drifting apart and producing nothing? Wächtershäuser maintained that the molecules formed around charged mineral surfaces that they would cling to, so they interacted instead of simply floating away. But Russell saw how his alkaline vents might trap organic molecules even more easily. The vents were made of chambers whose thin walls were perforated by pores. The tiny chambers were ideal to concentrate molecules, just like cells.

Russell saw another advantage to his vents. They had the same metals that Wächtershäuser believed served as catalysts, and they also had something else—a lot of hydrogen gas. Russell and the biochemist William Martin claimed that this abundance offered the crucial key to the origin of life.

They argued that tiny differences between the concentrations of protons—charged hydrogen ions—on opposite sides of the membrane-thin chamber walls created electrical potential—energy that could be tapped to form organic compounds. Astonishingly, this looked remarkably similar to how our own cells create energy. Our cells rely on small circulating power packs called ATP (or adenosine triphosphate). Your average cell consumes ten million to one hundred million of them every second. What's more, in the flow of charged hydrogen ions through pores in the vents' thin chamber walls, the scientists recognized an electric current similar to the ones that our cells use to produce ATP. Russell and Martin concluded that these currents in the vents once provided the energy for chemical cycles to develop. They turned carbon dioxide and hydrogen into organic molecules. Their by-products and accidental combinations spawned new cycles, and as they increased in sophistication, they created the whole kit and caboodle: amino acids, RNA, and the full machinery of life. As Russell and Martin saw it, the first ghostly traces of life began with the same kind of electric current that powers us today.

Could the first cell really have hatched in this type of vent? Geologists believe that once alkaline vents form beneath the ocean, they survive for only about a hundred thousand years. To Russell, it's a no-brainer. In each of our cells, he explained, a million to a billion electrons are moving every second. If we travel down to the electrons' level, they don't measure time in years, days, minutes, or seconds. They move in microseconds and picoseconds—millionths and trillionths of a second. On that time scale, with the right ingredients present, even a hundred years is a tremendously long time for a chemical system to spring to life.

Many find Russell and Martin's theory persuasive. It's certainly poetic to consider that life sprang from a current that is still driving the activities of all living creatures. In the interaction of minerals, gases, and water rising from the depths of the Earth, they see the first traces of how inanimate molecules created energy and triggered a cascade of molecular invention. Like an Olympic torch relayed through countless generations, the energy source that set life in motion still flows within us. In short, Russell and his colleagues believe we can at last explain the secret of life.

As you may have noticed, however, when it comes to the origin of life, there's always another point of view. Since the 1990s, the number of competing theories has exploded and, in many of these, life arises not miles beneath the ocean, but back on the surface of Earth. If you search for recent headlines on the origin of life in the *New Scientist*, here is what comes up: "Ponds, Not Oceans, the Cradle of Life," "Russian Hot Springs Point to Rocky Origin of Life," "Volcanic Lightning May Have Sparked Life on Earth," "First Life May Have Been Forged in Icy Seas on a Freezing Earth," and "Clay's Matchmaking May Have Sparked Life."

If you talk with ten biologists, you are likely to hear eleven different opinions. Hydrothermal vents, tide pools, ponds, volcanic lagoons, a radioactive beach, Antarctic lakes—all have their backers. Some see life's birthplace in a patch of clay rather than a pool of water, because crystalline patterns in clay could have concentrated organic compounds and helped link them into longer chains. Hot springs or geysers, like the kind at Yellowstone, have their proponents because hot water from the depths of the Earth carries minerals similar to those at deep-sea vents. When the springs episodically dried up, organic molecules created there might have been concentrated and mixed together on the edges of the pools.

Other researchers focus on how RNA might have kick-started life. The formation of RNA has always been a conundrum. However, the British chemist John Sutherland has found a multistep pathway to turn hydrogen cyanide (which was abundant in comets) and hydrogen sulfide (which was common on Earth) into nucleotides, RNA, and even the precursors of amino acids and lipids. He envisions various molecules that formed in disparate environments meeting in a body of water, from which life emerged.

Others think that we need to combine several scenarios. Some have suggested that a comet brought organic precursors to Earth, and its crater became a warm incubating environment, while geysers of hot water that erupted in its fractures delivered other molecules to the party. "We have got to be open to all possibilities," the geochemist George Cody said. "There are as many different people in the field as different hypotheses. It's humbling. You have to keep an open mind."

Then there is the minority opinion: the suggestion that life did not arise on Earth at all. I first heard this theory from Jay Melosh, a well-respected geophysicist who surprised me by saying, "My feeling is that if we need a location, the obvious one for the origin of life is probably Mars." I presumed that this was a marginal theory, but it turns out to be alive and well, and even mainstream. It gained traction when snowmobiling scientists who were hunting meteorites in Antarctica discovered a four-pound rock. They named it ALH 84001 after the

Allen Hills where it was found. In 1996, a team of NASA scientists concluded that the meteorite not only came from Mars, but also appeared to have traces of fossilized bacteria, as well as grains of magnetized minerals that resemble those produced by bacteria. President Bill Clinton even touted the discovery in a White House briefing. That evidence for Martian life was debated, and still is; most scientists don't accept it. But it spurred Melosh to explore whether such a space voyage was even feasible.

Melosh, an expert on crater formation, had previously calculated that, indeed, it was possible that a large impact on Mars would not vaporize or melt all of the nearby rock. Instead, it would hurl some hunks into space, and toward Earth, like bits of pasta sauce launched into the air by a fallen meatball. Could life hidden in the cracks of these rocks have possibly survived? Perhaps. When the geologists Ben Weiss and Joe Kirschvink analyzed the Martian meteorite, its magnetism revealed that the rock had never experienced temperatures higher than 104 degrees Fahrenheit. That's cooler than Phoenix, Arizona, on a hot day—hardly enough to kill life. And a journey through the vacuum of space is not a deal breaker. Bacteria have survived a 553-day joyride on the outside of the International Space Station.

The strongest evidence cited for this surprisingly popular belief is the fact that life appeared on our planet so soon after the Earth formed. The Earth was born 4.5 billion years ago. Some believe that life was already thriving 300 million years later. Others suspect it arose by 3.8 billion years ago, and it is generally agreed that life certainly existed by 3.5 billion years ago. That is blazingly fast, especially since, during much of this time, the Earth's surface was being pummeled by a troubling bombardment of massive asteroids (the Late Heavy Bombardment). "We keep pushing back the date when life was present on the Earth, to amazingly early times," Melosh said. "The conundrum is: How could life, with all its complexity, get started in the small amount of time that was available?"

On Mars, Melosh believes, life would have had more time to evolve. Its surface was tranquil longer because it didn't suffer the

Moon-forming Big Whack. "There was lots of water on the surface then, and it was warmer," Melosh said. "There were hydrothermal systems, and it was a stable environment long before the Earth settled down to the point that things were watery and pleasant." On Mars, life could have evolved in any of the environments, from volcanic lagoons to deep-sea vents, that were present on Earth. Kirschvink has complex additional reasons for thinking that the surface of early Mars had a chemistry more conducive to the evolution of life.

Melosh points out that there are further grounds for believing that if our bacterial ancestors had sheltered in the cracks or pores of a Martian rock, they could have survived the perilous journey here. Researchers have discovered many genes that could enable bacterial spores to withstand deadly UV radiation and lack of water in the vacuum of space. "What they do when they get into inclement conditions," Melosh said, "is that they wrap their DNA in proteins that stabilize the DNA. Then they go to sleep. That is how they survive." If Melosh and Kirschvink are right, we are all Martians.

All in all, the evidence is so sparse that the question of exactly how or where our most ancient cellular ancestors arose remains unsettled. No one can say for sure if life on Earth came from Mars or from our own planet, if it was an improbably lucky break or if the process is so inevitable that life is common throughout the universe. Did life evolve fast or slowly? Are we life 2.0? Were one (or a number) of earlier lifeforms wiped out by fearsome impacts many millions of years before our ancestors colonized the globe? We don't know, although we can be confident that, so far, every form of life found in every nook and cranny on Earth arose from a single lineage. We all share the same distinctive fundamental biochemistry. We have the same nucleotides in our DNA and RNA, the same twenty amino acids in our proteins, the same method of using ATP molecules to create energy.

Amid all the competing theories and frequent sparring, it's easy to lose sight of how far science *has* come. In truth, no one knows if, barring the invention of a time machine, we'll ever have a definitive account of how life on Earth began. Still, many researchers have a strong sense that we are closing in on the most likely scenario. We now have detailed (although admittedly incomplete) theories of how membranes, amino acids, RNA, and DNA might have formed, and how the first metabolism and replication might have begun. The chemical origin of life no longer seems so unlikely or so completely incomprehensible.

Many believe the most probable scenario is that, once organic complexity evolved—somewhere on Earth or possibly on Mars—membranes captured a small number of molecules. Those membranes were permeable enough that some other molecules could enter, furnishing raw materials for replication and for fuel. When those proto-cells grew too large for their membranes to hold, they broke into two smaller cells that also grew and multiplied. Within some of these primitive bubbles of life, accidental copying errors in the assembly of their RNA (or proto-RNA) produced more efficient structures, and eventually proteins, DNA, and the increasingly intricate souped-up, tricked-out machinery of cells.

Gradually, Earth's surface filled with life. And much of it was not entirely unfamiliar. Whatever the very first life-form was, scientists believe that it evolved into two kinds of single-cell organisms: bacteria and similar-looking creatures called archaea. Bacteria, you're familiar with, of course, but you may not know archaea. They live in extreme places like hot springs, acid lakes, and your gut—where they play a part in digestion and produce some of your flatulence. We are the descendants of these microbes. As they spread, over billions of years they reshaped almost everything on the surface of our planet.

PART III

FROM SUNLIGHT TO DINNER PLATE

In which we discover the magic of photosynthesis, learn how
this transformation of cosmic energy terraformed our planet,
and learn how "intelligent" plants colonized the continents
and began producing the building blocks of us.

8

LIGHT
ASSEMBLY
REQUIRED

Discovering Photosynthesis

Food is simply sunlight in cold storage.
—*John Harvey Kellogg*

In the summer of 1779, a well-coiffed forty-nine-year-old Dutch physician and natural philosopher named Jan Ingenhousz rode in a carriage from London to a manor he had rented in the English countryside. Originally, he had planned to use his summer "retirement" to write a book on smallpox inoculations, his medical specialty, but at some point, a much more exciting plan took hold. As he and his manservant, Dominique, departed London, their carriage was laden with four tables, half a dozen knives, forks, linen, and an armchair cushion, as well as experimental apparatus that included a glass instrument to measure the quality of air. Ingenhousz had a suspicion he was onto something. He could not have known that he was about to discover an invisible process that scientists had not yet dreamed existed, which is a little surprising, because it's perhaps the most important biochemical process on Earth: photosynthesis.

Life made the greatest change to our planet's surface with a single advance: it was with photosynthesis that cells could begin to tap the energy of the Sun. You drink water and eat salt from the Earth's surface, but almost every other molecule inside you was made, or collected, by photosynthesizing plants (or by animals that ate those plants). Photosynthesis is a series of remarkable chemical reactions that punches far above its weight. As will be seen later, it wrought great changes on our planet. Not least among them: plants rework the primary product of photosynthesis—sugar—into a wealth of substances that create the greenery around us and make the existence of land animals like you and me possible. Photosynthesis is responsible for our wood, rubber, coal, gas, and oil. And photosynthesis laid down a critical stretch of the road that our molecules journeyed to become us. Yet this herculean process that profoundly transformed our planet is impossible to see by simply looking around. In fact, nothing about it was ever obvious. So how did scientists ever learn of its existence and the remarkable way in which it works?

Had the good doctor Ingenhousz entered a forest seeking evidence that plants conduct a mysterious operation that transforms our atmosphere, he could have found only a few subtle clues. The vegetation would display innumerable shades of vibrant green. If it happened to be fall, leaves would be fluttering to the ground, a sign that the trees had decided to halt their invisible activity and hibernate for the season. Little else around him would have revealed that one of nature's most astonishing tricks was silently transforming the world.

Well-mannered, erudite, sometimes pompous, Ingenhousz was too serious to be the life of the party, but he was plenty smart. At age sixteen, he had entered university in the Netherlands, already accomplished enough in Greek and Latin to startle his teachers. He established a successful medical practice in the small city of Breda, perhaps to bolster the fortunes of his father, an apothecary. In 1764, immediately after his father's death, Ingenhousz left for London to learn from the most skilled doctors of his time. He soon joined their efforts to combat one of the greatest killers of the day: smallpox, a scourge that condemned twenty to thirty out of a hundred of its victims to miserable disfiguring deaths. Ingenhousz helped develop a controversial new treatment—inoculation—that required courageous doctors to scrape live microbes off scabs and inject them into healthy people. (Our modern practice of using dead or weakened microbes still lay in the future.) About 1 percent of those who were inoculated died, but that was a lot better than 20 to 30 percent. Ingenhousz's skill brought him an invitation from Maria Theresa, the empress of the Austrian Empire, to inoculate the Habsburg royal family, despite skepticism from her royal physician. She had survived smallpox, but several of her children and a daughter-in-law had not. She was desperate to save the rest of her children. Ingenhousz's reward was a lifelong appointment as a royal physician and a generous income. From that came leisure, which he used to pursue his science.

Ingenhousz was the very model of an Enlightenment scientist. Inspired by Ben Franklin, he conducted experiments with electricity. When he moved to London, where Franklin was living, the two men began a friendship that lasted until the end of their lives. Ingenhousz championed Franklin's lightning rods, at a time when some clergymen railed against the thought that man would dare interfere with God's punishment of the wicked. Among other matters, in their correspondence, Ingenhousz and Franklin exchanged stories of severe accidental electric shocks they suffered in the course of their investigations.

One such jolt knocked Ingenhousz out—which may have been the genesis of electroshock therapy. "I feared I should remain forever an idiot," he wrote Franklin. But the next morning, he awoke feeling more alive, and his thinking more acute, than ever. So much so that he recommended to several "mad-Doctors" that they try using electroshocks to restore the mental faculties of their patients. Soon after, some London doctors did.

Ingenhousz conducted a wide variety of investigations, including attempting to replace the gunpowder in a pistol with an explosive mixture of hydrogen gas and air. In 1779, at the age of forty-nine, he took the entire summer off in the hope of making a different break-through.

He rented a secluded country manor a two-hour carriage drive from London. There, undisturbed by his usual flow of visitors, he would try to build on a remarkable discovery by Joseph Priestley. Priestley was a British natural philosopher and chemist, famed for a marvelous invention that puts us forever in his debt: soda water. He was a fiery political and religious radical, a founder of Unitarianism, and an inventive genius when it came to the scientific investigation of gases. He discovered that, strangely, if he covered a candle with a glass vessel, it soon burned out as if some substance in the air had been exhausted. Even more unexpectedly, if he placed a "sprig of mint" in the vessel, it revived the "bad air"—allowing the candle to burn again. Like a flame, a mouse also had difficulty surviving long in a covered vessel. But the addition of a mint sprig (or more likely a full plant) allowed the mouse to continue puttering along. Priestley seemed to have discovered that plants make the invisible air that we breathe agreeable by turning "bad air" in our atmosphere into "good air."

Yet Priestley was befuddled. Sometimes plants revived the air in his vessel, and sometimes they did not. He couldn't figure out why, and attempts by Swedish chemist Carl Scheele to replicate Priestley's experiments utterly failed. When he placed the roots of germinating peas in a jar of water and covered them with a glass bell jar, they did not improve the air. Priestley's claims, Scheele announced, were entirely unwarranted.

Ingenhousz was intrigued. Do plants truly "cleanse the atmosphere"? Ensconced in his country manor, complete with an extensive garden, he began his own investigation. Fearful he would run out of time, he worked at a frenetic pace. At first he placed glass jars over the leaves of plants on the ground and used an instrument called a eudiometer to detect any change in the quantity of "good air" in his vessel. After a while, he found it easier to analyze the air in jars that contained plant clippings submerged in water. He had a rich variety of plants at his disposal: apple, lime, pear, mulberry, willow, and elm trees. French beans, artichokes, potatoes, willow sage, deadly nightshade. Ingenhousz tested all these and many more, experimenting with leaves, roots, and shoots, and carefully monitoring his plants morning, noon, and night.

Not long after he began, he wrote, "I saw a most important scene opened to my view." Nature had revealed a secret almost as thrilling as if he had learned how to turn lead into gold. He discovered that leaves could transform "corrupted air" into "good air" in just a few hours—but *only* if they were exposed to sunlight. When he placed a jar containing a clipping submerged in water in the sun, he could actually see bubbles rise from beneath the leaf in a continuous stream. Heating the leaf in the jar by putting it by a fire failed to create the same effect. The process required sunlight.

Ingenhousz continued his exhaustive experiments dawn to dusk seven days a week to learn more and to make sure that the "salubrious" air he detected was not coming from some other source. After conducting five hundred experiments in less than three months, he was satisfied. That fall, even before he left his country house, he completed a book he expansively titled *Experiments upon Vegetables: Discovering their great Power of purifying the Common Air in the Sun-shine and of Injuring it in the shade and at Night, to which is joined a new Method of examining the accurate Degree of Salubrity of the Atmosphere.*

Ingenhousz was thrilled to have discovered the "secret operations of plants." He found that they silently breathe: that is, they inhale a gas, "bad air" (which we call carbon dioxide), and exhale "good air" (which his friend, the chemist Antoine Lavoisier, would soon name oxygen).

To be fair, Priestley had detected the phenomena first. But he failed to see that only the green parts of plants could "purify" air, and that the process is dependent on sunlight.

The acclaim Ingenhousz expected, however, did not follow. He swiftly published editions of his book in English, French, Dutch, and

German. Unfortunately, that was not sufficient to prevent his former friend, Reverend Priestley, the Dutch pharmacist Willem van Barneveld, and the Swiss botanist Jean Senebier from each claiming that they had discovered photosynthesis first. "When two dogs fight for a bone, a third will be around, ready to make off with it," an incensed Ingenhousz vented to his Dutch translator. Van Barneveld and Senebier were less well known, so he did not bother writing public rebuttals. But the claim by the famed Priestley, whom Ingenhousz had praised as an "inventive genius," would attract great attention and was far more troubling.

Hearing of Ingenhousz's concerns, Priestley wrote him that he had discovered the role of sunlight and communicated it to others at just the same time that Ingenhousz had. Nonetheless, Priestley promised, he would recognize Ingenhousz's work when he published a second edition of his book, *Experiments and Observations on Different Kinds of Air*. But two years later, when Ingenhousz perused the new edition, he found no acknowledgment of his experiments. Irate, Ingenhousz sounded off to a friend that Priestley, clearly embarrassed and envious, was "a sultan who did not tolerate a competitor to his throne." Over the years, Ingenhousz's frustration simmered as Priestley continued to release new editions of his influential book without mentioning Ingenhousz's breakthrough. Finally, Ingenhousz hotly challenged Priestley to reveal where he published first. "If you have realy publish'd this doctrine before me, I owe you the justice to aknowledge it publicely . . . and I will very readily quote the volume of your works and the page in which you will indicate to me this doctrine is clearly and explicitly to be found." Priestley never sent him proof. Nor did he publicly acknowledge Ingenhousz's work.

By this omission Priestley prevailed. Not only was he justly famous for his experiments, but he was also a prolific writer who was willing to campaign for public recognition. The shy Ingenhousz did not relish the limelight and was likely wary of feuding with such a fiery figure. He published rebuttals to Priestley's claim only in the second French edition of his book, and in the appendix of an English government report released seventeen years after his experiments. So it is no surprise that most accounts hail Priestley as the discoverer of photosynthesis. Until recently, Ingenhousz was largely forgotten, one of the most important scientists you never heard of.

Even after Ingenhousz discovered the existence of photosynthesis, however, its nature remained a mystery. How did plants turn "bad air" (carbon dioxide) into "good air" (oxygen)?

One part of the answer would come from a deceptively simple question that had long misled scientists: What do plants eat? It is obvious to anyone that plants don't dine as we do. They don't lunch on other creatures (the Venus flytrap and a few other species excepted). So where does the mass of a towering tree, which weighs tens of thousands of pounds, come from? In other words: What is a tree made from?

One hundred fifty years before, another Jan—the Flemish alchemist Jan Baptist van Helmont, who would be sentenced to house arrest by the Spanish Inquisition for heresy—was among the first to investigate this mystery. Van Helmont, the son of a nobleman, was by any standard an unusually fervent seeker of truth and enlightenment. In the 1590s, he studied logic, astronomy, and natural philosophy at the Catholic University of Louvain, but he scorned the explanations of the natural world that his teachers offered. Like Galileo, he had no use for the theories that Aristotle and other ancients had spun out of pure reason without close observation of the natural world. He was convinced that what he had been taught was worthless and that he still knew nothing. He was "seeking truth, and knowledge, but not their appearance," he recalled. Refusing to accept his degree, he entered a Jesuit college, where he studied alchemy and magic. His professor taught that there was no such thing as good "white magic," only demonic magic. Van Helmont left feeling no wiser.

Regrettably, his first publication, "Of the Magnetic Curing of Wounds," drew the attention of the Inquisition. The Church was not pleased by his suggestion that the curative power of saintly relics might lie in natural causes, which the proto-scientist called "magnetic effects." Nor did he endear himself by casting aspersions on the fitness of Jesuit theologians to study natural science. (Mind you, not all of van Helmont's own science was exemplary. His recipe for the spontaneous generation of mice, for example, was a bit suspect: "Put a pair of sweaty underwear in a jar with wheat. Wait. And, after a period of fermentation, adult mice will crawl out.") In 1623, the medical faculty at the University of Louvain condemned his publication as a "monstrous pamphlet." He was seized by the Inquisition. Once he repented, he was sentenced to house arrest, the very same year that Galileo was also imprisoned at home. It may have been then that van Helmont conducted an experiment with a tree that still enshrines his name in textbooks four hundred years later.

Van Helmont asked, where does the mass of a tree come from? To most others of a scientific bent, the answer was long obvious. Plants eat dirt. Their bulk must come largely from soil. Van Helmont decided that it was high time to test this theory. (He may have been inspired by the German scholar Nicholas of Cusa, who, a century and a half earlier, proposed a similar experiment.) Van Helmont carefully weighed dry soil in a large tub. Then he planted a five-pound willow tree in it and watered it faithfully. Five years later, he removed the tree and weighed it again. By now, it tipped the scales at 169 pounds, 3 ounces; yet the soil had lost just 2 ounces. To van Helmont, the conclusion was obvious. The great mass of the tree must have come from the water it received. Plants were largely made from water, not soil.

Over 150 years later, chemistry had become a more advanced science. By 1796, 17 years after discovering photosynthesis, Ingenhousz knew that the "bad air" that plants absorb was made of carbon and oxygen. (We call it carbon dioxide.) And he knew that plants contained a great deal of carbon. So it was clear to him that a plant's principal nourishment comes not from water, but from the air. Not long after, Nicolas de Saussure showed that water is the only other substance that contributes substantially to a plant's mass. And Jean Senebier determined that the "good air" exhaled by plants is oxygen. So by the mid-1800s, when scientists began learning molecular formulas, they had a basic understanding of photosynthesis. It was obvious that plants live on very little indeed. From carbon dioxide (CO_2), water (H_2O), and sunlight alone, plants make their own food: glucose ($C_6H_{12}O_6$), which they convert into amino acids, fats, and the two-sugar molecule that we crave, known as sucrose or table sugar. *And* plants discard as waste the oxygen we breathe. It's an impressive, almost miraculous, trick of nature.

How it was carried out remained unknown.

That's largely where our knowledge of this process, which makes the preponderance of the molecules within us, remained for another eighty-some years—not for lack of trying, but simply because researchers lacked the tools to learn much more.

They did make modest advances. They discovered that plants use the sugar from photosynthesis to create and store energy as well as to make fats and proteins. They also learned where photosynthesis occurs. They found that it happens principally inside leaves in small green structures called chloroplasts. Within these, it takes place in two distinct reaction centers, and the process begins there when the pigment chlorophyll absorbs energy from the Sun.

Yet although our entire food chain is dependent on photosynthesis, scientists were like children ignorant of how their parents made a living. They could tell you almost nothing about the chemical reactions that produce the sugar that fuels us and makes us. They knew that plants pulled carbon from the air and hydrogen from water, but what was the source of the oxygen in sugar? Was it from carbon dioxide (CO_2), or water (H_2O)? You can't see chemical reactions with microscopes. So photosynthesis was a black box. Scientists could detect what went in and came out, but its inner workings remained hidden in darkness.

Researchers did not suspect that particle physics, which had bailed out physicists who despaired of ever peering inside the atom, would also bail out biologists. The breakthroughs would come from two teams of scientists. The first would struggle to create an extraordinarily powerful new tool, but would be robbed of the opportunity to use it. It would be left to the inheritors of this tool to finally puzzle out how photosynthesis creates the foundation of our existence.

The initial great leap forward would be launched by Martin Kamen and Sam Ruben, a like-minded duo of ambitious scientists eager to make their mark. Kamen, a short dark-haired chemist, grew up in Chicago in the 1920s and spoke a mile a minute. A musical prodigy, he played the violin and cello brilliantly and forged lifelong friendships with Isaac Stern and Yehudi Menuhin. But in his first years at the University of Chicago, he saw the Depression rob his family of their fortune. It began to occur to him that a career in music or a degree in English might be a fast track to poverty. His father, meanwhile, had taken to searching for get-rich-quick schemes in the advertisements of magazines like *Popular Mechanics*. One day he showed Kamen an ad that read, "Be a chemist and make millions." Six years later, in 1936, Kamen had a PhD in nuclear chemistry, and no notion of what to do next.

He decided to risk most of his savings, a few hundred dollars earned from jazz gigs on Chicago's South Side, to buy a train ticket to San Francisco. His plan was to try getting a foot in the door at the Radiation Lab at Berkeley by offering to volunteer. In the years before World War II, the Lab was the biggest, baddest science project around, the forerunner of today's huge undertakings like the Human Genome Project and the Hubble Space Telescope. He arrived at the right place at the right time.

The endeavor was the brainchild of the physicist Ernest Lawrence, who had invented the cyclotron a decade earlier. This was an ingenious machine that could accelerate subatomic particles to unheard-of speeds and energies. Lawrence was already on his fourth version, each larger than the one before. His circular particle accelerator was the great-grandfather of Switzerland's CERN, the largest in existence today. But unlike CERN's awe-inspiring diameter of 1.24 miles, Lawrence's accelerators had grown from five inches to thirty-seven inches across. It was easily small enough to fit in a former engineering building, an old two-story wood frame structure renamed the Radiation Lab. When Kamen arrived, Lawrence was in the process of assembling a team, bolstered by volunteer graduate students, to use the cyclotron for research in particle physics. He also hoped to make new radioactive isotopes that might serve as new tools for medicine. He wanted to try using radioactive phosphorus, for instance, to destroy cancer cells.

On the dark rainy day that Kamen walked into the Radiation Lab, he volunteered to collaborate on research and to help with the high-powered cyclotron's maintenance, a grimy task and constant struggle. Six months later, Kamen was on cloud nine. On learning that his doctoral work was in nuclear chemistry and physics, Lawrence offered him a salaried position supervising the production of radioisotopes. Kamen now had an important job in a cutting-edge program.

Not long after his arrival, he met Sam Ruben, an energetic young chemist, who would become his scientific partner. Ruben, whose peaked eyebrows imparted a look of perpetual inquisitiveness, was the son of a Polish immigrant, a cap maker–turned–carpenter. Ruben grew up in Berkeley. He had been coached by Jack Dempsey at a youth boxing club, and had been a high school basketball star. He was, Kamen recalled, keenly intelligent, intellectually confident, even "outspoken, abrasive and unafraid of confrontations." Ruben needed that courage to hold his own in Berkeley's fiercely competitive chemistry department, where he was working on his PhD. At the time, most of Ruben's colleagues looked down their noses at biology; it was a second-rate field for second-rate minds. Yet Ruben proposed to Kamen that they team up to use radioactive isotopes to study biological processes. They could be the first to investigate biochemistry with a new tool—carbon-11, an isotope newly created at the Radiation Lab. Kamen would make carbon-11 in his cyclotron, and Ruben would direct the biological research that would use it.

This was not the ordinary carbon in our bodies, known as carbon-12, which has an equal number of protons and neutrons in its nucleus, six of each. Lawrence's crew at the Radiation Lab had discovered that by firing a stream of subatomic particles at boron (the element one step down from carbon in the periodic table), they could sneak an extra proton into boron's nucleus. Boron has five protons and five neutrons. By adding another proton they could actually transmute it into a new element—carbon-11—an isotope of carbon with six protons but just five neutrons. But that combination of protons and neutrons was unstable. After a while, it would lose a proton, decay back into boron, and release radioactivity. That, Kamen and Ruben realized, held promise. Their initial plan was to use their carbon-11 to solve a problem unrelated to photosynthesis. They wanted to learn how laboratory rats metabolize sugar. The concept was simple: Feed them sugar made with radioactive carbon. Then follow the carbon as it moved from one chemical compound to another.

In practice that meant that they first had to introduce radioactive carbon into plants to make radioactive sugar. Then they had to feed the "hot" sugar to an animal and attempt to locate the new compounds the animal made from it, all before the radioactive carbon decayed. They soon realized their plans were wildly overambitious. Within a few weeks, a dejected Ruben told Kamen that the technical difficulties were overwhelming.

"During a recital of these troubles," Kamen recalled, "Sam suddenly stopped, his eyes widened, and he blurted, 'Why are we bothering with the rats at all? Hell, with you and me together, we could solve photosynthesis in no time.'"

Suddenly they realized that they had a shot at solving one of the greatest mysteries of biology: how plants make the sugar that our existence depends on. Over a century and a half after Ingenhousz discovered photosynthesis, scientists still didn't have the slightest idea how its chemical pathways actually worked. Now, Kamen and Ruben recognized, they had a tool that might finally reveal them. All they had to do was to make radioactive carbon dioxide, introduce it into plants, and stop the reactions at different times to see which new compounds the radioactive carbon had found its way into. They were delighted.

They had a virtual monopoly on carbon-11, so there was no fear of competition if they moved fast. And they figured that the work was so straightforward, so simple, that it shouldn't take more than a few months to crack. They immediately threw themselves into their work, without the slightest inkling of the pain and heartache ahead.

To make the radioactive carbon they needed, Kamen could begin his work only after the Lab's physicists, whose research was deemed more important, went home. So sometime after nine p.m., Kamen would place boron in the cyclotron, which was encircled by an eighty-ton magnet. Then he would sit behind a control console and bombard the boron for hours with proton-neutron pairs. Once he had his radioactive carbon-11 in hand, he would race several hundred feet, through a dark labyrinth of alleys and staircases, to Ruben's lab in a ramshackle shingled building called the Rat House. And he did have to race. Carbon-11's half-life was just over twenty minutes, meaning that after twenty minutes it lost half of its radioactivity, and after an hour only about 10 percent remained. In the early morning, when Kamen made his dash, Ruben would be waiting impatiently with prepared chemicals, pipettes, blotters, beakers of hot water, and everything else that he needed to introduce carbon-11 into plants and then track its progress. Kamen recalled that anyone looking in on them frantically conducting their experiments would have had "the impression of three mad men hopping about in an insane asylum." In Ruben's view, anyone who did not work eighteen hours a day was a slacker. Excitement, not sleep, fueled their efforts.

They succeeded in developing new biochemical techniques. And they made modest discoveries here and there. Yet, after three sleep-deprived years and hundreds of early morning experiments, when they met in the Rat House to review their progress, they glumly agreed that they had made little headway in identifying even the very first chemical step of photosynthesis. The problem was that they were chasing ephemeral fireflies. In practice, carbon-11 was just too short-lived to allow them to track chemical reactions.

Kamen suspected that another isotope of carbon, carbon-14, might also exist. And it was possible that it was longer-lived. The theoretical physicist Robert Oppenheimer told him that the chances of this were just about nil. Nonetheless, Kamen dreamed of trying to make it—especially since Lawrence had just built a second, more powerful cyclotron with a diameter of sixty inches. But Kamen felt like a drowning man watching a rescue ship pass him by. He would need to monopolize one of the cyclotrons for long periods of time, yet their project was hardly considered important enough to warrant it. Once again, the pair had reached a dead end.

Just as Kamen left their meeting filled with gloom, he received a summons from Lawrence. Running up the physics building's three wide flights of stairs, he found the great man in his office, in a state of agitation. Lawrence told him that Harold Urey, who discovered naturally occurring deuterium (and would later inspire Stanley Miller to investigate the origin of life), was making great strides at collecting naturally occurring stable isotopes of carbon, oxygen, nitrogen, and hydrogen that could prove useful for medical research. So far, the Radiation Lab's attempts to make new useful isotopes of these elements had failed, and now Urey was loudly proclaiming that Lawrence's cyclotrons would never succeed. He claimed it was probably physically impossible for new longer-lived isotopes of these elements to exist. If Urey was right, Lawrence told Kamen, it would be a disaster. He wanted to raise funds to build much more powerful cyclotrons, but in the 1930s, there was no National Science Foundation or Atomic Energy Commission, no federal agency doling out large grants to physicists. (That would change after the Manhattan Project showed that physics could yield practical dividends, especially for war.) For Lawrence, the only substantial sources of funding were foundations interested in biomedical research. They had kept him afloat so far, but now Urey was going around claiming that cyclotrons would never make biologically important isotopes. Lawrence was desperate. Unless they could find a way to make them, he told Kamen, his plans of building larger cyclotrons would remain just that. You can have unlimited access to the cyclotrons, whatever resources you need, Lawrence said.

Just make long-lived radioactive isotopes of carbon, nitrogen, or oxygen as fast as possible.

Kamen left dazed, hardly believing his luck. A genie named Lawrence had just granted his greatest wish. He immediately began drawing up a list of every way he could think of to try making new isotopes. The first isotope on his list was carbon-14. In late September 1939, he began running experiments, dressed as usual in a lab coat blackened by oil and grease and oblivious to the radiation that clung to the zippers of his pants and the coins in his pockets.

This was a business that Jan Baptist van Helmont, being an alchemist, would surely have appreciated. Kamen wanted to transform one element into another. Just a few days after meeting with Lawrence, he sat behind the control board of the thirty-seven-inch cyclotron and began firing alpha particles (containing two protons and two neutrons) at boron. He hoped to add a proton and three neutrons to the ten protons and neutrons already in boron's nucleus, thus transforming it into carbon-14. But after two full days of effort, a test of his sample in a device called an ion chamber revealed that nothing of importance had emerged. Moving to the more powerful sixty-inch cyclotron, he tried again, this time firing deuterons (neutron-proton pairs). But again he made no progress.

Now he tried a different tack. Instead of adding subatomic particles to a lighter element, he would try to make carbon-14 by knocking a proton and a neutron out of nitrogen, which is just one step up the periodic table.

This approach was also a disappointment.

Then the larger, sixty-inch cyclotron went down for repairs. By now, it had been several months since he had begun his trials. Kamen was beginning to feel desperate. He was coming up against the maddening reality that all their efforts might have come to naught.

In a last-ditch effort, he returned to the smaller, thirty-seven-inch cyclotron and decided to try bombarding carbon itself. He smeared graphite, a soft solid form of carbon, onto a probe that he planned to strafe with deuterons in an attempt to sneak two more neutrons into its nuclei. To maximize the intensity, he poked the graphite directly into one of the cyclotron's ports, cranked up the power as much as he dared, and bombarded the graphite every night for a month. For the final push, he stayed awake three nights in a row. Blearily, in the early morning of February 15, after seventy-two hours without sleep, he scraped the charred graphite into a bottle, walked to the Rat Lab, and deposited the sample on Ruben's desk.

Sometime around dawn that morning, as rain pounded the Berkeley streets, police cruising the neighborhood spotted Kamen stumbling along. The suspect for a mass murder committed just hours earlier was still at large. Eyes red, chin unshaven, and clothes unkempt, Kamen looked the part. He claimed that he was a chemist, and that he had been up for many hours working on something called a "cyclotron." The cops hauled him in. He was examined by a hysterical survivor of the gruesome crime. But she didn't recognize him, so they let him go. Once home, Kamen instantly toppled into a deep sleep.

Twelve hours later, he awoke and phoned Ruben, who was in the lab. Finally, it seemed, there was encouraging news. Ruben thought he detected faint radioactive signals, but he wasn't sure. Kamen rushed over, elated, and now fully awake, but he was so radioactive after his marathon stint that Ruben forbade him to approach or help. Within a few days, Ruben and Kamen confirmed that they had made an isotope that Robert Oppenheimer had told them could not plausibly exist—radioactive carbon-14 atoms that are very long-lived.

Lawrence was home in bed nursing a cold when he heard the news. He jumped up, dancing for joy. They had proved Urey wrong. They had shown they could create valuable isotopes for biological investigations. In a week's time, he would receive the Nobel Prize for inventing the cyclotron and creating artificial radioactive isotopes. Now he had an even stronger case for building more powerful cyclotrons. At the ceremony, the chairman of the physics department stepped back from the podium, raised his arm, and dramatically announced that Kamen and Ruben had just found a new isotope—carbon-14. And its half-life was not minutes, like carbon-11, but thousands of years (5,730 years, as we know now).[*]

Carbon-14 would be key to fully working out the remarkable chain of chemical reactions in photosynthesis. Even before that, however, Kamen and Ruben, in collaboration with two colleagues, solved a small but crucial part of the puzzle. Using radioactive oxygen-18, they at last discovered the source of the oxygen that photosynthesis uses to make sugar ($C_6H_{12}O_6$). They found that the oxygen comes from carbon dioxide, not water, as many supposed. How it did this remained a mystery. But now it was known that when a plant splits water, it is interested only in the hydrogen. The oxygen we breathe? Plants discard it as waste.

Kamen and Ruben had discovered part of the path that our oxygen molecules must journey to reach us. And in the process, they found that van Helmont could not have been more wrong. Most of the dry mass of a plant comes from neither water nor soil; it is plucked out of the air. If you dry out a tree, 50 percent of its mass is carbon, and 44 percent is oxygen. That means that almost all of its mass— 94 percent—came from carbon dioxide in the atmosphere. You and I are not so different. About 83 percent of the dry mass of our bodies comes from carbon dioxide molecules that were once wafting through the air until plants seized them and enmeshed them in organic molecules that we later ate. Another 10 percent of our mass is hydrogen that photosynthesizing plants stole from water. Meanwhile, all the oxygen molecules we breathe were once in molecules of water before plants discarded them into the atmosphere.

Both just twenty-six, Kamen and Ruben were now riding high. They had cracked a crucial mystery of photosynthesis. With the world's only supply of carbon-14 in their hands, solving the rest of the

---

[*] A few years later, scientists discovered rare, naturally occurring carbon-14, which is created when cosmic rays strike atoms in our atmosphere. Atmospheric carbon-14 revolutionized the dating of organic materials and has become an incredibly powerful new tool for archaeologists, anthropologists, geologists, and many others.

"big problem" was finally within their reach. At last, they could identify the long-hidden chemical reactions that make sugar from water
and thin air.

That at least was their hope. On December 7, 1941, however, at the
moment they were poised to reap the rewards of their discovery, their
world fell apart. The Japanese attacked Pearl Harbor. All nonmilitary
work at the Radiation Lab immediately stopped. Lawrence would de-
vise plans to make uranium-235 for the first atomic bomb and oversee
production of the isotope. Kamen would help, while Ruben was assigned to chemical warfare research. It would be his undoing.

Throughout the war, debates raged about whether the use of chemical weapons would save Allied lives. Commanders were consider-
ing deploying poison gas to wipe out enemy troops on beaches before
sending in Allied soldiers. The best candidate for this was phosgene,
which had been responsible for over 80 percent of the poison gas
deaths in World War I—seventy-five thousand of them. Now the
military needed to know how long phosgene would take to disperse,
so they could determine if it was practical to use for beach landings.
Ruben was asked to supervise mock trials to find out.

Ruben was eager to complete them as quickly as possible so he
could return to his work on photosynthesis. In October 1943, after
working round-the-clock for several days, he was so tired, he fell
asleep at the wheel on his long drive home and crashed. He was lucky
to escape with just a broken hand. The next morning, with his right
arm in a sling, he was back in his lab and handling a vial of phosgene
gas. Suddenly, the glass cracked. Perhaps it was flawed, or perhaps
he was too impatient to cool it in ice water, and used liquid nitrogen
instead. Whatever the reason, a dense cloud of phosgene escaped.
Ruben and his two student assistants ran outside and lay on the grass,
their best defense against the rising gas. His assistants survived. But
for Ruben, it was too late. He died two days later. He was twenty-
nine years old.

Heartbroken, Kamen soon ran into more trouble of his own. His
marriage had recently dissolved, so he began spending more time

with new pals, many of whom were leftists. He was also a good friend of Robert Oppenheimer, the scientific director of the Manhattan Project, who was a suspected Russian sympathizer. That was enough to bring Kamen under surveillance by FBI and army counterintelligence. They feared that Oppenheimer or his acquaintances might leak atom bomb secrets. In 1944, the violinist Isaac Stern introduced Kamen to the Russian vice consul in San Francisco, who asked Kamen for an introduction to Lawrence's brother so he could inquire about a treatment for leukemia. As thanks, he took Kamen to dinner. To army counterintelligence and the FBI, which were both trailing him, that added fuel to the fire. The Radiation Lab was now a top-secret arm of the Manhattan Project with checkpoints, guards, and constant warnings about enemy spies and the dangers of speaking loosely. The authorities feared that the gregarious Kamen might easily let slip classified information to his leftist friends. Under orders from General Leslie Groves, Lawrence had Kamen immediately expelled from military work and the Radiation Lab.

Kamen was distraught. His reputation was besmirched, and he was blackballed. He managed to find a job as an inspector in the Oakland shipyards. In time, he returned to research at another department at Berkeley, then went on to an eminent career working with radioactive isotopes at other universities, but he had lost his chance to continue work on photosynthesis and achieve the breakthroughs that he and Ruben had long hoped for.*

Back at Berkeley's Radiation Lab, two other chemists, Melvin Calvin and Andrew Benson, abruptly inherited the reins of photosynthesis research. Calvin was a thirty-four-year-old energetic rising star in the department of chemistry. Like Kamen, Calvin had seen his father, a Detroit automobile mechanic, struggle to make a living. Calvin looked around for a profession with more security, and chemistry

---

* Kamen would later be dragged before the House Un-American Activities Committee and spend years denying the FBI's accusations that he was a spy before eventually clearing his name.

seemed to fit the bill. Admired for his keen, bubbling intellect, Calvin made Lawrence's acquaintance because, when he took lunch at the faculty club, he chose to sit at the physicists' table instead of the chemists'.

In mid-August 1945, just a few days after the emperor announced that Japan would surrender, Lawrence decided that scientists had done enough war work. Outside the faculty club, he approached Calvin. "Time to quit. Time to do something useful," Lawrence told him. "Now do something with that radiocarbon."

In an instant, Calvin found himself fully funded to assemble two teams. One would investigate how radioactive elements could be used in medicine. The other would try to exploit the techniques that Kamen and Ruben had pioneered to crack photosynthesis. For help, Calvin recruited the young organic chemist Andrew Benson. A skilled experimenter, Benson had assisted Ruben and Kamen before the war and was already familiar with their techniques. Lawrence had offered Calvin the use of the old building that housed the obsolete thirty-seven-inch cyclotron. Calvin asked Benson to design and set up a new lab there. Calvin would be the captain and Benson, the first mate.

Calvin suspected that anyone who solved the mystery of photosynthesis stood a good chance of winning a Nobel Prize. To him, though, the stakes were even higher. If we could discover how plants make food, he thought, it's possible we could produce unlimited artificial food and eliminate world hunger. Moreover, if we could learn how photosynthesis harnesses energy, we could replicate that process to solve the world's energy problem. Once the mechanisms of photosynthesis were revealed, Calvin believed, these advances were just a matter of time.

In their new laboratory, Benson, together with James Bassham and Calvin's other young hires, realized they could simplify their work by using algae rather than plants. They could easily grow and treat algae in ingenious circular flat glass vessels of Benson's design, which they called lollipops. More important, Benson found a way to refine a newly invented tool called paper chromatography, which Kamen and Ruben had lacked. This would speed up the work tremendously. Once they had introduced radioactive carbon-14 to algae, they could kill the algae, grind them up, and place a small drop of the slurry on a sheet of paper. Adding solvents would make its compounds migrate to different areas of the paper, thus helpfully sorting themselves. To make their work easier still, Benson realized that if they exposed photographic film to the sheet, the compounds containing radioactive carbon would reveal themselves as dark spots, making them easy to locate for further identification.

The detectives now had a lineup of suspects, but they still lacked their chemical IDs. Finding their identities was not easy. As they tried to isolate and identify the compounds, they also struggled to work out bugs in their techniques. At eight a.m. every day, Calvin, always sharp in a business suit, made the rounds to ask, "What's new?" and to talk ideas over. For a long time, frustratingly, their list of the carbon-containing molecules that played a role in converting carbon dioxide into sugar kept changing. Some molecules vanished as they refined their techniques. Then they found that radiation from a nearby cyclotron had thrown off their readings for almost a year. Making sense of the long list of molecules they were detecting was maddeningly difficult. Simply identifying molecules—such as $C_3H_6O_4$, $C_3H_8O_{10}P_2$, and $C_5H_{12}O_{11}P_2$—told them little about which reactions had created them or the order in which they had been formed. Still, the lab buzzed with excitement. Calvin provided oversight and insight. He was the public face, the idea guy, the decider. He brought in the funds, more staff, and collaborators. Benson developed the lab techniques and made many key discoveries.

Nevertheless, there was friction. A colleague recalled that Calvin constantly walked in with bright ideas he was sure would be revolutionary. "He would come tearing into the lab with this new idea, which you'd have to stop and listen to, and he'd pull those finger joints. . . . And then he would go away, and Andy [Benson], who would have listened to all this, said, 'Oh, that's his latest theory, is it? Well, it's nonsense, it won't work because of this or that. . . .' Andy could see reasons why something wouldn't work, and he would know very well that, in two days' time, there would be another rush of ideas that would come in. . . ."

That tension would grow.

In 1954, their team published a landmark paper. It revealed the full, ridiculously complex Ferris wheel of reactions that transforms carbon dioxide into the sugar that makes and fuels us. Their tour de force would be known as the Calvin cycle, or more commonly today the Calvin-Benson cycle. When they began a decade earlier, they could never have envisioned the process that every carbon atom in your body endured when photosynthesis grabbed it from the air. In the first step of the process, which might have happened yesterday in your garden or hundreds of years ago, a carbon dioxide molecule (CO_2) was added to a molecule of five carbons, creating an unstable six-carbon molecule that immediately split into two molecules of three carbons each. That was just the beginning. As the Ferris wheel continued to whirl, these were rapidly converted into intermediate chains of four, five, six, and seven carbons. Only after many turns of the wheel did glucose—$C_6H_{12}O_6$—finally emerge.

That same year, the tensions between Calvin and Benson finally boiled over. In the early days, Calvin had been fully immersed in Benson's research, but as the details of the Calvin-Benson cycle were falling into place, Calvin became engrossed in solving another problem. He wanted to work out the chemical reactions that enable light to power photosynthesis. Calvin prided himself on making informed guesses, on floating inventive theories even before he had enough data to support them. "He could make interpretations that nobody else would even think of," a colleague recalled. Getting it wrong never embarrassed Calvin, because he knew that eventually he would make another breakthrough. Now, he constructed an elegant chemical model that seemed so right that one eminent biochemist jumped to his feet with tears in his eyes after hearing Calvin describe it. On many other occasions, Calvin's willingness to guess had paid off handsomely, but not this time. Instead, for two years, it led him on a wild-goose chase.

As he was pursuing his theory, he stopped checking in on Benson, while Benson didn't bother telling him that he was working with a colleague to solve another key question. After months of work, they were thrilled to identify the large enzyme at the heart of photosynthesis. Mercifully, its inelegant name—ribulose-1,5-bisphosphate carboxylase oxygenase—was later shortened to Rubisco. It's the first molecule in most plants and algae that grabs carbon dioxide to begin making glucose. Benson was amazed to find that Rubisco is also the most abundant protein in the leaves of plants, including our salad greens. Rubisco lies at the center of every chloroplast in every plant on Earth. Every carbon atom within our own bodies was seized by this large enzyme that begins the process of turning carbon dioxide into sugar.

Calvin, perhaps threatened and insecure, and certainly angered that Benson had not informed him of this research, could not tolerate working together any longer. He simply said, "Time to go." With those three words, Benson was fired. Eight years later, Calvin deservedly received the Nobel Prize, but most of his colleagues believed that Benson should have been standing beside him on the podium. Benson went on to have a successful career in biochemistry, yet he never shook the sting of losing the highest honor a scientist can ever receive. Calvin barely mentioned Benson in his acceptance speech. In Calvin's account of his research in his autobiography, published over thirty years later, Benson's name did not appear once.

Since then, investigators have cracked many other aspects of photosynthesis. It turns out to be even more insanely complex than Calvin or Benson imagined. And why should we expect it to be otherwise? It does, after all, turn flavorless air and water into sweet molecules that store energy and can be used to make the building blocks of life. The Calvin-Benson cycle is just one part of the process. Biochemists have discovered that the entire crazy act is a jury-rigged Rube Goldberg operation that takes place in two different stages. In the first part of the process, beams of light excite electrons in chlorophyll, which is embedded in a special membrane. These electrons travel down a bucket brigade of molecules that gradually use their energy to split water (and liberate hydrogen and excess oxygen). Then, with a kick of energy from another light beam, they generate molecules called ATP and NAHDP. These leave the membrane and fuel the second part of the process—the Ferris wheel of reactions, which Rubisco initiates in the Calvin-Benson cycle that transforms carbon dioxide and hydrogen into sugar.

The entire process requires an inordinate number of reactions because it is, in technical parlance, "energetically difficult"; in other words, it's chemically really really hard. The first part—splitting water—is tough because the bonds between oxygen and hydrogen are extremely strong. Even heating water to 3,000 degrees Fahrenheit severs only a small percentage of the bonds. Photosynthesis had to find a complex series of chemical reactions to steal hydrogen from water. The second part of photosynthesis—forcing carbon dioxide to bond with other molecules—is also no picnic.

Photosynthesis has devised ways to pull these tricks off, but it takes over 160 steps, depending on how you count them, says the plant scientist Stephen Long, who broke them down to simulate the process in a computer.

The entire operation is not only ungainly, but inefficient. Rubisco, the enzyme that first seizes carbon dioxide, misfires about 30 percent of the time because, mistakenly, it often grabs a similarly shaped oxygen molecule instead. And it works lazily, a hundred times more slowly than most other enzymes. "Rubisco is a silly enzyme, a bad enzyme. Why nature invented it, I have no clue," the biochemist Govindjee said. "God must have been looking the wrong way." He went on to explain that Rubisco performs so poorly because it evolved when our atmosphere had no oxygen, just lots of carbon dioxide. It worked brilliantly then. Now that our atmosphere is full of oxygen and has less carbon dioxide, Rubisco just sputters along. Yet it still works well enough that it is one of the most common proteins on our planet. All the Rubisco on Earth combined would weigh about 700 million tons, about as much as 120 million African elephants, more than enough to encircle the globe.

By the way, Calvin's dream of mimicking photosynthesis to solve global hunger and the world's energy crisis hasn't died. Plant scientists aren't trying to re-create photosynthesis, but they are fine-tuning it to improve crop yields. They are tweaking genes to allow plants to absorb more light and to help Rubisco work more reliably. In his later years, Calvin had also hoped it would be possible to invent an artificial photosynthetic device to split water with sunlight to make limitless hydrogen fuel. Researchers are still pursuing this idea. Calvin's dream is just taking much longer than he envisioned to achieve.

Once photosynthesis entered the picture, life on Earth was no longer solely dependent on whatever chemical energy it could scrounge up on the Earth's surface. Life could draw vastly more energy from the fusion of hydrogen into helium, over 90 million miles away in the Sun. All the energy we expend during our existence was emitted as light from our star and stored by plants in chemical bonds in our food. Without photosynthesis, our continents would look rocky and dusty like Mars. Instead, they're green. You don't have to get all New Agey to appreciate the wonders of that arrangement. But if you would like to, you can savor the words of the visionary Russian geochemist Vladimir Vernadsky, who saw photosynthesis as a "region of transformation of cosmic energy."

As we've seen, roughly 90 percent of a plant's and 83 percent of an animal's dry mass comes from carbon dioxide. That means that in the end, all the food you eat, the cotton you wear, the trees you climb, and the friends you hug were largely pulled from the air and packed with energy from the Sun. Photosynthesis made the bulk of every living thing you have ever talked to, climbed on, loved, or eaten.

It's also the source of our energy. Photosynthesis uses the Sun's energy to produce sugar from carbon dioxide and water (while expelling oxygen). We reverse that process to generate energy by burning sugar and oxygen (while expelling carbon dioxide and water). We liberate the energy that photosynthesis stores in molecules in order to think thoughts, play tubas, and dance the Lindy. Photosynthesis powers every breath you make, every step you take.

And when did this process that transforms cosmic energy into life first appear on our planet? You might think that photosynthesis evolved first in plants, but it didn't. It appeared billions of years earlier, when bacteria and other microorganisms were our planet's only inhabitants. Once that crazy chemical reaction was unleashed, it would change everything. It would deal our planet some of the greatest shocks it has ever experienced. It would extinguish almost all life before making plants and us possible.

9

LUCKY BREAKS

From Ocean Scum to Green Planet

Today photosynthesis runs our planet.
—*Stjepko Golubic*

Four billion years ago, the Earth's continents looked awfully dull. They were nothing but barren black, brown, and gray rock. Volcanoes spewed poisonous gas into an oxygenless atmosphere. If you traveled there in a time machine, you would have asphyxiated immediately. The only life on Earth was bacteria and other single-cell organisms far smaller than the period at the end of this sentence. Yet if you fast-forwarded billions of years, to a mere 350 million years ago, oxygen levels approached the luxurious 21 percent that we're accustomed to today. The oceans seethed with large creatures darting this way and that. And plants had invaded the continents and invented the pathways that our molecules would take to reach us. What was responsible for terraforming Earth so dramatically from such an inhospitable wasteland to a blue-green oasis?

Of all the forces at work, one deserves the lion's share of the credit. Only beginning in the 1960s have we come to appreciate the many strange and surprising ways in which photosynthesis, like a powerful geological force, has remade our planet. And they have been strange.

Along the way, photosynthesis may have provoked a mass die-off so vast that its effects were once considered as extreme as a nuclear holocaust. It turned our planet into a monstrous snowball. It aided and abetted "impossible" evolutionary shortcuts that ratcheted up the diversity of life, ultimately making the arrival of plants, and us, possible. How did scientists learn of these immense upheavals that occurred so long ago? And how did photosynthesis create so much havoc?

The first hint of its great antiquity appeared in the late nineteenth century. At the time, no one had found any evidence of life before the Cambrian era, which we now date to about 550 million years ago. In the winter of 1882, however, in the depths of the Grand Canyon, a rock lover named Charles Walcott changed that.

Walcott, who would later become director of the Smithsonian Institution, grew up in fossil heaven. As a boy in Utica, New York, the gangly youngster collected fossils on his parents' farm and in a nearby quarry belonging to his future father-in-law. At age eighteen, he left school to become a clerk in a hardware store, but he nurtured his passion by reading textbooks, writing scientific papers on fossils, and corresponding with famous geologists. He also managed to amass one of the best collections in the world of the ancient sea creatures called trilobites, which he sold to Harvard.

Walcott's prospecting savvy eventually won him a position in the newly formed United States Geological Survey. In November 1882, its director, the explorer John Wesley Powell, asked Walcott to survey the hitherto inaccessible depths of the Grand Canyon. Previously, Powell had only been able to glimpse the lowest levels of rocks from a small wooden boat as he floated by. Camping in occasional "driving frozen mist and whirling snow," Powell supervised the construction of a steep horse trail from the rim to the warmer regions three thousand feet below. Then he sent the thirty-three-year-old Walcott down the makeshift trail with a crew of three men, enough food to last three months, and nine saddled and packed mules.

"So much snow will have fallen on the plateau above," Powell told him, "that you and the packers will not be able to get out of the canyon until Spring. In the meantime, I want you to work out the stratigraphic sequence and to collect all the fossils you can. Good luck to you!"

Walcott saw this as a golden opportunity. He had already found some of the oldest fossils known—his trilobites, which resembled strange crustaceans. And he knew that the lack of very primitive animal, plant, or bacterial fossils had caused Darwin acute embarrassment when he published *The Origin of Species* just forty years earlier. Darwin's critics took this as evidence that all species were divinely created. If skeptics asked him for evidence that, as he presumed, even simpler creatures once existed, he could only murmur that fossils were rarely made, the organisms must have been very small, and he hoped they would appear someday.

Walcott was well aware of Darwin's dilemma, and as he descended the steep primitive trail into the largely lifeless Grand Canyon, he kept his eyes wide open. Walcott loved this stark red-tinged world of canyons, cliffs, and nothing but "rocks-rocks-rocks." His companions—a fossil hunter, a cook, and a pack animal handler—did not always share his exhilaration. They inched along eight-hundred-foot cliffs. Today, part of their path, the Nankoweap Trail, is considered the most dangerous in the Grand Canyon. The banks of the roaring river were too steep to travel along, so they had to make their own trails at times to reach the deepest rocks. One mule died and two were badly injured. On at least one occasion, the ink in Walcott's pen froze and they were forced to pile ice around fires to melt water for their animals. Above all, it was quiet and lonely. After three weeks, that was enough to make Walcott's fossil-collecting companion so depressed he had to leave. But Walcott was thrilled to be there. He would not return for seventy-two days.

One day, clambering about, he was intrigued by layers of lines in some rocks. They looked like cabbages that had been sliced in half. These patterns seemed so unusual that he was sure they could only have been created by some kind of life—cyanobacteria, he would later claim (which were then known as algae). They brought to mind similar fossils that he had seen in New York State. Those were named *Cryptozoön*, meaning "hidden life," and came from the Cambrian period. But these apparent fossils in the Grand Canyon lay within more ancient rock, making them much older than any others ever found.

Walcott went on to discover similarly ancient *Cryptozoön* in Montana and elsewhere. Other paleontologists also found unusual patterns in Precambrian rock that seemed to be fossils. It appeared likely that evidence for the most primitive life-forms would be found in rocks older than the Cambrian era. Nonetheless, skeptics abounded, particularly since one long-disputed fossil proved to be nothing but a distinctive mineral deposit in volcanic limestone created by pressure and heat.

In the 1930s, four years after Walcott's death, Albert Charles Seward of the University of Cambridge, the most influential paleobotanist of his day, decided to weigh in. As the paleobiologist William Schopf put it, Seward proceeded to snatch defeat from the jaws of victory. In what became known as the "*Cryptozoön* controversy," he took a hard look at the evidence for Precambrian fossils and concluded that it was all wishful thinking. He noted that the purported fossils had no obvious relation to living species, nor did the larger structures reveal any evidence they were made of smaller cells. He argued that the ringed patterns in Walcott's *Cryptozoön* could just as well have been created by deposits of calcium-rich mud on the ocean floor. Furthermore, he declared, we could never expect creatures as small as bacteria to be preserved in fossils. In a scathing warning, Seward cautioned scientists to be wary of overeager prospectors who claimed to find fossils so old.

This admonition from such an eminent figure dissuaded geologists from even bothering to look for fossils in rocks more than about a half a billion years old. It seemed impossible that we could ever find them. For many, that turned into the belief that life was a latecomer to our planet: that during Earth's first 4 billion years, the first 90 percent of its existence, no life existed here at all. In fact, the microbiologist Stjepko Golubic recalled that many scientists used the term the Precambrian Era to mean before life. They had fallen into the "If Our Current Tools Haven't Detected It, It Doesn't Exist" bias. The lack of a discovery turned into certainty that tiny bacteria were never there.

Then, in the mid-1950s, two decades later, a young Australian graduate student, Brian Logan, and his geology professor, Philip Playford, explored remote Shark Bay, an isolated salty lagoon on the northwest coast of Australia. If you stood on the beach with the tide receding, the shallow turquoise waters revealed a sight so strange it seemed like a dream. Hundreds of rocky cylindrical towers up to three feet high were scattered just feet apart, like a colony of huge rough mushrooms hard as stone. As they investigated these Tolkien-like formations, it dawned on them that they had found the key to understanding Walcott's *Cryptozoön*. They were looking at living fossils, the answer to a riddle: What is both dead and alive? The living part was a mat of photosynthesizing cyanobacteria stuck to the top of each formation. The mats trapped sediments as the tides swept in and out. When the cyanobacteria died, the sediments were still locked in place—building up sponge-textured towers as new bacterial mats grew above them. Bacteria had created Walcott's *Cryptozoön* in the same way in primeval oceans. We now call them stromatolites, from the Greek *stroma* (layer) and *lithos* (rock). Today, stromatolites exist in only a few places like Shark Bay, where it is too salty for most other organisms to survive. But ancient fossilized stromatolites have been found all over the world.

At just the same time that Australian geologists stumbled on living stromatolites, two American geologists, Stanley Tyler and Elso Barghoorn, announced the discovery of other fossils that Seward had claimed could never exist. These were single-cell and multicellular microorganisms, including hairlike strands of cyanobacteria. And they were almost 2 *billion* years old. "It was shocking to many," said Golubic. "The assumption was that life exploded in the Cambrian, and before that there was nothing. The Cambrian was supposed to be the beginning." Today, the oldest generally accepted fossils are stromatolites and microbes that lived an amazing 3.5 billion years ago, only 1 billion years after the Earth itself was created. Darwin and Walcott would have been astounded.

Which bacteria built the most ancient stromatolites? No one can say for sure if these were photosynthesizing cyanobacteria or their ancestors. But cyanobacteria were in the oceans by at least 2.4 billion years ago.

You're probably already familiar with cyanobacteria. They are the irksome creatures that are turning our ponds murky green. But they are more than just minor irritants; they are the most subversive organisms in our planet's history. They were microbial Bolsheviks, the geologist Joe Kirschvink once said, because they completely overthrew the existing system. Their bacterial ancestors lived only where they could find minerals to eat, but photosynthesizing cyanobacteria fed on just water, air, and sunlight. They were free to spread far and wide, to colonize the Earth as no organism had before. Once unleashed, these unassuming revolutionaries would launch a million changes that made the rise of plants, and humans, possible.

The first to recognize the outsize impact of photosynthesis on our planet was a wiry, driven five-foot-three whirlwind—Preston Cloud, a geologist with a tinge of a Napoleon complex. "He was feisty," the paleobiologist William Schopf said. Some called him "the little general." When he headed a department at the US Geologic Survey, he placed his desk and chair on four-inch risers to look down on his employees. Cloud had been a bantam-weight boxing champion in the United States Navy Pacific Fleet and held down full-time day jobs during the Depression to finance night classes at George Washington University. Eventually, after holding a succession of jobs at Harvard, the USGS, and elsewhere, he became chair of the geology department at the University of Minnesota. By this time, Cloud, now in his fifties, had become interested in searching for clues to something few ever thought about: how life and planet Earth may have interacted and influenced each other. "He was ahead of the game in trying to consider life and environment in the same breath," the geobiologist Andy Knoll told me. "He gets the gold star for trying to make sense of things as an integrated package."

Cloud traveled to Ontario, Canada, to search for the undisclosed location of the rocks where Tyler and Barghoorn had found ancient bacteria. (They kept it a secret, as they still hoped to make more discoveries there.) As Cloud scrambled over rock formations, he puzzled over why they looked so strange. Like a huge layer cake, bands of black rock were interspersed with layers of red rock chock-full of iron. He knew that these iron-rich stripes had once been laid down as sediment on an ancient ocean floor. Similar formations were well known. Minnesota's "iron ranges" had red deposits so deep that, until the Second World War, 25 percent of all the iron production in the United States was excavated from a single pit. He was aware of other iron-rich layers throughout the globe, including in South Africa, Australia, and Greenland. He even had an estimate of their ages. The very thickest were about 1.8 to 2.3 billion years old. But what stumped him was this: Why did these colossal deposits of iron suddenly appear on the ocean floor worldwide, and then abruptly stop?

A light bulb finally went off when Cloud realized that the red bands were rust. And what rusts iron? Oxygen.

That seemed strange. Geologists had determined that the early Earth's atmosphere had no oxygen. Yet, as Cloud saw it, these red rocks were telling him that 2.3 billion years ago, the Earth underwent a sudden transformation. So much oxygen showed up that it rusted most of the iron particles in the ocean. And these proceeded to sink to the bottom where, over time, pressure turned them into sedimentary rock.

He could think of only one explanation for the dramatic appearance of so much oxygen: the birth of the first oxygen-producing microorganisms. "What else could it be?" asked Schopf, an acquaintance of Cloud's, who has discovered some of the Earth's most ancient fossils. "My God, those red beds of rocks are thick," Schopf said, by which he meant that it took an *awful* lot of oxygen to make them.

"And the only strong source of oxygen known to us are organisms." Cloud proposed that, incredibly, in the ocean, huge green mats of these tiny photosynthesizers, which we know as cyanobacteria, pumped so much oxygen into the atmosphere that they rusted the Earth's surface. First, oxygen rusted the iron in exposed rocks on the continents, which weathering, rain, and rivers swept into the ocean. Then, as the cyanobacteria continued to multiply, oxygen rusted the large amounts of iron deposited by underwater volcanoes and vents in the ocean.* Finally, once all the exposed iron on Earth had turned to rust, oxygen levels in our atmosphere began to rise.

Some geologists call this surge the Great Oxygenation Event. More sinisterly, others call it the Oxygen Catastrophe or even the Oxygen Holocaust, because, although we think of oxygen as life-giving, it's also toxic. Oxygen is so highly reactive—it so desperately wants to steal electrons from other atoms—that it can be a deadly poison. Just watch what happens when you add oxygen to a fire by fanning the flames. When an oxygen molecule eludes its handlers and enters a cell, it eagerly latches on to DNA, or an enzyme, or what have you, and gums up the works.

That's why we evolved antioxidants—molecules whose only job is to snare wayward oxygen before it can do harm. (While a few minerals like copper, zinc, and selenium are antioxidants, most, like vitamin C, are made by organisms.) Cyanobacteria also evolved defenses against oxygen. But the oceans were already filled with countless other microorganisms, and to them, oxygen was poison.

When huge colonies of cyanobacteria pumped up oxygen levels, they triggered our planet's first mass extinction. In one of the most significant events in Earth's history, cyanobacteria poisoned most other life (all still single-cell organisms at this point). "Many kinds

---

* We know now that some iron formations appeared earlier. These were likely created by bacteria that did not produce oxygen through photosynthesis. Instead, when they fed on iron floating in the ocean, they transformed it into another form of iron that would sink to the ocean floor.

of microbes were immediately wiped out," wrote the microbiologist Lynn Margulis, who once called it a "catastrophe of global magnitude," an "oxygen holocaust." Now, however, most don't think that the impact was nearly so sudden (and they cringe at the use of the term *holocaust*). Many species probably had time to adapt to new habitats. Nonetheless, "the rule would have been retreat or die," Schopf explained. Vast populations of bacteria must have died off, fled to hydrothermal vents on the ocean floor, or found refuge in locales, like mud, far from oxygen's lethal reach, where their descendants remain today.

That left cyanobacteria to rule the Earth's surface. They floated in the oceans, flourished in green bacterial mats on shallow seafloors, and built vast cities of stromatolites. Our planet wasn't a man's world or a plant's world. It wasn't even a fish's world. It was a cyanobacterium's world.

The next step on the way to a green planet was another catastrophe, one so odd that geologists would have a hard time imagining anything less likely. Cyanobacteria were responsible for the appearance of mile-high glaciers that many believe smothered the Earth's entire surface, entombing everything in an icy embrace.

Scientists might have remained utterly ignorant of this chapter in our planet's history had it not been for surprising evidence left by ancient magnetism. One of the first suggestions of this decidedly chilly turn of events came in 1986 when the geologist Joe Kirschvink at Caltech was asked to review a paper on ancient Australian rocks. Kirschvink, a distinguished geologist with a penchant for tossing off wry comments, also had a knack for dreaming up imaginative, daring theories, many of which are related to his passion: magnetism. The custom-built equipment in his Pasadena "maglab" was capable of measuring magnetic fields a billion times weaker than those from a hand magnet. Kirschvink had studied the magnetic iron crystals in birds' brains that help them navigate, similar grains in bacteria that help them orient, and iron grains in rocks. Still, how much could faint magnetic signals tell us about life on the ancient Earth? The startling answer is—a lot.

The authors of the paper Kirschvink was reviewing made the surprising claim that the magnetic fields they detected in Australian rocks revealed that the rocks had been formed at the equator. Their theory was sound. As some rocks form, tiny iron crystals within them orient in the direction of the Earth's local magnetic field like tiny soon-to-be-frozen compass needles. At the Earth's poles, they point toward the center of the Earth, while at the equator they point horizontally. In the Australian rocks, the crystals' orientation was horizontal, but that seemed about as likely as finding an igloo in the Sahara. The rocks were a distinctive type that had clearly been deposited by glaciers, yet geologists were quite sure that the climate of the tropics had always been moderate and balmy enough for a swim.

Kirschvink did know of a few dissenting voices. A handful of geologists had reported finding distinctive rocks (out-of-place boulders and jumbled mixes of rocky debris) that showed evidence of having been transported by glaciers. Their claims, however, had long been dismissed. In the 1960s, the new theory of plate tectonics offered a simple explanation for how glacial rocks appeared in the tropics. The rocks had been created in cold latitudes and then carried to the equator by continental drift.

Kirschvink had yet another reason for skepticism of the authors' claims. He knew that the orientation of the tiny iron crystals in the rock could have changed if, after the rock formed, it was buried so deep underground that it heated up again. Yet the question of whether the equator ever had glaciers continued to nag at Kirschvink because he realized that he could perform a more definitive test. He decided he would analyze another rock from the same location in Australia. If pressure had folded and heated its layers, its tiny remagnetized crystals would all point in the same direction. If, however, the rock had not been reheated, the crystals' orientations would have changed as the layers bent.

Kirschvink was quite sure he would find that the rock had been remagnetized.

But he didn't. The magnetic signals changed with the folds. Presumably this told him that the rock formed at the equator: which meant that the equator might once have had glaciers.

That was provocative, yet hardly conclusive. Then Kirschvink was startled to make an intriguing connection. He had seen glacial rocks of the same age, about 700 million years old, in Canada, and just as in Australia, they were accompanied by thick layers of red rock. These were just like the iron bands that formed 2.4 to 1.8 billion years ago, when oxygen rusted the ocean's iron.

Why had these red bands abruptly reappeared? It was then that Kirschvink realized that the entire Earth must have been covered in ice. "It meant you had to freeze the whole planet, smother the oceans," he said. If photosynthesis was virtually shut down, that would not stop deep-sea vents beneath the ocean from continuing to pump iron into the waters. And when photosynthesis returned, a burst of new oxygen would have rusted this iron, creating a layer of red rock.

Still, Kirschvink had reasons to be dubious. It so happened that in the early 1960s, the heyday of atmospheric atomic bomb testing, scientists feared that nuclear weapons could destabilize the climate of our entire planet. The Russian geoscientist Mikhail Budyko had even created a model. Alarmingly, it revealed that there actually was a scenario in which the entire Earth could be covered by ice. If the planet cooled too much, and glaciers crept too close to the equator, the Earth would be caught in a runaway cooling cycle. The white ice would reflect so much of the Sun's heat that the cooling would swiftly accelerate. Glaciers would creep even closer to the equator, reflecting more heat in a feedback loop, and before you knew it, the entire planet would be frozen over. Everything. Not just the continents, but also the oceans. Budyko called this the "ice-catastrophe," because his calculations showed that, if this event were to occur, there was no way out: our planet would remain encased in ice forever.

Kirschvink realized, of course, that an eternal ice age had never happened, yet he couldn't shake off gnawing questions. Was it possible, he wondered, that we should take the rocks at face value? If the equator had once been covered by ice, how did our planet ever escape its deathly cold grip? One night he dreamed that he was trapped in the ocean beneath an ice-bound world. "I was under water," he said, "worried about, 'My God, how can we get out of this?'"

When he awoke that morning, the answer came knocking. He realized that even glaciers wrapped around the entire planet could not have prevented volcanoes fueled by hot magma from erupting. "Volcanoes don't give a damn about the ice. They're sitting there and spewing, and they don't care if it's ice around Greenland or Iceland." Volcanoes would have expelled carbon dioxide, a greenhouse gas, which would have collected in the atmosphere. Eventually, in what can be considered an extreme case of global warming, the atmosphere would have grown so hot that it would have forced the tropical glaciers to quickly roll back to higher latitudes. So, Kirschvink realized, it might have been possible for the Earth to escape a deep freeze after all.

But what on Earth could have tipped our planet into a frozen state to begin with? Surprisingly, Kirschvink learned of evidence that a worldwide freeze happened not once, but at least three times. The last two were about 700 and 640 million years ago. But the first one was much more ancient—about 2.4 billion years ago—and he could think of only one possible explanation for why it began. It occurred suspiciously soon after the time that, as Preston Cloud discovered, cyanobacteria were pouring oxygen into the atmosphere. Could photosynthesis have sent the Earth into a global deep freeze?

Mulling it over, Kirschvink realized that the early atmosphere of the Earth contained a lot of methane, a greenhouse gas that is much more effective than carbon dioxide at trapping heat. When cyanobacteria began growing like gangbusters and pouring out oxygen, not only would they have rusted the Earth and killed or driven off most other bacteria, but their oxygen would have also converted methane into carbon dioxide and water, thus removing the Earth's insulation blanket. Oxygen could have sent Earth's climate into a freezing tail-spin.

Satisfied that his seemingly loony theory fit the facts, Kirschvink gave it a catchy name, Snowball Earth. He seemed to have clinched the case that photosynthesis provoked one of the most extreme events in Earth's history. Unfortunately, his first article on the subject was initially buried like a layer of sedimentary rock. He submitted it to a collection of papers on the early Earth that were only published as a 1,400-page book four years later.

By chance, however, in 1989, at dinner at the International Geology Conference in Washington, DC, he happened to mention his theory to the Harvard geologist Paul Hoffman. A few years later, as Hoffman was conducting fieldwork in Namibia, he began to think seriously about it. It dawned on him that he was seeing rocks characteristic of glacial deposits—dropstones and accumulations of rocky debris—in land that once lay close to the equator. And he was struck by another rock formation above them: thick layers of white calcium carbonate. Back at Harvard, Hoffman and his colleague, Daniel Schrag, scratched their heads and generally twisted themselves into pretzels in late-night brainstorming, trying to figure out how the calcium carbonate rock could have possibly formed. At last, they decided that one of their zanier ideas actually jibed with Kirschvink's theory.

They realized that, at the end of Snowball Earth, the withdrawal of the ice would have created some of the wildest weather Earth has ever experienced. Towering glaciers rapidly retreated beneath an intensely hot atmosphere rich in insulating CO_2, and that clash would have provoked savage hyper-hurricanes. The ocean would have been thrown into a frenzy. Waves as high as three hundred feet would have incited great quantities of carbon dioxide and water in the air to create acidic rain. This would have rapidly weathered rock, depositing thick layers of calcium carbonate in the ocean.

Convinced that the evidence was now overwhelming, Hoffman

set out on a university lecture tour to persuade his colleagues that if they had walked on the equator during an episode of Snowball Earth, they would have faced temperatures of negative 50 degrees Fahrenheit, cold enough to freeze exposed skin, and if they wanted to touch solid rock, they would have had to dig through ice a mile thick. You can guess the response: unbridled skepticism. Extraordinary claims require extraordinary evidence. Their peers had yet to be convinced of the evidence. However, they also struggled against the "Too Weird to Be True" bias (the same one that Einstein fell prey to when he rejected the possibility that the universe might be expanding). An entirely iced-over planet was just too hard to imagine. Yet today, Snowball Earth is largely accepted. The only question is whether glaciers covered every last bit of the Earth in an icy snowball, or if the Earth was a slushball that still had open patches at the equator. Either way, most of the molecules now in living creatures were once entombed in an ocean and rocks smothered by cold ice.

In fact, Kirschvink believes that cyanobacteria nearly sent us into an eternal deep freeze. "We're damn lucky that the Earth wasn't a little bit further from the Sun," he said, "because if we had been a little bit closer to Mars, we would never have come out of the snowball." Instead, over tens of millions of years, volcanoes spewed out enough carbon dioxide to rewarm the planet and break the ice's spell.

How did that catastrophe affect the course of evolution? That's hard to say. Did Snowball Earth simply slow evolution down? Or did it spur it on? Both arguments have been made. Certainly, while the Earth was frozen, most bacteria and other microorganisms were wiped out. Yet some survived at volcanic hot springs, deep ocean vents, and other remote places. And it's possible that their isolation and the harsh conditions they had to adapt to may have spurred genetic innovation. No one can say for sure . . . yet.

We do know that when Snowball Earth ended, cyanobacteria were free to range far and wide once more. And not too long afterward, at least by about 2.1 billion years ago, a remarkable new type of cell abruptly appears in the fossil record that would change the world again.

This new arrival, called a eukaryotic cell, represents a stupefying jump in complexity, greater than the leap from the tricycle to the Space Shuttle. All multicellular organisms, from cannonball trees to thorny dragon lizards, are built from them. Eukaryotic cells are typically fifteen thousand times larger in volume than bacteria. Almost all have many more genes than bacteria. And unlike bacteria who arrange their genes in a single loop that bumps around in the cytoplasm with everything else, eukaryotes wall their genes off for protection in a nucleus (hence they are named for *karyon*, the Greek word for "nut"). But eukaryotes are also crammed with many other revolutionary features. They include organelles—specialized factories that carry out some jobs more efficiently and on a larger scale than bacteria do—jobs like burning sugar and oxygen to make energy, disposing of waste, and, in algae and plants, photosynthesizing. Eukaryotes also have intricate cargo transport systems that, like trucks on throughways, shuttle molecules around. How did these cells make such a huge evolutionary leap?

The strange answer came from a freethinking microbiologist, Lynn Margulis, who relished the role of a revolutionary. She was long ridiculed and dismissed as a crackpot. Yet like a once-scorned prophet, she would later be recognized for discovering the surprising ways that cyanobacteria, and the oxygen they unleashed, helped bring the ancestors of plants, and the likes of you and me, into being.

Margulis was passionate, outspoken, and ahead of her time. She was capable of debating and besting the most brilliant minds in biology. Some (principally male) colleagues also considered her arrogant and quick-tempered. "Lynn was good as a needler," one of her friends, the microbiologist Fred Spiegel, told me. "She was always thinking outside of the box. She would always push and push and push. She would tick people off. They would go out to prove her wrong, and a lot of times they couldn't." A quotation by General George Patton on her refrigerator read, IF EVERYONE IS THINKING ALIKE, THEN SOMEBODY ISN'T THINKING. "She liked to start trouble," her son, the writer Dorion Sagan, wrote. "But it wasn't trouble without a purpose."

Perhaps it was Margulis's precociousness that set her on the road to trouble. She was born Lynn Alexander. Her mother ran a travel agency in Chicago; her father was an attorney and businessman. Her parents drank and fought a lot. Lynn escaped into books and learning about nature and blossomed into an intellectual.

In 1952, at age fourteen, without bothering to tell her parents, she took an entrance exam for the University of Chicago. She was admitted. Two years later, on the stairs of the math building, she literally ran into a handsome "big shot" on campus named Carl Sagan. He was an articulate twenty-year-old graduate student studying physics, and his irresistible passion for science encouraged and cemented her own. A week after she graduated, they married.

It was at the University of Wisconsin–Madison, in a master's program in genetics, that she first encountered an idea she would be associated with for the rest of her life. She became fascinated by two organelles: chloroplasts, which carry out photosynthesis, and mitochondria, which burn sugar in plant and animal cells to create energy. What captivated her was the statement by one of her professors that these organelles resembled free-living bacteria so strongly that he suspected they once had been bacteria themselves. He thought they might even have their own DNA, which would be evidence that their ancestors once lived independently.

In 1960, Margulis and Sagan moved from Wisconsin to the University of California, Berkeley, where she enrolled in a doctoral program. Despite her thesis advisor's skepticism, she insisted on looking for evidence that the chloroplasts in *Euglenas*, single-cell photosynthesizing microorganisms, contained their own DNA. Most geneticists wondered why anyone would bother. One professor told her it would be like looking for Father Christmas. It had long been a central tenet of genetics that if a cell had a nucleus, all its genes were sequestered inside it—and these alone determined the organism's heredity. If by some fluke a chloroplast contained DNA, these genes must have escaped from the nucleus and no longer have any significance.

Yet Margulis read about earlier scientists from the 1880s through the 1920s who claimed that mitochondria or chloroplasts descended from bacteria. They included a German, Andreas Schimper; a Frenchman, Paul Portier; a Russian, Konstantin Mereschkowski; and an American, Ivan Wallin. Wallin even went so far as to claim that he could remove mitochondria from cells and culture them as free-living organisms. (He couldn't.)

Most scientists dismissed this speculation that mitochondria and chloroplasts were once independent bacteria. It seemed so unlikely. Surely they evolved from scratch inside cells, just like all the other structures there. Researchers had yet another reason for skepticism. Until the 1960s, bacteria were best known for spreading nasty diseases like anthrax, the plague, tuberculosis, and syphilis. Why would anyone think that the mitochondria in our cells came from germs? It seemed positively distasteful. Once again, the "Too Weird to Be True" bias swayed scientific minds.

One day as Margulis sat reading in the library, she had an epiphany. Could the title of Wallin's 1922 book, *Symbionticism and the Origin of Species*, have said it all? Was symbiosis—mutually beneficial cooperation between organisms—the cause of many of the greatest jumps in evolution? Perhaps symbiosis explained the origin of mitochondria, chloroplasts, and other huge evolutionary leaps. The idea, she recalled, hit her like lightning. But could she find proof?

Sadly, at this time, one of her other passions was fading. She and Carl Sagan both had scientific careers, and he was very supportive of her work—so long as it didn't impinge on the time he needed for his own. Like most 1950s husbands, Sagan, "who never changed a diaper in his life," expected her to cook, clean, mind their two kids, and do the bills. To add to their difficulties, as brilliant and charismatic as Sagan was, Margulis felt overwhelmed by his constant need for attention. (Her marriage, she wrote later, was "a torture chamber.") She left him, reunited with him when he moved to Harvard in 1963, and then left him again within a year. She soon married a crystallographer in Boston and adopted her new partner's last name, Margulis.

Although still without a doctorate, she was eager to champion the idea that mitochondria and chloroplasts, as well as cilia—tiny hair-like structures that propel cells like sperm—evolved from formerly proud independent bacteria. And by now, her former professors at Wisconsin, Hans Ris and Walter Plaut, had strengthened the case. With their electron microscope, they glimpsed DNA strands inside chloroplasts, and these resembled the DNA in bacteria. Moreover, two scientists in Sweden had also spotted DNA inside mitochondria. Yet neither group expended much effort trying to convince their colleagues that these organelles once lived independently. Margulis would.

While teaching at Brandeis University, working other jobs, and tending her kids, she managed to squeeze out the time to write an exhaustive paper that wove together evidence for her thesis from many different fields. That, in and of itself, was unusual. At Berkeley, she had been shocked by what she called "academic apartheid." She rarely saw scientists stray from their narrow lanes. Cell biologists didn't talk to geneticists, much less to geologists or paleontologists.

Margulis talked to everyone. She created a narrative of ancient evolution that she bolstered with evidence from microbiology, bio-chemistry, geology, and paleontology.

This was the mid-1960s, and Preston Cloud had just made his claim that about 2.3 billion years ago, photosynthetic bacteria suddenly introduced oxygen into the atmosphere. Margulis argued that, sometime after, one type of ancient bacteria became spectacularly efficient at using the oxygen to make energy. Fatefully, one of these was somehow ingested by another cell and, strangely, the enveloped cell was not kicked out or digested for dinner. Instead, it survived, and the cell and its captive came to a surprisingly happy accommodation. The engulfing cell provided food to its talented new cellmate, who could burn sugar with oxygen to produce energy more efficiently for both of them. It was a win-win. The captives, Margulis contended, evolved into energy-producing factories—mitochondria. And the descendants of this union became the first eukaryotic cells, the ancestors of all plants and animals, including us. "At root," the microbiologist John Archibald said, "she was claiming that even an individual human cell is a merger."*

Margulis went on to claim that in the same way, sometime later, a type of cyanobacterium that was extraordinarily skilled at photosynthesis found its way into a eukaryotic cell. The descendants of that cyanobacterium became chloroplasts, and the cell created by their union was the ancestor of all plants.

For years, she submitted her forty-nine-page paper to journals without success. Her grant applications fared no better. "Your research is crap. Don't ever bother to apply again," one reviewer wrote. She stubbornly refused to give up. In 1967, her article, "On the Origin of Mitosing Cells," was finally published after fifteen rejections.

It attracted much attention, yet most of her colleagues discounted and even ridiculed her work. It was considered fringe science. One reason was that when Margulis was self-confident, fierce, and passionate, some of her peers dismissed her as an abrasive woman with a fiery temper. But the nub of the conflict was that many scientists could not swallow the idea that evolution would take such a shortcut; it almost seemed like cheating. Biologists had long agreed that the *only* drivers of evolution were incremental mutations of genes. Margulis's theory was "retrogressive," one critic charged, because "it avoids the difficult thought necessary to understand how mitochondria and chloroplasts have evolved as a result of small evolutionary steps." It

---

* Most scientists now believe that this merger occurred between two major groups of microorganisms. A single-cell organism known as an archaeon swallowed an ancient bacterium, and each brought something different to the new cell they created.

seemed laughable to claim that some of the most important jumps in evolution occurred because one kind of single-cell creature was eaten or enslaved by another.

Margulis parried: How do you explain why mitochondria and chloroplasts have DNA of their own? And why, as biologists had long known, do both these organelles replicate independently of their cells? Evolution, she insisted, can be spurred by cooperation, not just by red-in-tooth-and-claw competition. William Schopf, a paleobiologist whom she consulted as she was writing her groundbreaking paper, told me, "It seemed like a harebrained idea, frankly, because that ain't the way evolution was supposed to work. She had a lot of courage. There were a lot of naysayers." For years, she remained defiant, and in the wilderness.

Then in 1975, just as it seemed that the issue would be debated from now to kingdom come, a new tool—genetic sequencing—appeared. Researchers could actually compare the RNA of chloroplasts and cyanobacteria. They turned out to be remarkably similar. In 1977, a comparison of the RNA of mitochondria and bacteria revealed that they too were related. Margulis was jubilant. It was a moment of triumph.

She was wrong about one major point, though. Expecting to find more examples of symbiosis, she also claimed that cilia—waving appendages in some eukaryotic cells—arose from a union with bacteria that could swim about. That theory did not hold up, but she was spectacularly right about the origin of chloroplasts and mitochondria.* Her idea that symbiosis explains two of the most important leaps in all of evolution has stood the test of time.

The implications were profound. Bacteria were "the greatest chemical inventors in the history of Earth," according to Margulis. They created chloroplasts, which would make plants possible. Bacteria and their microbial relatives, the archaea, invented the essential chemi-

---

* Today we know that we inherit our mitochondria only from our mother's egg. That makes mitochondrial DNA particularly useful for tracing maternal ancestry.

cal processes in our cells: They learned to make sugars, nucleic acids, amino acids, proteins, fats, and membranes. Huge colonies of the former bacteria known as mitochondria even produce our energy. "It may come as a blow to our collective ego," Margulis wrote, "but we are not masters of life perched on the top rung of an evolutionary ladder." Instead, "beneath our superficial differences, we are all of us walking communities of bacteria." Some like to sum up our relationship to microbes another way: "Germs are us."

Like Einstein and the astrophysicist Fred Hoyle, after making a great breakthrough in the face of fierce skepticism, Margulis leaned ever more on her intuition in her later years. "As her career progressed, she became less and less in touch with the data generated by frontline researchers. She preferred to work from first principles and generate her own ideas," said the microbiologist John Archibald. That, and her utter fearlessness, led her to champion questionable scientific theories like Gaia—the idea that Earth is a self-regulating organism. Margulis also promoted much more dubious ideas, such as the claim that a syphilis bacterium caused AIDS. Yet there is no denying that she stimulated a tremendous amount of fruitful research—and she opened our eyes to how the evolution of complex cells suddenly leaped forward.

Fossils unearthed in Australia, China, and elsewhere back her up. They reveal that by 1.7 billion years ago, if not earlier—after our atmosphere had oxygen and our planet was no longer encased in ice—eukaryotic cells first appear in the fossil record.

Although not everyone agrees, the biochemist Nick Lane and the microbiologist William Martin argue that the complex features in eukaryotic cells could never have evolved had one cell not swallowed another. It was one of the most momentous steps in all of evolution, and they think it happened only once. The ingested bacterium could generate much more energy than its host. As Lane and Martin see it, the bacterium's descendants—mitochondria—were cheap to maintain. One human mitochondrion has to support only about thirty-eight genes, while its ancient ancestor may have had three thousand of them. All the other genes our mitochondria need are now in the cell nucleus. So, if a cell needs more energy, making new mitochondria doesn't cost much. That allows our cells to support huge colonies of them without a lot of overhead. Lane and Martin argue that once mitochondria supercharged cells with energy, they could finally afford to create many more genes and use them to invent elaborate new structures and processes.

You take about twenty thousand breaths a day just to supply these former bacteria with the oxygen they need to burn sugar to produce energy for you. If you spread out all your mitochondria side by side, they would cover two basketball courts. Altogether, about a quadrillion of them keep you moving.

If a free-living bacterium, the ancestor of mitochondria, had not found a home inside another cell, would evolution have stalled? Would we all still be single-cell organisms? Lane and Martin think so, because bacteria and their unicellular relatives, the archaea, seem to have remained largely unchanged for the last 3 billion years. Lane and Martin, and the planetary scientist David Catling, suggest that, if we ever find life elsewhere in the universe, it is likely to be boring. It will look like microorganisms—unless evolution elsewhere also created cellular complexity with a similarly daring symbiotic leap.

Fossils of ancient algae, like those found in 1990 on Somerset Island in the Canadian Arctic, reveal that Margulis was also right about the next great leap forward. Their age suggests that by about 1.25 billion years ago or earlier, a eukaryotic cell swallowed a cyanobacterium that was skilled at photosynthesis. The consequences of this rare, "pretty freaky" event, as Nick Lane put it, were profound. The cyanobacterium evolved into chloroplasts, and the descendants of that merger evolved into algae, the ancestors of plants. Without that symbiotic union, plants (and humans) would likely not exist on Earth.

This new understanding left scientists scratching their heads over a bewildering new question. Eukaryotic cells—the ancestors of plants, animals, and us—were happily floating around in the ocean by at least 1.75 billion years ago. By around 1.25 billion years ago, algae, the more recent forebears of plants, had appeared. Moreover, we know that evolution can move very quickly; a mere 70 million years separates us from our small mammalian ancestors who were fearfully scurrying beneath the feet of dinosaurs. Yet large fast-moving animals in the oceans didn't show up until about 540 million years ago. And plants didn't appear on the continents until even later, roughly 500 million years ago. What took them so long? Why was there such a vast span of time in which so little seemed to have happened that some call it the Boring Billion?

In the 1990s, one reason for evolution's seeming laziness became apparent. You can't move fast without oxygen. As geologists dreamed up new techniques to tease clues about oxygen levels from ancient rocks, they made a surprising discovery: until about 700 million years ago, our atmosphere contained considerably less than 1 percent oxygen—not the 21 percent we luxuriate in today. It was not until about 540 million years ago that oxygen levels reached 5 to 10 percent in the atmosphere and were finally high enough in the ocean to support large active creatures. Before then, they simply wouldn't have been able to breathe.

Why did oxygen levels on Earth stay so low for so long? Researchers are still juggling a confounding number of competing theories. To understand one of the current favorites, you need only look at the bothersome green scum that clogs up our ponds and lakes—blooms of algae and cyanobacteria. They multiply rapidly when fertilizers rich with phosphorus and nitrogen drain into water. All organisms need these two elements to make membranes, proteins, and DNA. Phosphorus plays a starring role as well in the production of small molecular power packs that fuel the activities in our cells. Nothing can grow without phosphorus and nitrogen.

Now, finding nitrogen in the ocean a billion years ago may not have been a great problem. Cyanobacteria had figured out how to pull it from the atmosphere long before.

Researchers realized, however, that phosphorus may have been much harder to find. That's because, as the Earth cooled from a molten ball billions of years ago, phosphorus rose and hardened into patches of very lightweight rock that was found only on the continents. Phosphorus made its way to the ocean, where life first evolved, only as debris, as runoff from the weathering of these rare rocks. In 2016, a team of scientists from the Georgia Institute of Technology, Yale, and the University of California, Riverside decided to study ancient phosphorus levels. They painstakingly analyzed more than fifteen thousand samples of ancient oceanic rocks from around the globe, and they found that, until 800 million years ago, phosphorus levels remained quite low. Without more phosphorus, the existence of more photosynthesizing organisms, higher oxygen levels in the atmosphere and oceans, and the evolution of larger, more active animals were all on hold. It's possible that the shortage of one crucial ingredient delayed the appearance of big fast-moving creatures for many hundreds of millions of years. They would have to patiently wait for the appearance of the phosphorus locked in the Earth's crust.*

So what finally liberated this crucial mineral from the land? Most likely nothing less than a sandpapering of the continents. Lava flows may have brought some phosphorus to the surface where it was released by weathering. A recurring episode of Snowball Earth 770 million years ago, and a final one about 635 million years ago, also pitched in. As mile-high glaciers flowed toward and then away from the equator, rocky debris at their bases ground the poor mountains beneath them like sandpaper, shredding large quantities of phosphorus and other minerals and sweeping them into the ocean. Like rich fertilizer, these deposits encouraged the growth

---

* Many scientists warn that sometime, possibly in a few hundred years, we too will face a phosphorus crisis. Most of the phosphorus in our fertilizer comes from deposits that we are rapidly depleting. At some point in the future, we may be scrounging for enough phosphorus to supply the world's burgeoning population with food.

of green algal and cyanobacterial blooms on the ocean surface so immense that they may have hiked up the amount of oxygen in the atmosphere—at last making the evolution of large active oxygen-guzzling animals possible.*

Incidentally, the planetary scientist David Catling argues that if intelligent beings exist elsewhere in the universe, they also breathe oxygen. That's because the only other molecules capable of creating as much energy are fluorine, which makes organic matter explode, and chlorine, which destroys organics. Thus he thinks that any contemporaries elsewhere in space must also be fueled by oxygen created when water is split by photosynthesis. That means that the old science fiction films could be right. If aliens stepped out of a spaceship, they would feel right at home.

On Earth, after a billion years of relatively little happening, the Boring Billion was suddenly over. The first animals arose in our oxygenated ocean hundreds of millions of years before plants evolved on the continents. They appeared 700 to 800 million years ago—around the time of Snowball Earth and when phosphorus and oxygen became more plentiful—although whether the appearance of animals was directly linked to any of these is a matter of spirited debate.

Who were our first animal ancestors? On that intriguing question there is disagreement. We know that they were creatures who couldn't make their own food (as photosynthesizers did from air, water, and sunlight), but they were happy to dine on those who could. A number

---

* Geologists will want you to know that the rise in oxygen did not simply come from more photosynthesis. That's because, globally, photosynthesis and respiration and decay exactly balance out. Photosynthesis releases an oxygen molecule for each carbon dioxide it pulls from the air. The exact opposite happens in respiration and decay: an oxygen is consumed and a carbon dioxide released. So oxygen levels in the atmosphere could rise only if creatures made of carbon dioxide, like cyanobacteria, were prevented from decaying and returning their oxygen to the air. Luckily for us, that happened. Millions of years ago, great quantities of organic matter sank to the ocean floor, were buried, and became deposits of oil and gas. Unfortunately, when we burn these fossil fuels, we are returning carbon dioxide to the atmosphere, and that warms the Earth—an inconvenient fact we are facing today.

of scientists believe that we descended from comb jellies, organisms something like jellyfish. But I like the more generally accepted theory that our first animal forebear was a sponge. Every time I take a bath, I pause for a moment to savor how far we have come.

Sponges didn't have to chase prey, so they didn't need a lot of oxygen. But roving animals did. So, after the final chilly episode of Snowball Earth, when the oceans had more oxygen, it was pedal to the metal, evolutionarily speaking. Sometime about 575 million years ago, larger animals show up. "That is when the first card-carrying animals appear," the geologist Tim Lyons explained. "Bigness and motility were luxuries that came with high oxygen levels." That oxygen allowed them to make collagen as well, a strong, flexible protein that could bind shells, skeletons, and tissues together. We owe a lot to that innovation; about 30 percent of our protein is collagen (a word derived from the Greek word for "glue"). Most of it is in our cartilage, tendons, bones, skin, and muscles. Oxygen gave animals an infusion of energy and a way to keep themselves together. At last, about 510 million years ago, ancient fish were flicking their fins as they stalked other fish. Carnivorous fish feasted on smaller swimmers who dined on creatures that supped on photosynthesizing algae and cyanobacteria. The marine food chain that supplies much food to us was finally in place.

But photosynthesis was not done with our planet. Plants would arrive next. Their appearance was eased by the fact that photosynthesis had already transformed Earth in another profound way. When photosynthesis introduced oxygen into the air over 2 billion years ago, ultraviolet rays high in the atmosphere began splitting some of the O_2—creating a thin layer of ozone (O_3).* As luck would have it, that ozone cast a sunshade over the Earth that shielded its surface from 98 percent of the ultraviolet rays that would otherwise shred organic

---

* In the nineteenth century, scientists surmised the existence of this layer as they analyzed the continuous spectrum of light from the Sun. A narrow band of wavelengths was missing, and it corresponded to the wavelengths that ozone reflects, so they realized that the Earth must be surrounded by a layer of ozone that absorbs deadly ultraviolet rays.

molecules like razors. That made it safer for any enterprising life to leave the oceans. Sometime about 500 to 700 million years ago, algae accepted that challenge and began to invade the rocky continents. Eventually they would evolve into primitive plants such as mosses and liverworts—and then land plants.

Once plants spread across the continents, they sent oxygen levels shooting skyward. It was this last extraordinary infusion into our atmosphere about 300 to 400 million years ago that made it possible for fish to wriggle out of the oceans, and for their fast-moving oxygen-hungry descendants, like us, to live on land. Oxygen levels rose from about 10 percent to a staggering 30 to 35 percent, before falling to 21 percent, the amount to which we are accustomed today. Every year, photosynthesis pumps hundreds of billions of tons of oxygen into our atmosphere, of which you consume about thirty-six thousand gallons. About half the oxygen you breathe was produced by various forms of algae and cyanobacteria. For the other half, you can send a love letter to land plants.

Over more than 2 billion years, photosynthesis terraformed our planet from a world with Mars-like continents and oceans harboring only single-cell organisms into a green-blue planet filled with all kinds of exuberant life. It's hard not to wax rhapsodic over the magnitude of the changes that photosynthesis wrought. The toxic oxygen that cyanobacteria unleashed rusted the Earth and killed or drove away the cyanobacteria's competitors. They spread widely and introduced vast amounts of rocket fuel—oxygen—into the atmosphere. Suddenly, new souped-up cells crammed with mitochondria appeared. They could produce more energy and build more genes and proteins, and with that, life exploded in complexity. Some of these cells, with the help of photosynthesizing factories, or chloroplasts, jacked up oxygen levels even higher. Thereafter, ferocious predators and dazzling ecosystems appeared in the oceans, while photosynthesizing plants began turning the continents green.

Meanwhile, as our atmosphere was oxygenated, the principal atoms in living creatures entered the whoosh of a great Mixmaster. Carbon, nitrogen, phosphorus, sulfur, and oxygen traveled through the atmosphere and waters, were incorporated into organisms, sank to the bottom of the ocean, were thrust deep into the Earth by plate tectonics, and were expelled again in volcanic eruptions and as plates were pushed around, before finding themselves in plants and other life-forms once more.*

Of course, the ancient green plants that swept over the continents ultimately made our existence possible in one last way. While humankind obtains about 15 percent of its protein from fish, most of us are largely built from food that came from land plants. They make the building blocks of our bodies. Without the nutrients that plants assemble, you would not be here.

But before plants could make those nutrients, they would have to carry out mission impossible: they would have to conquer the continents' hard bare rock. How could they manage a challenge that sounds about as easy as making bread from concrete? Scientists seeking to solve that mystery would find that part of the answer is that plants turn out to be more clever than they let on—perhaps, some would say, more intelligent.

---

* It was the appreciation of this remarkable balancing act between life and the chemistry and geology of the Earth's surface that inspired the British scientist James Lovelock and Lynn Margulis to propose the controversial Gaia hypothesis. Few scientists accept Lovelock's claim that the Earth itself is a living superorganism. Nonetheless, virtually all agree that self-correcting feedback loops involving photosynthesis, respiration, mineral cycles, and plate tectonics have created unique conditions on our planet that make Earth habitable.

PLANTING
THE SEEDS

How Greenery and Its Allies Made Us Possible

> Shall I not have intelligence with the earth?
> Am I not partly leaves and vegetable mould myself?
> —*Henry David Thoreau*, Walden

On September 10, 1867, at the annual meeting of the Swiss Natural History Society in Rheinfelden, Switzerland, a placid town of salt baths and picturesque towers, a mild-mannered botanist named Simon Schwendener dropped a bombshell. Schwendener, thirty-eight years old, was a kindhearted, bewhiskered bachelor, a writer of poetry, a sensitive soul. Yet he didn't shy away from fiercely defending the ideas he believed in. In the same year that Karl Marx published *Das Kapital*, Schwendener proposed a theory that, to lichenologists, appeared equally revolutionary. It set off shock waves of indignation and outrage. Yet, we now know, his radical idea that lichens were not what they seemed helps explain how plants pulled off a truly astonishing feat.

You could say that plants rule the world. Although we often pay them little mind, they make up about 80 percent of the Earth's biomass, while land animals make up less than a tenth of 1 percent. (Single-cell organisms and fungi constitute most of the rest.) Plants blanket all of the Earth's surface that is not frozen or too dry. Yet, until 500 million years ago, there were no plants at all. Period. The mountains, valleys, and plains of our vast continents were cold hard rock: stark, windswept, and dull. Somehow, from this stone, air, and water, plants conjured up a vibrant tapestry of life. How did they pull off such an absurdly difficult feat that wholly transformed the continents?

Fittingly, Simon Schwendener, who would help answer this question, first intended to cultivate plants for a living. His father, a not-so-prosperous farmer, encouraged him to find a comfortable job in the civil service, but Schwendener was drawn to science. He followed his passion, although, he later confided to a colleague wistfully, the meager researcher's salary he received for years kept him from marrying. After earning his doctorate, Schwendener began to work with one of Switzerland's master microscopists. Microscopes were opening unknown vistas and Schwendener wanted to use them to plumb the deeper mysteries of biology. It was then that he began to study the humble organisms known as lichens.

Lichens are painfully slow-growing creatures that resemble small dry seaweed fronds. They happily subsist on bare rock, gravestones, and other terribly inhospitable places. Botanists considered them ancient plants—"primordial vegetation"—yet when Schwendener focused his lens on them, he saw only confusion.

To his inquisitive eye, they appeared to be something entirely different. They seemed to be two different types of creatures locked in a strange embrace. He saw thin white fungal threads encircling colonies of plump green algae, and the algae appeared to be trapped like a spider's prey. The "master is a fungus," he concluded, ". . . a parasite, accustomed to live upon the work of others. Its slaves are green Algae, which it has gathered around itself, at any rate, holds onto, and forces into service."

Schwendener's claim provoked an uproar within the passionate community of lichenologists. According to the well-established Linnaean classification system, an organism could belong to only one species, not two. Nor did fungi and algae even seem like they would be a congenial pair. "Destructiveness is a character of fungi," the author of *Fungi of West Cornwall* protested. "Whatever they feed upon they disease or destroy. . . . Yet, contrary to all experience, algae, when fed on by 'lichen-forming fungi' are supposed to flourish and grow." To the writer of *The Lichens of the Environs of Paris*, Schwendener's suggestion was "an assertion either of pure fantasy or a slander." The author of *A Monograph of Lichens Found in Britain* mocked Schwendener's sensationalized "Romance of Lichenology," with its tale of an "unnatural union between a captive Algael damsel and a tyrant Fungal master." Schwendener's theory, the nature writer Edward Step wrote, "met with the ridicule it deserved."

Once again, the "Too Weird to Be True" bias had exerted its powerful influence. Even in the 1950s, more than eighty years later, at least one eminent lichenologist still dismissed Schwendener's claim. Yet Schwendener turned out to be right in all respects, except perhaps one. It remains debatable whether the lichen's fungi and algae are locked in an exploitative master-slave relationship or enjoy a cozier partnership beneficial to both. Although the algae give up some of their sweet, photosynthesized sugar, in return the fungi give algae some of the minerals they eke from rock.

The fungi can do that because they possess a talent that's a little hard to believe: they can eat rock. Fungi, a unique form of life that has been around about a billion years, are distinct from bacteria, plants, and animals. They come equipped with two powerful tools. They excrete rock-dissolving acids. And they can push their tiny filaments into small cracks, bond firmly, and then create high pressure to make the rock crumble. That's how they get their minerals. The rest of their nutrients come from whatever dead or living organic matter they manage to find. An ancient alliance between fungi and algae who exchanged nutrients turned lichens into ridiculously hearty organisms.

They were photosynthesizers, acid excreters, supergluers, and pressurized crowbars all in one.

About 500 million years ago, primitive plants like low-lying leafless mosses and liverworts evolved from single-cell algae. However, they faced great challenges. Their algae forebears could afford to loll about in water while they waited for currents to obligingly deliver vital minerals to them. Living on land was not as cushy. There were some bacteria, algae, and fungi scattered here and there that began gradually building up a thin proto-soil—a thin crust of organic matter and minerals. But Schwendener's revolutionary discovery explained one reason that plants could have any real hope of invading the continents. Lichens were able to travel much more widely because they could both photosynthesize and find minerals. They paved the way for plants by helping make soil from rock.*

Even so, the earliest plants faced stiff odds. Any soil, where it existed, was still a sparse crust. Yet plants needed a continuous supply of minerals, such as phosphorus to make DNA, RNA, and proteins; magnesium and manganese for their chlorophyll; calcium to strengthen cell walls; and potassium, iron, and sulfur to make enzymes. The first plants lacked roots, and it wasn't as if they could carry pickaxes. So how could these would-be pioneers expect to wrest the copious amounts of minerals they would need from rock?

In 1880, when Albert Frank, a distinguished plant pathologist from Berlin, began rooting about in soil at the base of trees, he wasn't interested in that question. In fact, he wasn't studying ancient plants at all. He was after the beloved savory fungi called truffles. The Prussian minister of agriculture, domains, and forestry had commissioned him to determine if farmers could cultivate truffles. As Frank dug into the rich forest soil, he was astonished to discover thin threads

---

* A recent genetic study suggests that modern lichens evolved after plants. Nonetheless, explained the paleobotanist Paul Kenrick, older, now-extinct lineages of lichens most likely existed that helped create the first thin crust of soil. Some of these ancient lichens may have been partnerships of algae and fungi, while others were alliances of cyanobacteria and fungi.

extending from the truffles all the way to the tips of nearby tree roots. The threads wove a covering there so dense that it covered the root tips like a delicate glove, preventing the ends of the roots from even touching the soil.

What were the fungi doing? Others had seen them before and concluded they were merely parasites, but Frank looked more closely. He observed the fungal threads only on living trees, not dead ones. And he found them on trees both young and old. Surely if the fungi were parasites, the older trees would have suffered: yet they showed no apparent harm. Instead, he decided, he was seeing a surprisingly different kind of relationship that had a familiar ring. Frank, a general botanist, had not been offended, as so many lichenologists had been, by Schwendener's demotion of lichens from the plant kingdom. In fact, Frank coined the word *symbiotismus*, meaning "living together," to describe their relationship. Now, it seemed to Frank that he had stumbled on another startling alliance, this one between fungi and trees. The fungus, he claimed, was a "wet nurse" that feeds minerals and water to trees, which reciprocate with gifts of sugar.

One of Frank's many critics complained that his theory was "calculated to try our patience and credulity." But, like Schwendener, Frank was right. He named his fungal threads *mycorrhizae*, after the Greek words for "fungus" and "roots." Although individual strands are thirty times thinner than a thread, their webs are powerful force multipliers. They vastly increase the ability of a tree to absorb minerals. Today a cubic foot of soil may contain hundreds of miles of mycorrhizal fungi. About 90 percent of plant species carry on affairs with them.

In 1912, a fossil discovery revealed the ancient roots of this relationship when 407-million-year-old fossils from Rhynie, Scotland, were found to contain astonishingly clear images of primitive, rootless plants—and branching out beneath them are structures that look just like mycorrhizal fungi. The first plants were able to colonize the continents only because, in addition to getting the help of lichens, they created a mutual admiration society with webs of mineral-mining fungi. It was mycorrhizae, thin fungal threads, that first pried many of the minerals living in your body from hard rock.

Even then, however, any primitive plants who were contemplating taking over the land might still have hesitated. Through a quirk of chemistry, there was one crucial mineral that no fungi could help them find, and without it, plants could go nowhere. They needed nitrogen to make their DNA, RNA, and proteins, yet nitrogen was extremely rare in rock.

To add insult to injury, nitrogen was in the air all around them. Our atmosphere is 78 percent nitrogen; 500 million years ago, there was certainly plenty of it floating about. But a molecule of nitrogen gas consists of two nitrogen atoms. And the triple bonds between them are so strong that, like lovers clinched in an embrace, their atoms only have eyes for each other. They have no interest in bonding with anyone else. Nitrogen gas is virtually inert. That's why the great quantity of nitrogen you take in with every breath is simply exhaled again. And that is a good thing. Breaking the bonds between nitrogen atoms releases a lot of energy: think nitroglycerin or TNT. Yet for scientists, this created a maddening mystery. How on Earth did plants manage to find the nitrogen they needed? It would take astute sleuthing and two discoveries forty years apart before a "revolutionary announcement," as the journal *Nature* put it, revealed the surprising answer.

Mind you, the question was not just academic. It was the most pressing scientific question of the day. That's because Europe in the early nineteenth century was not able to feed itself. When harvests were poor, its burgeoning populations suffered devastating food shortages. This was something that the French chemist Jean-Baptiste Boussingault knew all too well. A shopkeeper's son, Boussingault grew up in a dark, miserably poor quarter of Paris. Many of his neighbors were ill-clothed and could find work only as ragpickers. "Their children suffered from hunger and cold," he recalled. "They would come to beg for bread and the kitchen left overs, of which there was so little at our house; then the father or the mother would fall ill, followed by privations." Children were frequently orphaned.

So years later, when he was visiting the coastal plain of Peru, one particular sight left a deep impression on him. Farmers were turning their sandy clay soil into fertile fields by adding just a single ingredient—guano—otherwise known as bird and bat poop.

Boussingault was intrigued. He knew that guano contained lots of nitrogen-rich ammonia (NH_3). Now he began to wonder, could nitrogen be as vital to a plant as oxygen to a fire?

Throughout his life, Boussingault specialized in doing things for which he had never been trained. Certainly his start in life did not appear promising. In elementary school, he recalled, "we passed from class to class like iron bars in a rolling mill." His teachers treated him like a blockhead. He understood little. Feeling hopeless, he dropped out at age ten and found employment helping a friend clean the laboratory of a famous chemistry professor, until he was sacked because of his young age. Luckily, his parents abandoned their hopes that he would become a pastry chef or pharmacist and allowed him to pursue his own interests, which generally meant reading science books, such as the four-volume chemistry text his mother bought him. At age fourteen, he began attending university science lectures, which were open at the time to visitors willing to stand in packed halls.

At sixteen, Boussingault gained admission to a school of mines. For two years, he studied geology, chemistry, and other relevant subjects. He worked briefly in a mine in Alsace. Then he received an unusual break: an offer to teach at a mining school in Peru. It was the best thing that ever happened to him. The famed naturalist Baron Alexander von Humboldt urged him to follow in his own footsteps and use the opportunity for extensive exploration and research. With von Humboldt's encouragement, Boussingault traveled widely in Latin America and, despite little formal education, wrote numerous letters and scientific papers on the continent's geology, geography, meteorology, and the customs of its indigenous peoples. Ten years later, he sailed back to France, married the daughter of a well-to-do Alsace farmer, and found himself in charge of an extensive estate.

At this point, he was in a position to begin another pursuit for which he had never been trained. He turned himself into an agricultural chemist and made his father-in-law's farm the first agricultural research station.

In his open-air laboratory, Boussingault was eager to return to the questions concerning guano that had so intrigued him in Peru. Was nitrogen crucial to the growth of plants? If so, by what hidden process did they obtain it? He began meticulous agricultural trials to find out. In the small lab he built on his estate, he measured the nitrogen content of many fertilizers, such as manure and straw. He demonstrated that the most powerful fertilizers were the richest in nitrogen. Once this was widely recognized through Boussingault's work and others', farmers looked for the cheapest source of nitrogen they could find. They began importing vast quantities of guano from South America, fueling a lucrative guano boom.

In 1836, Boussingault began to investigate a related question: Why was it, he wondered, that farmers who stuck to the old ways and planted only grains eventually exhausted their soil, while farmers who rotated grains with peas, clover, or other legumes had land that remained productive? In a large-scale five-year experiment, Boussingault rotated his crops while keeping track of exactly how much nitrogen was in the fertilizer he applied. In one impressive trial, his yield of sugar beets, wheat, clover, and oats contained 105 pounds more nitrogen than he had added to the exhausted soil. That was astonishing. Moreover, he discovered a clue as to why. He found that grown stalks of wheat contained no more nitrogen than their seeds. But clover, a legume, mysteriously increased its nitrogen content by a third. It was as if legumes were pulling nitrogen out of a hat, or rather, out of thin air. Yet where that nitrogen came from remained a frustrating enigma.

Decades later, in the 1880s, two patient Hermanns, the agricultural chemists Hermann Hellriegel and Hermann Wilfarth, revisited this mystery. They wondered if something in the soil was helping plants obtain nitrogen. At their well-financed agriculture research station in Prussia, they decided to plant two groups of peas in sandy soil that lacked nitrogen. The two groups were identical in all but one respect: the soil of one had been sterilized by steam.

They found that the peas in the unsterilized soil flourished and had bumps on their roots, which they, like others, suspected stored excess nitrogen. Yet for some reason, the peas grown in sterilized soil were sickly.

Now they were baffled. What secret ingredient could make unsterilized soil so much more productive? Grasping at straws, they mixed some of it with water to make a slurry. Then they added a small amount of it, just over a tablespoon, to sterile soil in which they planted pea seeds.

Their results were not what they expected. The peas now thrived. And the researchers discovered that bumps that they and others had seen on the roots of legumes were packed with helpful guests— bacteria.

Unexpectedly, Hellriegel and Wilfarth had stumbled on another cooperative arrangement. Plants can't pull virtually inert nitrogen from the atmosphere, but they can use the nitrogen found in ammonia (NH_3). Bacteria, it turns out, are the only life-form that has found a way to make ammonia from nitrogen gas. So bacteria give ammonia to their legume hosts and in return receive sugar for the energy-guzzling operation of pulling nitrogen from the air.[*]

In 1886, Hellriegel presented their findings at a conference in Berlin. Cries of "bravo!" filled the room. Suddenly Hellriegel and Wilfarth were rock stars of the agricultural world. In a boon to farmers everywhere, it was finally clear how "highly gifted" legumes like clover, peas, and beans improve farm yields immensely. Soon after,

---

[*] We now know that a legume's roots even initiate an underground conversation. They release a chemical that attracts nearby bacteria, who in turn send a chemical signal back asking for permission to enter the roots.

scientists discovered that free-living cyanobacteria also add nitrogen to soil. Plants would not exist without the nitrogen they ultimately obtain from bacteria.

That process that bacteria pull off is not just crucial to life, but extraordinarily difficult. In 1908, the German chemist Fritz Haber attempted to invent a method of making ammonia fertilizer (NH_3) in a factory. (Yes, the same Haber who later directed the first use of chemical warfare.) Several decades after South America's guano boom began, little guano remained. Scientists warned that Europe would face the specter of mass starvation again unless they could learn how to manufacture fertilizer. Haber, a German Jew, heeded as well the warnings of the Kaiser's generals, who feared that without a secure supply of nitrogen, Germany would not have enough explosives to win its next war. He mixed nitrogen and hydrogen gases in a tank that contained a metal, a catalyst that would encourage them to react. He found, however, that the nitrogen atoms would not loosen their grips on each other unless he also subjected them to infernal temperatures and pressures: on the order of 800 degrees Fahrenheit and 250 times atmospheric pressure. In 1918, Haber was awarded a Nobel Prize for a process that arguably exceeds the importance of radar, the personal computer, and sliced bread combined. Much of the nitrogen within you was likely extracted from the air and embedded in ammonia in hot pressurized tanks. Today, crops grown with factory-made fertilizers feed billions of people. Without manufactured ammonia, the Earth's population would be almost 50 percent smaller, and you and I might not be here.

Our bodies each contain about four pounds of nitrogen, much of it in our DNA and proteins. Some was extracted from nitrogen gas in factories. The rest, as Hellriegel and Wilfarth discovered, bacteria pulled from the air.

Hundreds of millions of years ago, when small earth-hugging primitive plants like mosses and liverworts evolved, they were able to begin spreading over the continents because, like ancient tribes that adopted hunting dogs, they had allies. They had the close companionship of bacteria and fungi. Together they joined forces to liberate the atoms within us from rock and air.

Theoretically, at least, once plants had the means to find nutrients, they were free to take over the continents. But only if plants could grow larger and upright. And how was a limp low-lying moss to do that? Only by coming up with many ingenious inventions.

Among them, they would have to develop backbones—or rather stems—to stand up. And for that they needed stronger building materials. Luckily, their algae ancestors already made a fiber called cellulose that they could use to strengthen their cell walls. By weight, a strand of cellulose can be many times stronger than steel. A tree is 40 to 50 percent cellulose, which now is the most abundant organic compound on our planet. Yet cellulose has a drawback. It weakens when it gets wet. So plants used another substance—lignin—to waterproof and cement together cellulose in their cell walls. Lignin is the second most abundant organic molecule on Earth.

With these fantastically strong building materials, plants grew stems and began their improbable climb into the sky. To make more energy, they began ferrying water and nutrients through long veins to their versions of solar panels: leaves lined with chloroplasts to catch the rays of the Sun.

But plants still couldn't rise very high if their weight threatened to topple them over. And they needed more water and minerals to feed their leaves high above the ground. So they deployed one of their most amazing inventions of all—roots. While the thickest part of a root may look impressive, it is largely just a conduit, drill, and anchor. Most of the crucial action takes place just above the root tips, where tiny silk-thin appendages forage for nutrients. These are the root hairs. By one estimate, a single rye plant may have fourteen billion of them. "They suck up all the nutrients. They are mining machines," the botanist Simon Gilroy said. In addition to pulling in water and any minerals dissolved in it, they also have other tricks up their sleeves. Minuscule nanomachines called proton pumps expel hydrogen ions into the dirt. When soil particles latch on to that hydrogen, they loosen their holds on other minerals that other pumps and channels swiftly drag inside. (Incidentally, while we lose minerals through sweat and urine, plants never let minerals go. That means, as the botanist Jim Mauseth explained, "the way plants are set up is that they take minerals in from the soil and trap them, and that's how we get our minerals. We don't go out and eat dirt.")

In roots' ceaseless search for minerals and water, they also learned to venture far and wide. The groundbreaking ecologist John Weaver, the author of many books and articles, such as "The Wonderful Prairie Sod" and "Who's Who Among the Prairie Grasses," was one of the first to discover just how deep plants will dig. Over four years, beginning in 1918, Weaver and his strong-backed students wielded shovels, ice picks, and toothpicks to expose the intricate structure of the roots of more than 1,150 prairie plants. The deepest-digging plant they discovered, alfalfa, burrowed thirty-one feet down, deeper than a two-story house. Some plants do better. The roots of a South African fig tree listed in the *Guinness Book of World Records* took advantage of a cave system to penetrate four hundred feet into the ground.

Hundreds of millions of years ago, armed with stems, leaves, roots, and other brilliant innovations, plants spread across the continents. In the process, they performed a feat akin to making stone soup. Their leaves plucked carbon and oxygen from the air above, their roots excavated minerals and water from the ground below, and with these, plants began creating an in-between layer of life.

Even then plants faced unnerving challenges. In fact, if you stop to think about it, you may wonder how they survive at all. To make it in this world, they have to begin their lives by taking a stand. They must commit to living out their days in a single spot, with no option to change their minds, much less flee. They are forced to cope with constant changes in light and the seasons. Though wind, rain, snow, and hail lash them, and water turns to ice, they can't run for shelter. They must confront drought, floods, and competitors for minerals and light. Then ravenous creatures arrive. Yet, like Spartan warriors, plants remain rooted to the patches of ground they've claimed. They survive only through their wily defenses and their remarkable ability to eke out nutrients from within their reach. In other words, to create the world from which we arise, plants needed to develop an extraordinarily important quality, one that would prove as crucial as any other to their success: they had to become insanely adaptable to whatever fate throws at them.

They learned to do that, first, by turning into biochemical geniuses. Unlike animals, plants produce hundreds of thousands of complex molecules that are not for internal use. They deploy them to fend off competitors, attract pollinators, communicate, and scare creatures that want to devour them. In plants' fight for survival, chemicals are their weapons of choice. (They are particularly good at poisoning animals.) That's why they make so many of our drugs, such as salicin, a relative of aspirin (found in the bark of willow trees), the cancer fighter Taxol (in yew trees), and the malaria drug quinine (in the Andean cinchona tree). To mess with the brains of predatory insects, plants make dopamine, acetylcholine, GABA, and the serotonin precursor 5-hydroxytryptophan: all neurotransmitters also found in our brains. To drive off insects and other animals, plants synthesize nicotine, caffeine, morphine, and opium. "Why do plants make cocaine?" the biochemist Tony Trewavas asked rhetorically. "Can you imagine what it's like for an insect to chew those leaves? What you find, of course, is that most insects decide *not* to chew those leaves." And the spices that we use to flavor our food? Plants produce most of them, as well, to keep animals and microbes at bay.

How could plants have learned to make so many compounds? In the early 2000s, scientists found one part of the answer. Researchers were decoding the genomes of organisms for the very first time by sequencing and counting their genes. Many expected that because humans are so sophisticated and intelligent, we would have at least

100,000 genes. They were shocked to discover that we have far fewer, only about 24,000 (the latest count is lower still). When a similar effort by twenty international institutions first decoded the genome of a plant, geneticists expected to find that "simple" plants have far fewer genes than we do. But in the first plant genome they decoded, belonging to a small, fast-growing weed called thale cress, they counted 25,498 genes. A ginkgo tree has 40,000. A Golden Delicious apple has 57,000, more than twice as many as us.

Incidentally, scientists were also startled to find that we have so many genes in common. About a third of your genes have counterparts in bananas. That is, the genes have similar functions and the proteins they code for have similar features, all of which strongly suggests a common ancestry.

Plants were able to turn into such brilliant chemical innovators because errors in reproduction doubled their chromosomes. This condition was usually lethal, but any rare survivors were left with an abundance of extra genes. While most of these duplicate genes lost their functions, sometimes a plant's descendants could adapt some of them for new uses.

In fact, plants evolved so many sophisticated adaptations that some researchers suggest they are downright smart. Such claims have set off one of the most heated furors in contemporary biology. Is it possible that plants—the creatures that make all the building blocks within us—succeeded so well because they are actually intelligent?

That suggestion has a surprisingly impressive pedigree. In the last paragraph of Charles Darwin's book *The Power of Movement in Plants* (1898), he wrote, "It is hardly an exaggeration to say that the tip of the radicle [the seedling root] . . . having the power of directing the movements of the adjoining parts, acts like the brain of one of the lower animals. . . ." After Darwin, other scientists also occasionally observed plant behavior that appeared intelligent.

However, in the 1970s, the publication of a book, *The Secret Life of Plants*, tarred the field of plant physiology with the taint of pseudoscience. The book profiled experiments by a former CIA interrogation expert who hooked plants up to polygraph machines and claimed that they responded to human emotions. His experiments were never replicated, and the book horrified most respectable biologists. It made any further suggestion that plants could be intelligent appear to smack of parapsychology.

Then, in 1981, two Dartmouth scientists, Jack Schultz and Ian Baldwin, attended a conference of the American Chemical Society in Las Vegas and heard a remarkable presentation. Their colleagues David Rhodes and Gordon Orians reported that when caterpillars attacked the leaves of willows, neighboring trees began producing toxic chemicals to repel insects, as if they had been warned.

Could trees really communicate? Rhodes and Orians's idea was heretical and easy to dismiss. They conducted their experiment outdoors, making it difficult to eliminate many other potential explanations of their results.

But immediately after listening to their presentation, Schultz told me, he and Baldwin looked at each other and said, "Well, *we* could actually answer that question in a much more controlled fashion." They realized that they could conduct a similar experiment in their laboratory where it would be possible to rule out other causes. In their greenhouse, they placed poplars and maples in plexiglass boxes and simulated a caterpillar attack by tearing their leaves. The researchers discovered that the wounded trees responded by wafting chemicals into the air (tannins that keep leather from rotting); neighboring trees, connected only by an air duct, detected these and also began making compounds to deter predators.

The press loved it. Schultz and Baldwin made *People* magazine and the front page of the *New York Times*. The *Boston Globe* enthused, "Injured Trees Sound the Alarm."

Many of their colleagues, however, were dismissive. It didn't help that the *National Enquirer* had been the first paper to run their story. "The headline was something like 'Scientists Show Trees Talk,'"

Schultz recalled, with a mixture of humor and regret. "A couple months later, they published a follow-up with a banner that said, 'You read it here first.'" Schultz and Baldwin's colleagues were not amused by even the hint of the suggestion that plants communicate. "It seemed too woo-woo," one botanist observed.

"I made enemies. Scientists hated the idea," recalled Schultz. "There were a couple of papers published, fortunately mostly in not very popular places, accusing us of BS. I spoke at the International Etymological Congress in Vancouver about that time. And someone in the audience actually stood up and said, 'You know, this absolutely cannot be true. Plants cannot do this.' Which is very unusual in the scientific community to have someone stand up like that. This was a big audience, a thousand people or so, and if someone accused you of making up a story or something, that's really quite unusual, and for a young scientist pretty scary."

It wasn't until two decades later that the distinguished plant biochemist Anthony Trewavas decided to break this taboo of modern biology. Trewavas, a member of the Royal Society, Britain's premier scientific organization, is tall with a long face and gray bushy eyebrows. He has a commanding presence and speaks in long, polished paragraphs. In 2003, his lengthy paper called "Aspects of Plant Intelligence" argued that, well, plants are intelligent. "The major problem," he wrote, "is a mind-set, common in plant scientists, that regards plants basically as automatons." He meant to change that.

Trewavas's definition of intelligence is simple. "When an organism is placed in threatening or highly competitive circumstances and modifies its behavior to increase its chances of survival, it is exhibiting intelligent behavior," he wrote in an e-mail. "Plants respond to certain signals by changing their structure," he explained. "That is a way they can improve the probability of their survival. How are they doing that? What are they assessing? Well, if it is called intelligence in animals, it ought to be called intelligence in plants because the characteristics of the behavior in biological terms are identical."

"How was your paper received?" I asked.

"I think a lot of people thought I have probably gone round the bend," he replied.

This was not the first time that Trewavas had proposed an unorthodox idea. He began studying plant hormones in graduate school in 1961. Some years later, he stirred up a hornets' nest by suggesting that these hormones work in much more complex ways than his colleagues appreciated. He had always read widely in science, and in 1972 he read a book called *General Systems Theory* by the theoretical biologist Ludwig von Bertalanffy, which profoundly influenced his thinking. Trewavas came away from it convinced that in the complex systems found in individual cells, ecosystems, and even plants, multilayered networks interact and function in much more sophisticated ways than most scientists recognized. Complex outcomes and emergent properties arise that are not easily understood as the sum of their parts. Yet most scientists clung to the reductionist belief that, if they could just identify every simple component in a plant, they would understand how the plant worked as a whole. To Trewavas, that was as likely as expecting that by studying the structure of a country's branches of government, you could predict how its leader would respond to next week's events.

In 1991, influenced by this framework, Trewavas performed experiments that transformed his view of plants even more radically. In a small, dimly lit room in the botany building of the University of Edinburgh, he and postdoc Marc Knight placed a plant seedling inside a luminometer. This microwave oven–size box had a tube inside that could detect extremely low levels of light. Trewavas and Knight knew that, in animals, the mineral calcium plays many roles in cell-to-cell signaling and nerve transmission. They wanted to know if it also played a role in sending signals between plant cells. But tracking minute real-time movements of calcium in tiny cells seemed impossible, until Trewavas realized he might go about it in a peculiar way. A colleague told him about an unusual protein in jellyfish—nicknamed the burglar-alarm protein—that actually glows when it detects changes in very low levels of calcium. Trewavas and Knight decided to try inserting the gene that makes this protein into a plant. They grappled with technical challenges for a year until, at last, they had made a transgenic plant. Now they placed it in their luminometer. If they touched a tiny spot on a leaf, would that trigger calcium signals to cells farther away? They pushed a wire through a tiny hole and gently poked the leaf.

The plant glowed.

We all know that plants move slowly. They just stand around and don't do much. Typically, if you apply a stimulus, you won't see a plant respond for days or weeks. So Trewavas was stunned to see his seedling light up instantly when he touched it. The cells sent calcium signals to their neighbors in milliseconds. Graphs of the luminescence revealed spikes that to Trewavas looked just like action potentials—rapid changes in electrical charge that travel along our nerves. And the signals spread through entire seedlings, not just a few cells. "It made such a difference to my perception of what was going on," Trewavas said. "Here we were getting what looked like a nerve cell signal, and yet these are not nerve cells. These are plants."

In just a few weeks, they discovered that calcium signals are extremely sensitive and are set off by almost everything that plants respond to in the wild: touch, wind, cold, heat, different wavelengths of light, even invading fungi. This suggested to Trewavas that plants may constantly monitor and transmit all kinds of information about their environment. And that made him wonder if plants and animals might be more similar than scientists thought. He began thinking more broadly about how plants might use all that information—and how it might influence their behavior. As far-fetched as it sounded, Trewavas was undaunted. "When I see a problem, I am one of those people who tries to get a solution."

It's common for botanists to lament that their colleagues who study animals receive the lion's share of publicity and attention. Plants just don't get much respect. We're mesmerized by suspense, stalking, the hunt, the kill; growth of an inch a day, not so much. Researchers speak of plant blindness: our tendency to overlook plants in a landscape. If we are fish, they are the water. To put it another way, they are an unobtrusive green backdrop on our stage.

Even as Trewavas was conducting his experiments, other studies were also challenging the assumption that plants' behavior was simple. Time-lapse cameras and investigations into plant senses, genetics, hormones, signaling, and the behavior of roots were revealing unappreciated sophistication. All of this was food for thought for Trewavas as he began to wonder: Are plants smarter than we give them credit for? In the wild, he recognized, they face enormous challenges just to stay alive. "Plants have been around the same length of time as animals," he concluded. "There is absolutely no reason that they shouldn't have developed as good and cunning methods for survival."

In 2003, Trewavas published a lengthy provocative article on plant intelligence. It opened the floodgates. A number of chemists, biophysicists, and botanists had been harboring similar thoughts. Now they found encouragement to voice them. Led by Stefano Mancuso of the University of Florence and František Baluška of the University of Bonn, they founded the Society for Plant Neurobiology to discuss their new vision of plants. In 2005, at their first conference, in Florence, Italy, they discussed interplant communication, plant information processing, plant senses, and memories. Mancuso and Baluška were among those who also spoke of plant neurotransmitters and neuronlike cell behavior.

The reaction from the greater community of botanists to the suggestion that plants have neurons, synapses, or anything at all analogous to an animal's nervous system was fierce. In a scathing joint rebuttal, thirty-six plant biologists asked, "What long-term scientific benefits will the plant science research community gain from the concept of 'plant neurobiology'? We suggest these will be limited until plant neurobiology is no longer founded on superficial analogies and questionable extrapolations." In 2009, in fact, the Society for Plant

Neurobiology changed its name to a more acceptable one, the Society of Plant Signaling and Behavior.

The change did little to moderate the ideas of some of its members. They point out that plants have over fifteen senses, including all of ours except hearing—and even then, plants respond to vibrations, such as those of a caterpillar chewing, so in a limited sense, you could say that they hear as well. Plants smell chemicals in the air and taste others in the ground. They sense gravity and touch. They detect neighboring plants with photoreceptors sensitive to infrared light. They constantly integrate a tremendous amount of information about their environment, and they use calcium, proteins, and hormones to send internal signals about it. Some cells also send action potentials— the nervelike spikes of electrical activity that Trewavas observed are propagated by calcium and other chemicals. Some of these signals' functions have been discovered, like telling leaves to make insecticides or telling the trap of a Venus flytrap to close. Others remain unknown. Plants even send a different kind of electrical signal (called a slow wave potential) that is still less well understood.

The botanists also noted the remarkable flexibility of plants' behavior. Each time an insect or microbe attacks, plants emit a cocktail of airborne chemicals that are specific to that type of threat. (Schultz suspects that these warning signals are likely intended for other parts of the plant, but eavesdropping neighbors pick up on them as well.) Plants also retain "memories" of harmful events like drought, which allow them to cope better with future ones. And they adapt their growth to suit their circumstances. Plants in windy conditions grow shorter with thicker stems and smaller leaves. "If you grow plants in really nice, pleasant conditions—no wind, no harsh temperatures, no animals crawling over them—and you put them outside, they don't do well," the plant geneticist Janet Braam said. "They don't put energy into getting tough." That's why gardeners harden greenhouse seedlings by gradually introducing them to the outdoors, and Japanese farmers tread over wheat and barley seedlings in their fields.

Scientists have also discovered that trees in forests are connected by a Wood Wide Web—a remarkably broad underground network of roots, bacteria, and mycorrhizal fungi. Trees use the web to communicate and share nutrients. Sugar that is made in the leaves of a beech tree can end up high in a neighboring spruce.

Of course, plants don't have centralized brains like us. That's a clever strategy on their part. If we lose our head, or a limb, we're in trouble. But if plants lose something, they just grow another one. Plants possess a different kind of intelligence, one that distributes their "decision making" more broadly and democratically. For instance, they have no fixed blueprint for growth. Instead, they choose the best angles and heights to grow new stems, branches, and leaves. Those critical decisions are made by an inner layer of their shoots called the cambium. Trewavas believes that a plant's cambium constantly monitors the productivity of its branches and decides which should receive growth hormones and nutrients, and which to limit or even cut off. Overall, he suspects the internal signals that plants process "are likely to be as complex as those in a brain."

Of all the parts of a plant, its roots appear particularly intelligent. They are brilliant at finding the minerals that they, and we, need. Baluška, at the University of Bonn, and Mancuso (who founded the unapologetically named International Laboratory of Plant Neurobiology, near Florence), take this argument the furthest. They think that a region near the root tips, called the root apex transition zone, behaves even more like a command center or a brain. They like to cite an observation by the famed nineteenth-century biochemist Justus von Liebig: "Plants search for food as if they have eyes." Plants easily snake around stones as they prospect for water and minerals. When water is scarce, they seek it out more actively. They are also eerily skilled at detecting nutrients like phosphorus and nitrogen and following their gradients like trails of crumbs. Once they sense a rich patch of nutrients, they move toward it with "explosive growth." Then they pause, grow dense rootlets, and mine that area before moving on. Plants can also be nasty. Their roots sometimes release chemicals to keep their competitors' seeds from germinating.

Roots and shoots communicate through a two-way hotline to share information about the availability of water and nutrients and to determine how to respond to sensory information. Provocatively, Mancuso and Baluška argue that this communication includes some kind of rapid electrical signaling that roots also use to decide how they will grow.

Even more controversially, they claim that plants are not just aware of their environment and "purpose driven," but may also be conscious and have a sense of self. For that reason, they suggest, it may be time for us to rethink our own place in the world. "Therefore we should be aware," they write, "that any living unit equipped with complex sensory systems and organs is 'constructing' its own worldview, which might be radically different, but principally not better or worse, from our human-specific worldviews."

Trewavas marvels that "there is so much we didn't know even fifteen years ago. One thing that has always struck me is that so many scientists' attitudes are built on the fact that we don't know about things. And we assume that, if we don't know about something, it doesn't exist." Trewavas won't fall for the "As an Expert, I've Lost Sight of How Much Is Still Unknown" bias. Nonetheless, on the question of plant consciousness, he remains circumspect. "You aren't going to get an answer because you can't ask a plant anything. If you look at how it behaves, that's as far as you can deduce. It could be conscious. That doesn't say it is. And that's as far as you can go. But that is still an important change in understanding."

Nonetheless, most researchers still balk at any suggestion that plants could be conscious, never mind intelligent. "I think that the complexity and dynamic nature of root behavior is so impressive," Schultz said, "that people have trouble imagining it without an integrated system like ours. But you can describe everything that roots do on the basis of simple responses to environmental signals." Still, like most botanists, he believes that plants are exquisitely sensitive to their environment, and they respond to it by behaving in remarkable ways that we are just beginning to understand. "If plants were simpleminded," Braam told me, "if they didn't have all this complicated stuff going on, we would have been able to figure them out. I'm a plant biologist, and they constantly confuse me."

With their bacteria and fungi brethren, brilliant innovations such as leaves, shoots, and roots, and remarkable adaptability, if not intelligence, plants began their reign of the continents. They began creating the final tangled pathways that our atoms traveled to reach us. About 500 million years ago, as primitive fish swam in the oceans, plants began to conquer every part of the continents that were not too arid, salty, or iced over. And animals were not far behind. Previously in the ocean, the first animals, possibly sponges, didn't have to bother making their own food because they could feed on photosynthesizers like cyanobacteria, which did. Small animals soon offered tempting targets for larger ones that looked like tasty snacks to even toothier ones. Before photosynthesizers in the ocean knew it, they were supporting an entire ecosystem of animals. That story would be repeated. As plants later appeared on the continents, they offered filling meals to any animals who could leave the ocean behind. One of the creatures who would come from the sea was a fishy four-finned crawler. Its descendants, who evolved lungs to absorb oxygen from the air, spawned amphibians, reptiles, birds, and mammals and us. We rely on plants for our sustenance as much as fish rely on photosynthesizers in the ocean.

It bears remarking that plants, in partnership with fungi and bacteria—their constant companions—made their own beds to lie in. They created their own soil, ecosystems, and ever larger pools of nutrients. Although plants die, their atoms don't. They are recycled into soil, the oceans, sedimentary rock, the atmosphere, and other life. Thus, although your soul may or may not have been reincarnated, the atoms inside you had past lives in many other organisms large and small. Some of the nitrogen in the fingernail of your right thumb may have once drifted in the air, before it was pulled into the root of a clover, embedded into ammonia by bacteria, and then used to make a protein in a leaf that was eaten by a moth, which decomposed next to a mushroom that you ate three weeks ago in a salad.

While other top predators like lions turn up their noses at vegetables, the creatures they hunt still eat plants. And that means that plants ultimately make or collect the building blocks of virtually everything (except water) that you are made of.

Incidentally, if you've ever wondered why we don't make our own food—in other words, photosynthesize—you need only look at a large tree. If photosynthesizing chloroplasts generated all our energy, they would take up as much space as a tree canopy. Simply covering our skin with chloroplasts wouldn't provide us with enough power to walk around, much less run, stalk, or catch prey. Eating plants or other animals gives us a fast, concentrated shot of energy. That allowed our hunter-gatherer ancestors to range over many miles to find our next meal. But, as free as we may feel, make no mistake, we are dependent on plants. Put another way, if *we* disappeared, plants would do just fine. If *plants* disappeared, we would vanish from the Earth within weeks or months.

It is through plants that our atoms, which came from the Big Bang and stars, finally arrived at our front doors, or rather, mouths. Almost every single atom within us, water and some salt excepted, reached us through plants. But how are those mere atoms reorganized within us to create life? For many years, scientists had no idea at all.

PART IV

FROM ATOMS TO YOU

In which we are astonished to discover what we must eat to survive, the most eminent experts' teachings are overturned, and we uncover many astonishing secret mechanisms within our cells that turn the food on our dinner plates into us.

SO MUCH
DEPENDS ON
SO LITTLE

What Do You Need to Eat to Survive?

Imagine all the food you have eaten in your
life and consider that you are simply some of that
food, rearranged.
—*Max Tegmark*

Think of your dinner plate. But don't think of that pizza, chicken masala, cassoulet, or dim sum as food. Think of them as atoms bound into molecules. We now know where those atoms came from, how they were made, and how they ended up at our dinner tables. But once eaten, how do those molecules create a living person? How are they organized to create a master blueprint in our cells and the astonishing molecular machinery that animates us? Before scientists could solve these mysteries, which we will ultimately get to, they needed to answer a more fundamental question: What are the substances that plants make and gather that we have to eat in order to build ourselves?

The story of how scientists discovered them begins with a "fiery and impetuous" investigator, Justus von Liebig.

Until the German scientific revolutionary Justus von Liebig boldly strode onto the scene, chemists could say almost nothing about how our cells or bodies worked, much less what we are made of. In the nineteenth century, that ignorance meant that millions were suffering and dying from malnutrition. In portraits, Liebig's determined gaze and visage bore a certain resemblance to Napoléon Bonaparte; like the French general, he was ambitious and brilliant, and he knew it. He did not shy away from proposing radical ideas or picking nasty quarrels with competitors. As one admirer observed, Liebig "built up new kingdoms on the ruins of empires overthrown."

In the year 1840, he decided it was time to apply his great knowledge and intellect to a problem most scientists were sure would always lie beyond our comprehension. He wanted to understand how our food is refashioned into us. And the first step, naturally, was to find the identity of the molecules we build our bodies from. In Liebig's day, this question had a decidedly practical relevance as well. He was asking what we needed to eat to survive. Liebig charged into his inquiry brimming with confidence, yet rarely have so many great advances been inspired by a theory so wrong.

Liebig's interest in chemistry was born in the workshop of his father, a merchant in Darmstadt who prepared dyes, paints, varnishes, and boot polish for sale. Liebig found school, especially the required memorization of Greek and Latin, painful. His classmates, following the example of the assistant headmaster, called him *schafskopf*—sheep's head. His headmaster informed him that he was "the plague of his teachers and the sorrow of his parents." He dropped out at age fourteen. His father then apprenticed him to an apothecary. Liebig, however, decided he wanted to be a chemist, so he withdrew, choosing instead to spend his time engaged in two favorite activities: reading chemistry books and experimenting in the workshop. Two years later, he enrolled in a university. His promise was soon recognized. After a year, he was awarded a scholarship to study in Paris with the celebrated chemist Joseph Gay-Lussac. Liebig recalled that time fondly, especially because Gay-Lussac insisted that they dance jubilantly together around the room whenever they made a new discovery.

At the age of just twenty-one, he received a teaching appointment at a small university in the medieval town of Giessen. Scientifically speaking, it was a "provincial backwater." The professor of anatomy did not believe in the circulation of blood. The university had just one building, and the other chemistry instructor refused to share his small lab space in the Botanical Garden. Ever bold and ambitious, Liebig persuaded the Duchy of Hesse-Darmstadt to transform a guardhouse at a nearby abandoned military barracks into a laboratory. He pitched it as a school to train pharmaceutical chemists, but he soon had bigger plans. He observed that, in Germany, chemistry was regarded as a mere adjunct subject, and that stuck in his craw. It was usually taught by professors of pharmacy or medicine, or worse, as a craft, which reduced it simply to "rules useful for making soda and soap or manufacturing better iron and steel." Liebig's vision was much loftier. As he saw it, "Alles ist Chemie," everything is chemistry. "The consciousness dawned on me," he recalled in his autobiography, that the mineral, vegetable, and animal kingdoms were all united and connected by the same chemical laws. So by rights, chemistry should stand equal to the other sciences, if not above them.

He would in time turn his guardhouse into a marvel, one of the most advanced laboratories of the day. An illustration of his legendary lab shows young men, many in frock coats and top hats, working at crowded lab benches in a well-appointed room with spacious windows and modern gas lamps. Visitors admired the cupboards filled with flasks, condensers, and other apparatus conveniently placed within easy reach. A central furnace drew noxious gases out of fume hoods, also a rarity.

More significantly, he had an inspiration that would profoundly transform universities everywhere. It occurred to him that he could accomplish more work with the help of students than he could on his own. So he began to assign them challenging research projects that he merely supervised—an innovation that would be recognized by any graduate student or postdoc today who is swilling coffee in a laboratory late at night.

In quiet Giessen, Liebig made many advances. He invented a tool that allowed chemists to determine the carbon content of organic molecules much more precisely than before. It turned an exasperating and tricky daylong process into an easier one that could take just an hour. With his apparatus, chemists could now make rapid progress in identifying the composition of organic molecules. He made other fundamental contributions to the emerging field of organic chemistry (the chemistry of living organisms). As a result, between 1830 and 1850, over seven hundred students from around the world flocked to his laboratory. Many of them would go on to become leading lights in the field.

His influence increased when he founded the first journal devoted to organic chemistry. As its editor, he used his bully pulpit to castigate and insult those who disagreed with his ideas. One of Liebig's unlucky targets lamented, "Storming and raging, his pen flies over the paper, mad with fury against everyone who dares to make a single remark upon his opinions." In a typical bon mot, Liebig argued that one colleague's theory "has arisen out of a complete ignorance of the principles of true scientific research." Moreover, referring to this theory, he added, "It is the curse that lies in a false opinion that carries in itself the seed of new errors, which are brought forth into the world as . . . miserable creatures that die when exposed to a healthy atmosphere."

Liebig's fame grew with the publication of a book, lauded for its bold imagination, that explained how chemistry could be applied to agriculture. He was a contemporary and rival of Jean-Baptiste Boussingault at a time when Europe was struggling to feed its population. Liebig's "law of the minimum," his claim that the growth of plants can be limited by the lack of any key mineral and that farmers should consider this when adding fertilizers, was an eye-opening revelation. By the 1840s, he was acclaimed as perhaps the leading chemist of his generation.

Yet Liebig was still frustrated. Despite organic chemistry's many advances, almost nothing was known about our own chemistry. Medicine had been dominated for two thousand years by the theory of the Greek physician Galen. He had believed that diseases were created by imbalances of four essential fluids or "humors": blood, black bile, yellow bile, and phlegm. That's why bloodletting to restore the balance of humors was common and why one of Britain's first medical journals was named *The Lancet*. In Liebig's day—the age of the steam engine—physiologists had also begun to think about the body in a new way: as a mechanical machine. The heart was clearly a pump, for instance, and until relatively recently physiologists believed that the stomach agitated or kneaded food to digest it. Still, it seemed inconceivable that chemistry could ever hope to explain the miraculous transformations that take place in living things like a seed in the ground or a baby in the womb. Something else was going on that just couldn't be explained. Although the atoms in our bodies bond to form molecules, just as they do in rocks, that was where the similarity ended. "In living Nature," the great chemist Jöns Berzelius observed, "the elements appear to obey laws quite different from those they obey in dead nature." Similarly, the American physician Charles Caldwell, who steadfastly opposed introducing chemistry into medical curricula, argued that "the principles of chemistry and vitality are universally acknowledged to be different from each other." It seemed that the mystery of life could be explained only by the presence of an inexplicable "vital force."

Liebig didn't buy it. Drawing on his unparalleled knowledge of organic chemistry and his by now limitless self-confidence, he decided to confront the baffling question of how we make use of the chemicals we eat. What kinds of molecules are in our food? And which of these are transformed into muscle, flesh, and energy?

Liebig began, promisingly enough, by building on work that the London physician William Prout had conducted several decades earlier. Prout, like Liebig, had faith that chemistry would reveal much about our bodies. This led him to search for chemical signs in urine that could diagnose diseases. His boundless enthusiasm for research inspired his more general interest in animal secretions, and so he spent a great deal of time analyzing the feces of boa constrictors and other animals, in addition to the contents of their stomachs. One of his great breakthroughs was discovering that their stomachs, and ours, produce hydrochloric acid to help digest food. (We make about six tablespoons a day.) Prout was also the first to propose that our food contains three essential substances, which he called "the saccharine, the oily, and the albuminous." We know them as carbohydrates, fats, and proteins. As proof, Prout observed that all three are found in a mother's milk. If we did not need them, it was evident to Prout, God would not have put them there.

Liebig agreed with Prout that our food is made of carbohydrates, fats, and proteins. So far so good. But then, emboldened by his past successes, he waded swiftly and decisively into murky waters. An apparent breakthrough by a colleague made a profound impression on him. Just two years before, the Dutch chemist Gerhard Johan Mulder, who was busily analyzing the composition of organic substances, believed he had discovered that the proteins found in plants and in our blood are virtually identical. In fact, he named this albuminous substance protein, from the Greek *protos*, meaning "the first."

From this apparent breakthrough, Liebig drew a sweeping conclusion. If these proteins are identical, and ultimately we eat a plant-based diet, we must obtain all our protein from plants. "Vegetables produce in their organism the blood of all animals," he wrote. In short, animals don't make their own protein. This was, in fact, consistent with everything known, but in his enthusiasm, he had fallen into the error of accepting an unconfirmed experimental finding. That was his first mistake.

Liebig's next leap was even more questionable. Knowing that carbohydrates and fats contain only carbon, hydrogen, and oxygen, while proteins also contain nitrogen and sulfur, he analyzed the composition of wild animals. He observed that they are always lean, and he failed to find carbohydrates or fats in their muscles and organs. He concluded that their bodies, and ours, are made entirely of protein.

Protein, he decided, is the only substance we eat that serves as a raw material for building our cells and bodily tissues.

He knew of one other observation that seemed to back him up. Our urine contains a great deal of a molecule called urea, which, like proteins, is full of nitrogen. So it appeared obvious to Liebig that when our muscles do work, they must break down some protein to create energy, and this must be the source of the nitrogen we excrete as urea.

In this tidy scheme, that left only a supporting role for fats and carbohydrates. We consume them, he argued, for only one reason: to burn them like coal to heat ourselves and stay warm. As Henry David Thoreau wrote in *Walden*, "According to Liebig, man's body is a stove, and food the fuel which keeps up the internal combustion in the lungs."

In 1842, Liebig presented this groundbreaking theory in his book, *Animal Chemistry: Or Organic Chemistry in Its Application to Physiology and Pathology*. He revealed that the only nutritional substances we need are proteins, carbohydrates, fats, and a few minerals. Ultimately, he declared, they all come from plants. And protein is king. It is the only material that our muscles and tissues are actually made of. (No doubt this was a great disappointment to his Bavarian compatriots who, at the time, believed that they needed to drink beer to build strong muscles.)

Liebig's ideas were, at first, widely embraced. In a foreword to his book, a Scottish chemist wrote that he "experienced the highest admiration of the profound sagacity, which enabled the author to erect so beautiful a structure on the foundation of facts, which others had allowed to remain for so long a time utterly useless. . . ." A British physician gushed that a conversation with Liebig "filled me with admiration and appeared like a new light where all had been confusion and incomprehension before." Although some questioned his conclusions, the great respect with which many held the "living scientific pioneer" helped enshrine his theory as textbook wisdom. In other words, many were influenced by the "World's Greatest Expert Must Be Correct" bias.

Now, to further advance the science of chemistry, Liebig set about applying his knowledge to problems of nutrition. In the process, he would make himself quite wealthy. One of his inventions—Liebig's Soluble Food for Babies—was the first scientifically designed baby formula. He also invented a beef bouillon paste made from boiled meat, which Liebig's Extract of Meat Company later marketed under the trade name OXO. Despite Liebig's claim that bouillon was a stimulant to the body, it had little nutritional value. No matter, it remains a tried-and-true flavoring to this day.

Just about the time that Liebig's products began to make him rich, telling critiques of his theory began rolling in. In 1866, it occurred to the Swiss scientists Adolf Fick and Johannes Wislicenus that they could easily test Liebig's theory. They needed only simple equations of physics and an invigorating hike. At five a.m. one cool misty morning, they set off on an eight-hour walk up a Swiss mountain peak. They had stopped eating protein the previous day and faithfully collected their urine during the climb and for six hours after. Then they analyzed its nitrogen content. From that they would determine how much protein their muscles should have consumed to make energy. A few more calculations and they had their answer. Knowing their body weights and the height they had climbed, it was easy for them to figure out how much work they had performed. Was the energy released from the protein their bodies had presumably broken down enough to power their climb?

Not by a long shot. It amounted to only about half of what they needed. Instead, it was evident that they must have produced energy another way—not by breaking down protein, but by burning carbohydrates and fat.

Two of Liebig's former students, Carl von Voit and Max Joseph Pettenkofer, conducted an experiment that turned out to be equally damaging. They built a sealed room, "a respiration chamber," that was large enough for a person to live in for days, and furnished it with a table, bed, and chair. But they didn't ask their experimental subjects just to lounge around. They were to work by turning a heavy hand crank 7,500 turns over 9 hours. All the while, von Voit and Pettenkofer precisely tracked every substance going into and out of their subjects' bodies to prove Liebig's theory. They fed them a no-protein diet and measured the nitrogen content of their urine and feces. Meanwhile, an analysis of the air coming out of the chamber told them how much carbon dioxide their subjects exhaled, which was a measure of the carbohydrates or fat they burned. Their results were a disappointment. When the subjects turned the hand crank, their nitrogen production remained unchanged. They didn't seem to be consuming protein to produce energy, as Liebig claimed they would. Instead, a spike in their carbon dioxide production suggested just the opposite—they were generating energy by burning carbohydrates or fat. Liebig was wounded. He and his students came up with convoluted arguments to salvage his theory, but by now the handwriting was on the wall. Liebig had been spectacularly wrong.

Still, you have to give the guy credit. His kick in the pants pushed scientists to consider questions they never thought it would be possible to explore. For the first time, scientists began to learn what humans are made of. We are more than just protein. Some structures in all living cells, including our muscles, tissues, and bones, are also made of carbohydrates and fats.

Liebig was also mistaken to conclude that we can't make proteins ourselves. Proteins are long folded strings made of twenty types of amino acids (each composed of about twenty atoms). Any proteins you eat, in fact, are disassembled into amino acids; then your cells link these up again, like beads on a necklace, to make new proteins.

Liebig also had our fuel source wrong. Normally our energy comes from burning carbohydrates and fats. In trying situations, like famine, we do consume some protein from our muscles, as Liebig suspected, but our bodies resort to this desperate measure only if they have no other choice.

That's not to say that all of Liebig's ideas have vanished. One of them is still alive and kicking. If you find yourself thinking that you need to eat a high-protein diet in order to build muscles, you can thank Liebig. In popular culture, that idea has never gone away. Nonetheless, once you have enough protein, eating more won't help you build more muscle—just more fat. Sadly, the only way to make more muscle is to expend more energy.

Despite many missteps, Liebig remained highly respected, deservedly so for his many innovations. One of his core teachings became universally accepted—we need four types of molecules in order to build our bodies.

Liebig correctly deduced that the first three, the "dietetic trinity," are made by plants. Plants transform carbon dioxide and water into sugar. Then they transform sugar into the three substances that, if you exclude water, make up roughly 90 percent of our mass: proteins, fats, and carbohydrates (chains of sugars). Liebig also identified a fourth ingredient that we need to build our bodies—a few minerals, such as sodium and potassium. These four kinds of molecules form the scientific basis of the ingredient list in Liebig's Soluble Food for Babies, "the most perfect substitute for mother's milk." Unfortunately, no one suspected that his list was incomplete, which would explain why babies raised solely on his formula did not thrive. It turns out that we have to eat one more type of molecule to assemble ourselves.

Unhappily, a lack of this last kind of substance was responsible for four exceptionally gruesome diseases. In the seafaring age between 1500 and 1800, scurvy killed about two million sailors, many more than those who died in battle. Throughout Asia, a pernicious disease called beriberi sporadically paralyzed and killed millions. Pellagra, memorably known for its four Ds—dementia, dermatitis, diarrhea, and death—afflicted the poor in Europe and America, particularly many in the American South who primarily ate bacon, cornbread, and molasses. Rickets deformed the bones of the children of rich and poor alike. Growing up in Arkansas during the Great Depression, my own mother-in-law's sisters were stricken by it. Until scientists could discover the reason for these inexplicable ailments, countless victims would suffer and die hideous deaths.

Some clues, however, had long been visible, including a particularly promising one that appeared half a century before Liebig was even born. In 1747, a thirty-one-year-old British naval surgeon named James Lind stood one day on the rolling deck of the HMS *Salisbury*, a three-masted ship of war outfitted with fifty cannon. As they patrolled the Bay of Biscay, off the coast of France, Lind relished the fresh air, a welcome relief from the stagnant hold below and the vexing mystery he faced there.

It had been only eight weeks since they'd left port, and already forty of the three hundred sailors on board had contracted scurvy. The men limping to Lind's sick bay had putrid gums and red, blue, or black spots resembling bruises on their skin. They were lethargic and losing the strength to walk. He knew that, if the disease grew too advanced, he would have to cut away their grossly swollen gums just so that they could swallow their food.

In the British navy, this was hardly unusual. Scurvy was common on longer voyages. Lind was all too aware of the single worst incident, as it had happened just seven years earlier. The navy had dispatched a squadron of eight ships under the command of Sir George Anson to attack Spanish galleons in South America. Three and a half years later, Anson returned with a treasure so vast, he needed thirty-two wagons to haul it to the Tower of London. But only about 400 of his 1,900 men returned with him. Most had died of scurvy.

It was not that the navy completely ignored the disease. The problem was, there was no agreement on how to cure it.

Yet, this knowledge had once been known, at least by some. Two hundred years before, many ship captains could have told you that scurvy breaks out on long voyages that deprive sailors of fresh fruit and vegetables. The writer Stephen Bown observes that in the seventeenth century, captains made mad dashes from port to port in an attempt to outrun the disease. It was also known that lemon juice could prevent or cure it. In his 1617 textbook, *The Surgeon's Mate*, John Woodall recommended lemon juice daily. The Dutch East India Company even established plantations in the Cape of Good Hope and Mauritius to provide lemons for their crews.

Over time, unfortunately, the knowledge of lemon juice's beneficial properties somehow vanished. The reasons were many, including simple complacency. When the incidence of scurvy grew worse again, there was resistance to citrus. Lemon juice was expensive and some shipowners suspected that merchants touted the imaginary medicinal powers of lemons just to drive up the price. At the same time, physicians were peddling a confusing variety of many other supposed cures. As author David Harvie observes, there were even "anti-fruiters," who claimed that lemons hurt rather than helped sailors on some expeditions.

Lind had seen relatively little scurvy himself until, on their ten-week voyage the previous summer, eighty of his crew had been laid low. As he cast about for an explanation, he noted that the rainy cold weather they encountered had made it hard for the crew to dry out and fostered stale air in the hold. Lind wondered if this bad air was the culprit. He also contemplated the possibility that the lack of a proper diet was to blame. Yet that seemed unlikely. "They had been afflicted by scurvy," he would write, "even though the captain supplied the crew with mutton-broth fowls and meat from his own table." On Lord Anson's ships, Lind noted, scurvy had broken out in spite of a plentiful supply of what he believed to be adequate provisions and good water.

Despite Anson's staggering loss, the brass in the British Admiralty displayed a disastrous lack of urgency. There was a great difference of opinion about its cause. Was it overcrowding? An excess of salt? Bad air? Some believed that only sluggish and lazy sailors succumbed to it. Moreover, even if they were to accept that for some strange reason lemons helped prevent it, carrying large crates of lemons on long voyages would entail great expense and was impractical besides, because lemons and lemon juice spoil. Perhaps most important, scurvy usually passed over the officers and higher-ranking seamen. So it simply seemed more expedient to replace casualties by pressing more unwitting men into service (often through trickery or kidnapping) than it was to shoulder the burden and expense of trying to prevent the disease.

Lind, newly promoted to ship's surgeon, was horrified by scurvy.

Having a sound scientific mind, he requested permission from his captain to search for a remedy by conducting an experiment that is considered by some to be the first clinical trial in all of medicine. Lind divided twelve sailors suffering from scurvy into six pairs and lodged them in hammocks in the ship's forehold. He doled out a different remedy to each: either cider, sulfuric acid, vinegar, seawater, or oranges and lemons. The unfortunate sixth pair received a formulation that one of Lind's colleagues recommended: an unappetizing paste of garlic, mustard seed, dried radish root, a tree resin known as balsam of Peru, gum myrrh, and for good measure, an occasional dose of barley water with tamarind along with cream of tartar to purge the system.

After a week, he ran out of fruit and had to end his trial. It was by now evident that only two of the remedies had any effect. The cider appeared to help a little bit, while, incredible as it seemed, the citrus largely cured the disease—so much so that one sailor returned to duty, and Lind put the other to work nursing his companions.

You might think that Lind would immediately jump up and down yelling "Eureka," because he had just proven, for all time, that something in citrus fruit cured scurvy. Not a chance. The unfortunate Lind was mired up to his hips in intellectual quicksand—the confusing medical theories of his day.

Lind gave himself time to make sense of his work. He retired from the navy, earned a medical degree in Edinburgh, and established a practice as a physician. Then he settled down to review many accounts of scurvy by others, before finally and conclusively explaining it.

In 1753, six years after his landmark experiment, Lind published a 456-page opus. Although the results of his experiment may seem clear-cut, his conclusions could have been, well, more conclusive. This is the point in our story where one wants to say, "Wait, wait! Can't you see?" After perceptively reviewing fifty-four other works on scurvy, he only gets around to his own trial a third of the way through the book—and devotes just five paragraphs to it. He was confident he had shown that citrus could cure scurvy, yet he struggled to explain the malady's cause. Concepts of disease at the time were a complete mess.

They were dominated by Galen's idea that sickness resulted from an imbalance of bodily humors. So Lind concluded that on ships, a combination of poor diet and moist cold air blocked perspiration, and this trapped putrid unwholesome humors inside the body. He explained that citrus could open up the skin's pores, but in a later edition he conceded that other medicines could also do the same. "I do not mean to say," he opined, "that lemon juice and wine are the only remedy for the scurvy. This disease, like many others, may be cured by Medicines of very different and opposite qualities to each other, and to that of lemons." As the author Frances Frankenburg observed, "If there was ever a researcher who doubted his own findings, it was James Lind."

On the bright side, Lind did recommend that sailors use lemon juice to prevent the disease. But he followed that sound suggestion with an uncharacteristically sloppy error. To prevent the juice from rotting, he suggested it should be heated to make a syrup—little suspecting that heat destroys the juice's curative powers. To add to the confusion, many distinguished physicians championed other cures that were entirely ineffective. One sea surgeon wrote sourly, "Dr. Lind reckons the want of fresh vegetables and greens a very powerful cause of the Scurvy; he might with equal reason, have added fresh animal food, wine, punch, spruce beer, or whatever else is capable of preventing this disease." Lind's critic went on to recommend rice as a remedy, or a mixture of one-fourth brandy and three-fourths water. Scurvy rampaged on, unabated.

In 1756, three years after Lind published his treatise, the Seven Years' War broke out between Britain and France. Of the 184,899 sailors who enlisted or were pressed into the Royal Navy, only 1,512 were killed in action. Another 133,708 expired from disease—primarily scurvy. Scurvy continued to hamstring the British navy during the American Revolution that followed soon after. If the Admiralty had provided lemons to their crews, some argue, the British might have prevailed against the colonies, or at least held off France's navy and negotiated a more favorable settlement.

It wasn't until 1795, a year after Lind's death, that the Royal Navy began issuing lemon juice to sailors. For a time, scurvy actually ceased to be a problem. But after taking one fruitful step forward, the navy leaped two steps back. Eighty years later, they switched to limes, which they could buy more cheaply from plantations in the British West Indies. Henceforth, British sailors were, of course, known as limeys. But regrettably, limes were much less effective at preventing scurvy, and this cast doubt on the value of any citrus juice as a cure. Even in the early twentieth century, when doctors agreed that fresh fruit and vegetables could treat scurvy, they still could not agree on the disease's cause, which is why, in 1912, scurvy plagued the British explorer Robert Scott's meticulously planned expedition to the South Pole. His conviction that bacterial food poisoning was to blame likely hastened his own demise. After hundreds of years, scurvy's cause still remained a mystery.

It was a Dutch war of conquest that brought to light the decisive clue to scurvy's true nature. In the late nineteenth century, the Netherlands added a Muslim sultanate in northeast Sumatra to the Dutch East Indies, now known as Indonesia. Their invasion provoked a ferocious guerrilla war, and a horrid malady called beriberi contributed to the carnage. In 1885, it afflicted 7 percent of Dutch troops and many more of their native soldiers. Elsewhere in the Dutch East Indies many hospital patients were also dying of the disease.

The government appointed an eminent pathologist with the appropriately Dutch name of Cornelis Pekelharing to find out what caused beriberi. He in turn recruited a neurologist, Cornelis Winkler, to help.

The two physicians had every reason to think they would swiftly succeed. Little more than a decade earlier, Louis Pasteur had become a French national hero when he demonstrated that virtually invisible enemies, called germs, transmit disease. Pasteur had shown that bacteria transmit deadly anthrax, which was periodically devastating European livestock. Just several years later, the German physician Robert Koch identified the bacteria that were responsible for tuberculosis and cholera. Now the race was on to find the germs responsible for many other diseases.

In 1886, Pekelharing and Winkler traveled to Berlin to visit Koch and pick up some pointers. As the story goes, at the elegant Café Bauer, they asked for a Dutch newspaper to peruse over their coffee. They were directed to a young man with an exuberant mustache who was already reading it. On approaching his table, they were delighted to find themselves speaking with another Dutch physician, Christiaan Eijkman. The sad-eyed twenty-nine-year-old had already served in the East Indies, seen the effects of beriberi, and was equally keen to discover its cause. A bout of malaria had forced him to return home—after claiming his young wife. Nevertheless, he was uncowed by the prospect of returning to the tropics. He had come to Koch's lab to study bacterial diseases, and so he eagerly signed on as an assistant to Pekelharing and Winkler.

In February 1887, the three physicians arrived at the scene of the fighting, Aceh province on the northern tip of Sumatra, where beriberi was endemic. With a hospital laboratory at their disposal, they began their investigation. Soon they determined that beriberi affected the nervous system. It induced a surprising variety of symptoms, such as swollen legs, difficulty walking, paralysis, heart problems, and an appalling lack of sensation. "I found my legs and feet perfectly numbed and swollen, and the space around my mouth, reaching nearly to my eyes, felt numb," one sufferer recalled. Eijkman was shocked to learn how quickly troops could fall ill. A soldier who hit a bull's-eye at target practice in the morning might lie dead that same night.

Something else made beriberi particularly strange. It rarely appeared in native villages, yet ran rampant in the army, in hospitals, and in prisons. Incarcerating a prisoner for several months to await trial could be tantamount to a death sentence.

In their new lab, Pekelharing, Winkler, and Eijkman immediately searched for the bacterial culprit to blame, but this was more difficult than expected. At first they failed to find any bacteria in the blood of sick soldiers; then they did detect bacteria, but they also found them in the blood of healthy troops, which, they concluded, meant that the germs must spread through the barracks in almost no time. Yet when they injected bacteria cultured from beriberi sufferers into dogs, rabbits, and monkeys, the animals seemed unaffected, unless they were injected many times. This seemed odd.

Nonetheless, after eight months, Pekelharing and Winkler decided that their task was largely complete. They concluded that the germs responsible were probably inhaled. They must be endemic in tropical regions where heat and humidity fostered their growth. And the bacteria spread extraordinarily swiftly. So the physicians recommended sterilizing infected buildings from top to bottom—although, they admitted, if the soil outside was also contaminated, it might be easier simply to move the inhabitants to a new location to prevent reinfection. Satisfied, Pekelharing and Winkler returned to Holland, leaving the thirty-year-old Eijkman to mop up, to find the identity of the bacteria that were responsible.

Thus in 1887, Eijkman found himself in charge of a laboratory, on the grounds of a military hospital full of beriberi sufferers, in Batavia (now known as Jakarta), the capital of the Dutch East Indies. His space consisted of two sizable rooms entered through a covered veranda, which he furnished with a couch and an icebox to entertain visitors. Batavia itself was swiftly changing. The golden flicker of gaslights now lit up recently dark streets. Steam-powered trams, which could run at the awesome speed of 10 miles an hour, were replacing horse-drawn carriages. Even diets were changing. Newly introduced steam mills were churning out gleaming white rice that seemed more appetizing than the less polished brown rice milled by hand. Yet beriberi still killed.

As Eijkman searched for the germ responsible, however, the bacteria once more refused to cooperate. When he again injected bacteria cultured from beriberi sufferers into rabbits and monkeys, the animals did not sicken. Settling in for a long haul, he switched to working with chickens, probably because they were cheaper to keep in large numbers. That decision would be a stroke of blind luck. When Eijkman injected chickens with bacteria, they fell ill. They developed an unsteady gait and couldn't walk, symptoms remarkably similar to those of beriberi. Then, just as things were going swimmingly, confusion returned. The control animals that he housed in a different location came down with the same symptoms. Once again, it appeared that the disease could spread extraordinarily rapidly. Soon, however, even more confoundingly, all of the chickens suddenly recovered, for no apparent reason. It was enough to drive anyone mad.

It was then that Eijkman learned of a strange coincidence. His assistant told him that he had changed the food he fed the chickens. When they fell sick, he had been giving them the cheapest food he could find: leftovers—scraps of cooked white rice that were donated by the cook at the hospital kitchen. But the cook had been replaced and, as Eijkman put it, "his successor refused to allow military rice to be taken for civilian chickens." So Eijkman's assistant fed them uncooked brown rice, and soon after, they began to recover.

Louis Pasteur famously observed that "chance favors only those minds which are prepared." Eijkman, after his disappointments and setbacks, was prepared. He knew that white rice was an innovation. Steam-powered mills, introduced just two decades earlier, could husk rice much more thoroughly. Machine-milled rice could be stored much longer before it grew rancid, and most people preferred these gleaming white kernels to brown hand-milled rice that retained a thin layer of bran.

Eijkman fed some chickens a diet of cooked white rice, and he was surprised to see that they came down with beriberi-like symptoms in three to eight weeks. At last he seemed to be getting somewhere.

At which point, Eijkman himself also fell ill, most likely from malaria. Nonetheless, he persisted, launching a flurry of experiments to identify the mysterious ingredient in white rice that made chickens sick. Was a particular variety of white rice poisonous? Did white rice spoil faster than brown rice? To stay healthy, did chickens need protein or salt that was found only in the rice's outer layer? None of these theories panned out.

After five years of experiments, to Eijkman's incisive mind, only one logical possibility remained. He concluded that the white part of the rice contained a toxin, and the brown bran surrounding it, an antitoxin. Or alternately (because many scientists in his day believed that bacteria in our guts can create toxins) he hypothesized that, when bacteria in our stomachs feed on white rice, they release a toxin that is counteracted by an antitoxin in the rice's outer layer.

Eijkman was delighted to find strong support for his theory from his friend, Adolphe Vorderman, the medical inspector of the East Indies' prison system. Vorderman analyzed the incidence of beriberi in 101 prisons holding almost 250,000 inmates. He discovered that fewer than 1 in 10,000 suffered from beriberi in the prisons that served brown rice. In those doling out white rice, however, the incidence was 1 in 39, and it was even higher, 1 in 4, among those incarcerated for a long time. All of this was strong proof, Vorderman agreed, that the outer bran layer of rice that milling removed contained an antitoxin that neutralized a poison in white rice.

In 1896, Eijkman's illness forced him to return to Europe, and there he faced withering criticism. Memorably, one British physician charged that beriberi had as much to do with eating rice "as eating fish had to do with leprosy or chewing tiger's flesh with the production of courage."

In Asia, however, the idea that beriberi might be linked to food was beginning to spread. A deadly epidemic in a Malaysian "lunatic asylum" prompted a British physician to conduct a trial with his inmates. To his surprise, he discovered that brown rice prevented and even cured beriberi. He too faced ferocious skepticism from no less than the director of the Malay Institute for Medical Research. Yet the evidence was growing. In Japan, the physician Takaki Kanehiro erroneously believed that a protein deficiency caused beriberi. Nonetheless, the changes in diet he recommended to the Japanese navy virtually eliminated the disease. By the 1910s, many physicians in Asia were convinced that eating polished rice caused beriberi, but they were at a loss to explain how.

Meanwhile, in Europe most scientists continued to insist that bacteria were the culprits. It was as if they were looking at an optical illusion, a drawing that contained two images and they could see only one of them. It would take an entirely different kind of experiment for them to suddenly see the picture in an entirely new light.

That experiment would come from Frederick Gowland Hopkins of Cambridge University, who had begun his career working on famous poisoning cases and would later be called the father of British biochemistry. Hopkins hadn't set out to study disease. He was trying to refine our knowledge of our nutritional needs by concocting artificial food from scratch. He measured out proteins, carbohydrates, fats, and minerals—all the essentials that Justus Liebig identified sixty years before—and fed them to young rats. To his surprise, he found that their growth was stunted—unless he also gave them a tiny drop of milk.

Hopkins was mystified. Was Liebig wrong? Do we need to eat the tiniest amount of some other substance that Liebig had not identified? That seemed hard to believe. "So much careful scientific work on nutrition had been carried on for half a century and more," he wrote. "How could fundamentals have been missed?" But after some time, he wondered, "Why not?" He called the mysterious substance found in milk an "accessory factor." Going out on a limb, in 1912, he suggested that the lack of tiny amounts of accessory factors, not the influence of germs, may be responsible for two diseases: scurvy and rickets.

Meanwhile, a shy Polish scientist named Casimir Funk was hotly pursuing the same quarry at London's Lister Institute. The director alerted him that researchers in Eijkman's old lab in Batavia were racing to identify a substance in rice bran that cured beriberi. Funk decided to try to beat them to it. He began by feeding pigeons a diet of white rice. Sure enough, they exhibited beriberi-like symptoms: their necks craned up, their wings and legs weakened, and they had difficulty walking. Next, he tried to isolate an active portion of the polishings of brown rice that cured the disease. Working alone at "full blast," often late into the night, he attempted to isolate a curative substance in rice bran—toiling laboriously through a great many steps that included mixing it with alcohol, filtering and evaporating the liquid, pressing the residues, and adding other chemicals. He would feed his final extract to ailing pigeons and, if they recovered, he would try to isolate an even smaller portion of it. Eventually, from two thousand pounds of rice polishings, he extracted a spoonful of an active substance. In tiny doses, it had his pigeons standing and walking again in three to ten hours. Even then, skeptical colleagues still questioned the validity of his "cure," because it only lasted for seven to ten days.

In 1912, in a landmark paper, Funk proclaimed that dietary deficiencies of very small amounts of unknown substances were responsible for scurvy, rickets, and two other dreaded illnesses: beriberi and pellagra. Shrewdly, Funk came up with a much sexier name than "accessory factors." He called them *vitamines*, from *vita* (Latin for "life") and amine (a nitrogen-containing compound he erroneously believed they were made of). The name stuck. It simply lost its *e*. At long last, the idea spread that if we want to avoid dreadful diseases, we might need something else in addition to Liebig's proteins, fats, carbohydrates, and minerals.

Looking back, you have to wonder: For heaven's sake, why did it take so long? The evidence was there long ago. Hundreds of years earlier, many ship captains knew that lemons cured scurvy. Yet this knowledge was misinterpreted, disregarded, and forgotten. The link between beriberi and white rice was also strong. A decade before Hopkins and Funk published, Eijkman's successor in Batavia, Gerrit Grijns, and Eijkman's old boss, Cornelis Pekelharing, both concluded from further experiments that the lack of an unknown substance caused beriberi. Yet their papers, dropped in Dutch journals, created as little stir as pebbles in the ocean. Most of their colleagues refused to believe them, suspecting instead that the experimenters had not ruled out the presence of highly contagious, invisible germs. Only a completely different type of experiment—Hopkins's discovery of the inadequacy of his synthetic food—began to convince scientists that they might have been missing something all along.

Why did they seem so willfully blind? To begin with, we have to admit, of course, that they are human. It is as hard for a scientist as it is for the next person to shake off the teachings of one's revered teachers. It is also a rare person who easily admits being wrong. Then there's another obstacle, a thinking trap—the "You Look For and See the Evidence That Matches Your Existing Theory" bias, commonly known as confirmation bias, the tendency to look for and accept only information that confirms what we already believe. That baked-in wiring that we all walk around with is important. It helps us quickly make sense of the world. You don't want to take a long time debating whether a thin long brown object on a path in front of you is a snake or a tree branch. But that bias also has drawbacks. Lind, Eijkman, and so many others searched for an imbalance of humors, or germs, or a toxin—for evidence that fit their existing understanding of disease. They were unwilling to take a mental leap to a radically different concept, until overwhelming evidence forced them to hunt for a new explanation.

From a young age, we are taught to avoid spoiled food and drink. In Victorian England, it was often safer to drink beer than water. Everyone knows you'll get sick if you eat the wrong thing, but the quirky concept of vitamins appeared to turn that idea upside down. As the biochemist Albert Szent-Györgyi put it, "a vitamin is a substance you get sick from if you don't eat it." So it is not surprising that long after Hopkins and Funk made their discoveries, a great many scientists still regarded vitamins as "a mere name"—especially since no one had identified the chemical composition of those presumed substances or had any idea of how they were supposed to work.

Yet now there were some as well who sensed that a great change was coming. At the University of Wisconsin in 1913, Elmer McCollum and his young volunteer Marguerite Davis also concocted a synthetic diet for rats. They found that pups stopped growing unless their food contained minuscule amounts of two substances: a "factor" they isolated from fats and a water-soluble one found in wheat germ. They became known as vitamins A and B, but they might have just as well been named X and Y. Researchers knew almost nothing about them.

This was the starting pistol for a scientific gold rush. After much work, biochemists would discover that a vitamin they christened B$_1$

was the mysterious factor in brown rice that cured Eijkman's beriberi. A deficiency of vitamin C, later called ascorbic acid (which, by the way, consists entirely of three of our favorite elements: carbon, hydrogen, and oxygen), caused James Lind's old nemesis, scurvy. Vitamin B_3 cured pellagra, which was common in the American South. And vitamin D cured rickets, a scourge of many infants living in dark city tenements that had no sunlight.

Despite their staggering potency, researchers also discovered that vitamins are quite small. They are molecules of only about 12 to 180 atoms.

Not surprisingly, it wasn't long before a vitamin craze swept the world. Cure-all claims abounded. "Scientists Find Indication of a Vitamin Which Prevents Softening of the Brain," enthused a 1931 headline in the *New York Times*. It appeared that vitamins could do it all: increase vitality and pep, improve sexual drive, and prevent cancer. Though many rosy assertions were vastly overblown, foods newly "fortified" with vitamins drastically reduced some diseases. Among them, the addition of vitamin A to margarine eliminated night blindness, while vitamin D added to milk and margarine helped eliminate rickets.

In 1941, vitamin-enriched foods received another boost. Just six months before Pearl Harbor, President Roosevelt convened the National Nutrition Conference for Defense in Washington, DC. At this gathering of nine hundred physicians and experts on the nation's food supply, eminent speakers rang alarm bells. They warned that America might be about to field vitamin-deficient troops against well-nourished German soldiers. The FDA lost no time in recommending that millers and bakers fortify white flour and bread with the nutrients that industrial milling stripped away. Manufacturers voluntarily added vitamin B_1 (thiamin), vitamin B_2 (riboflavin), and vitamin B_3 (niacin) to flour. "You're in the Army, too!" a 1942 *Good Housekeeping* column told readers. "It's your patriotic duty to give your family these health values by using enriched bread and flour." Foods like Wonder Bread, proudly fortified with vitamins, acquired the aura of superfoods. For many, these health foods of their day turned a whole class of diseases into a distant memory.*

Today, it is generally agreed that we need a total of thirteen vitamins if we don't want to become swollen, discolored, paralyzed, and subject to a plethora of other agonizing indignities. Our bodies expect us to feed them the vitamins A, C, D, E, K, and the eight B vitamins. (Incidentally, we don't need the others; the letters F to J and some B vitamins were once assigned to discoveries that never panned out.)

What do all vitamins have in common? "Vitamins are another name for molecules we can't make," the BBC interviewer Melvyn Bragg quipped. That's true, with a caveat: we do actually make three of them—vitamins B_3, D, and K; we just don't always produce enough of them. While we use heaping servings of carbohydrates, fats, and proteins as raw materials to build cells and produce energy, our vitamins are, for the most part, small essential tools that simply help encourage chemical reactions. They are like the grease and oil in a car. Without them a car remains intact, but after a while it won't be going anywhere. Nor do you need a lot of them. You need only 2.5 millionths of gram a day of vitamin B_{12}, for example—about one-thirtieth the weight of a grain of salt.

All of which raises an obvious question: If vitamins are so essential, and shortages of them so perilous, why didn't our bodies evolve to just go ahead and make them ourselves? One simple explanation is that we are lazy; we don't because we don't have to. Take vitamin C. Our primate ancestors made it, and most vertebrates, including our cats and dogs, still do. That is why we don't have to feed our pets broccoli. But around 60 million years ago, mutations disabled that gene in our ancestors. We still carry it around; it just doesn't work anymore. "Since we have to eat every day," the biochemist Chris Walsh explained, "we took

---

* Even today, a significant number of Americans still suffer from at least one vitamin deficiency. In the developing world, these deficiencies are even more acute. That's one reason that researchers are trying to genetically engineer crops such as Golden Rice, which could prevent vitamin A deficiency.

a gamble that we would always find enough vitamins in the food we eat." Luckily for our ancestors and us, plants made vitamin C in spades.

Don't let the small amounts that we need fool you. Vitamins assist in many of our cells' most basic functions. For instance, vitamins A, C, K, and the B vitamins are coenzymes. These Lilliputian molecules help their massive big brothers, the enzymes, speed up chemical reactions. Enzymes—long, folded strings of amino acids—can make a reaction that would otherwise take place only once in a million or billion years happen instead many times a second. Enzymes encourage molecules to react by seizing and precisely positioning them to tolerances of a few billionths of an inch. But to do that, some enzymes need the help of tiny sidekicks with different chemical handles. In other words, they need coenzymes: some of our vitamins. Just as construction workers need power drills, in addition to cranes and bulldozers, to assemble buildings, so our cells need vitamins to carry out many basic tasks.

Not surprisingly, we are not the only creatures that appreciate them. All living organisms, whether *E. coli*, mushrooms, or platypuses, thirst for them. The work of vitamins is so fundamental that the biochemist Harold White suspects that they evolved in some of the very earliest cells. Recall that some researchers believe that life first evolved from RNA—long before DNA and proteins had been invented. RNA could replicate, *and* it could speed up chemical reactions. But RNA may have needed the help of coenzymes, and as the saying goes, if it ain't broke don't fix it. So we may have been stuck with them ever since.

By now, vitamins have been coopted for so many other functions that it's hard to keep track. Vitamin A helps make rod-shaped cells in our eyes that detect dim light. Without it, you would bump into walls in the dark. You also need it to manufacture cells for your skin, bones, teeth, nails, hair, and immune system. On the other hand, without vitamin C, you wouldn't be able to walk. You need it to produce collagen—the elastic stuff that adds flexibility to your skin, bones, tendons, and muscles. At least 30 percent of your protein is collagen.

If you skimp on vitamin C, your gums and legs will swell up like, well, a limey in the British Royal Navy. Vitamin D's major claim to fame is that it helps cells absorb calcium, which means, of course, that without it you would have no skeleton. But your muscles and nerves need it too. So if you are deficient in vitamin D, your body will rob Peter to pay Paul. It will pull calcium from your bones to keep your muscles and nerves going. If you are young, the resulting bow in your legs will make you instantly recognizable as suffering from rickets.* Some vitamins, such as A, D, and E, also work as tiny brooms. They are antioxidants that sweep up dangerous charged molecules called free radicals before they can gum up machinery in our cells.

Over a century ago, plants and occasional fungi, like mushrooms and yeast, would have made all these vitamins for you (with one exception: only bacteria make vitamin B_{12}). But today many of our vitamins are factory-made. They're delivered in pills and added to our pasta, orange juice, and breakfast cereal. As Catherine Price notes in *Vitamania*, some synthetic vitamins we swallow may have been made from nylon, acetone, formaldehyde, and coal tar. As unappetizing as that sounds, the synthetic versions work just fine. Manufactured and plant-assembled vitamins have identical (or almost identical) chemical structures. Nonetheless, many biochemists suspect that we are much better off getting our vitamins from whole foods. They readily admit that there is much we don't know about nutrition. Researchers are still learning the nutritional roles of many compounds we are finding in plants, such as the powerful antioxidants called flavonoids. Vitamins may work with these and other nutrients in whole foods in synergistic ways we don't yet appreciate. Eating your broccoli may be better for you than you think.

By the way, you may be wondering at this point if you should just take daily vitamins as a cheap insurance policy. Well, that all depends.

---

* The good news is that you make vitamin D yourself in your skin. But you need sunlight to do that. Twenty to thirty minutes at midday a few times a week is probably enough, less if you wear a loincloth.

If you are pregnant, for instance, it may make good sense. Or if you are elderly. As we age, our ability to absorb vitamins D and B$_{12}$ declines. Moreover, in some countries, the abundance of inexpensive rice, wheat, and maize has supplanted nutrient-dense foods like beans, lentils, and peas, creating vitamin deficiencies, such as modern beriberi. Even in America, a surprising number of people may be deficient in one or more vitamins. However, if you are among those who eat a healthy well-balanced diet, vitamins are plentiful enough that you can freeload off plants and bacteria to get what you need. And once you've eaten enough vitamins, downing more from health food stores won't do you any good, and can even do your body (and your wallet) harm. As Gerald Combs Jr., who coauthored the 612-page text *The Vitamins*, put it, "Americans have the most expensive urine in the world." My skeptical father-in-law shared the same sentiment. "Vitamins are reverse alchemy," he liked to say. "They turn gold into piss."

Justus Liebig may have missed vitamins, but he got the rest of our ingredient list right. Remember how he said there was a fourth substance in addition to proteins, fats, and carbs? There is, though even he didn't fully appreciate it. And no offense, but the truth is that, without it, you would hardly amount to much. Liebig taught that, in order to survive, we also needed a few minerals. His list rightly included iron, phosphorus, and the highly prized sodium and chloride in salt.

Of course, our hunger for salt is so great that food doesn't taste as good without it. Among its many roles, salt helps maintain blood pressure, send nerve impulses, and contract muscles. It is so valuable, it was once part of Roman soldiers' wages—hence the question "Are you worth your salt?" Many battles have been fought over it, including raids in the American Civil War, when Northern generals attacked the saltworks of Saltville, Virginia. They hoped to sap Southerners of the strength to fight by depriving them of salt.

Beginning in the 1930s, scientists who concocted artificial food for lab animals discovered that we crave small amounts of other minerals.

We need magnesium, manganese, copper, zinc, and small traces of others, including vanadium, selenium, and chromium, the metal that puts the gleam in chrome.

It's difficult to overestimate what your minerals do for you. And it would be hard to decide which is most critical. Calcium and phosphorus, your most abundant minerals, strengthen your bones. Calcium weighs in at about 1 percent of your dry mass, while phosphorus is about half of that. You need sodium and potassium to think or walk. You use them to create differences in electric charge in and outside of your cell membranes. And as we'll see, this brilliant arrangement allows you to send electrical impulses along your nerves and muscles. On the other hand, if iron entered the contest for the Most Critical Mineral of All, it could campaign on the slogan: "No iron, no energy." When iron in the hemoglobin in your lungs rusts, it snares oxygen, and iron molecules haul it through your bloodstream to every cell in your body. But you also need iodine to make thyroid hormones that regulate your metabolism. Without enough iodine, you will come down with goiter and your eyes will bug out. Your body also craves nickel, zinc, manganese, and cobalt. And who knew how important selenium is? A deficit may produce hair loss, fatigue, mental fog, weight gain, heart problems, goiter, a weakened immune system, and physical and mental disabilities.

Then there is the poison arsenic—the faintest amount of it may be essential. But don't go taking more of it. Its role is not yet understood, and in larger amounts it's certainly toxic.

Finally, you walk around with tiny traces of minerals that you have no use for, but found their way into you anyway: minerals like yttrium, tantalum, strontium, niobium, gold, and silver. "Anything that's in the soil, anything that's in the dirt, gets into the human body," said the mineral nutrition researcher James Collins. That helps explain why, although our bodies contain about sixty elements, about half are just along for the ride. Only about twenty-five are considered essential.

But think about what life would be like if you had to regularly run around trying to collect them all. Where would you even look for molybdenum or vanadium? You would go crazy. Fortunately, many common foods, like tomatoes and peas, contain molybdenum, while pepper, dill, and grains contain vanadium. You can be thankful for the hardworking plants, bacteria, and fungi that first pried many minerals free from rock and continue to do the gathering for you. They are our mineral-delivery services.

Ironically, some mineral and vitamin deficiencies that torment humans, like anemia, scurvy, beriberi, and pellagra, are largely modern diseases that less commonly plagued our hunter-gatherer ancestors. Before the agricultural revolution, our forebears ate a varied mix of plants, fruits, nuts, and meat. It was only beginning about twelve thousand years ago, as civilizations started to rely heavily on domesticated wheat, corn, and rice, that we unknowingly risked falling victim to nasty nutritional deficiencies. And only in the last hundred years, thanks to the voracious curiosity of Liebig, Lind, Eijkman, and many others, have we learned how to avoid those gruesome ailments.

The list of ingredients that we need to build our bodies is now complete. We assemble our cells from five types of molecules: proteins, fats, carbohydrates, vitamins, and minerals.* Virtually all of them come from plants (with a handful from bacteria and fungi).

This list, however, raises one of the most baffling mysteries in all of science. How do we manage to make a living, breathing person from a collection of sliced and diced nutrients? How do they create life within our cells? And the first question is: Where does the knowledge live within us that tells us how to assemble a human being from the trillions upon trillions of atoms that we consume at every meal? Where is our instruction manual? That question once seemed entirely incomprehensible. Until a clue emerged from a scientific investigation of . . . pus.

---

* The fats include one other type of molecule that we can't make ourselves: the essential fatty acids omega-3 and omega-6. They play many roles in our bodies, most notably in the proper functioning of our brains.

HIDDEN IN PLAIN SIGHT

The Discovery of Your Master Blueprint

*Exploratory research is really like working in
a fog. You don't know where you're going. You're just
groping. Then people learn about it afterwards and
think how straightforward it was.*
—*Francis Crick*

In the fall of 1868, a young Swiss physician named Friedrich Miescher strode through the grand stone arch entrance of an imposing castle high above the medieval German city of Tübingen. Just twenty-four, shy and introverted, the medical school graduate was on his way to his future laboratory, the former kitchen of the castle. Miescher once intended to follow in the footsteps of his father and uncle, both distinguished physicians. But after a typhus infection damaged his hearing and made it difficult to use a stethoscope, he'd decided on a career in research and had come to Tübingen to work under the great pioneer of biochemistry, Felix Hoppe-Seyler.

Within just six months, Miescher would uncover a vital clue to a question that can easily make your head spin: How can atoms that arrived here helter-skelter from space direct the fantastically complex activity within our cells? In other words, where is the molecular instruction manual, the blueprint, that tells your cells how to build and maintain you? Miescher would come surprisingly close, within a hair's breadth, of predicting the answer just thirty-some years after Darwin published *On the Origin of Species*, and long before James Watson and Francis Crick were even born.

When Miescher arrived in Tübingen, so little was known about our bodies that Hoppe-Seyler's goal was to identify the chemicals found in different kinds of cells. He hoped that analyzing the proteins in them would shed light on how cells operate. Miescher would work on white blood cells, which they believed were among the simplest cells of all.

Conveniently, the young researcher had no trouble finding a ready supply. In the days before antiseptics and the germ theory of disease, it was commonly believed that dead white blood cells—pus—helpfully rid the body of poisonous "humors." Physicians were encouraged by the presence of copious pus at a wound and regarded the chief role of bandages as simply to absorb it. Miescher had no trouble getting his hands on plenty of smelly pus-filled bandages from a nearby hospital full of soldiers.

Under the vaulted ceiling of his spacious laboratory, Miescher extracted a "cloudy, thick, slimy mass" from the bandages and broke the cells open with salt solutions. Then he got down to the tricky part, identifying the chemicals within them. As he expected, he found proteins and fats, but he also detected another kind of molecule. This one had phosphorus in it. That was surprising. Proteins, fats, and carbohydrates don't contain phosphorus, so he began to wonder if he had found an entirely new type of molecule in our cells. The hunt was on.

It appeared to him that in some tests he had isolated nuclei and that these were the source of his new molecules. If he wanted to prove it, he would have to fully isolate nuclei from cells, something never done before. He attacked the problem with his characteristic zest for focused work. (On his wedding day, years later, a friend would have to drag him out of his lab because he had forgotten the date.) The process Miescher developed necessitated visiting a butcher to obtain the odiferous lining of a pig's stomach from which he extracted the digestive enzyme pepsin, which he would use, along with alcohol and hydrochloric acid, to isolate the nuclei.

After many months, he extracted a white substance from nuclei that contained phosphorus. This, he was sure, was a groundbreaking discovery. He had found a new type of molecule that must play a unique role in the cell. It might even rival proteins in importance.

Hoppe-Seyler was reluctant to believe that his young researcher had made such a fundamental discovery. He was not willing to place Miescher's paper in the biochemical journal he edited until he could replicate the results himself, even though, as a year passed, Miescher desperately needed this paper to obtain a lectureship and feared that someone else would publish first. His paper, "On the Chemical Composition of Pus Cells," finally appeared in print after two years, with a note from Hoppe-Seyler explaining that it had been delayed for "unforeseen circumstances."

Miescher called his new molecule nuclein. We know it as DNA.

Given Miescher's interest in nuclei, it is not surprising that he also took an interest in heredity. Nuclei and heredity seemed to be linked, although scientists had only vague clues about how heredity worked. Hundreds of years earlier, when the English natural philosopher Robert Hooke peered into his microscope, he was astonished to see that his slice of cork was divided into many tiny compartments. They brought to mind small living quarters in monasteries, so he called them cells. By the 1850s, more-powerful microscopes revealed that every living creature is composed of cells. Nonetheless, most scientists agreed that new life could arise only from dust, dead flesh, or organic matter through spontaneous generation. Then, even *more*-powerful microscopes showed that cells divide, thus revealing that every cell comes from another cell. Moreover, they revealed that when a cell divides, its nucleus also divides. And in the fortuitously large transparent egg of a sea urchin, they could observe that an embryo is formed by the fusion of two nuclei, one from a sperm and the other from an egg.

Miescher speculated about the kind of molecule in the nucleus that might be able to transmit heredity. The chemical techniques of the day were too crude to provide an answer, yet he was on the verge of making one of the greatest predictions in all of biology. In 1874, he suggested, "If one . . . wants to assume that a single substance . . . is the specific cause of fertilization, then one should undoubtedly first and foremost consider nuclein." Later, in 1892, in a remarkable letter to his uncle, he presciently suggested that, just as languages with alphabets of only twenty-four to thirty letters can express an unlimited number of words and thoughts, so molecules with a similar number of features might tell a cell how to reproduce. This was astonishingly close to the mark. DNA contains a code for twenty different amino acids. And this code ultimately controls most of our cells' activities.

Ironically, however, over twenty years after discovering nuclein, Miescher was more taken by proteins. Proteins were known to be large and complicated, and he just couldn't see how nuclein could have the complexity necessary to transmit heredity. Because of his change of mind and his reasonable but unfair slight of DNA, he missed making one of the greatest predictions in science. He died at age fifty-six. Overwork weakened his immune system, and tuberculosis did him in. His contribution was largely forgotten.

By the turn of the century, however, a number of biochemists disagreed with Miescher's questioning of nuclein's potential. They suggested that it was nuclein, not protein, that transmitted heredity. Nonetheless, that nascent idea was firmly stamped out, largely by the work of one man—Phoebus Levene, the esteemed head of the chemistry department at the Rockefeller Institute for Medical Research in New York. Levene was the undisputed expert on nuclein. In fact, he renamed it DNA—deoxyribose nucleic acid (because, if you care to know, its sugar—ribose—lacks an oxygen atom, unlike the ribose in ribonucleic acid, RNA). He knew that DNA's distinguishing feature was that it contained four small chemical bases: adenine, guanine, cytosine, and thymine. The approximate measurements he could make at the time suggested that DNA contained the same proportion of each base. So, it seemed most likely to Levene that DNA was a simple molecule with its four bases repeated in the same fixed order.

It was a reasonable speculation, but somewhere along the way, it solidified into unquestioned wisdom. Soon, everyone agreed that DNA was extraordinarily boring. They had fallen for the "Because It Seems Most Likely, It Must Be True" bias. Once everyone agreed that DNA was simple, the shakiness of the assumption underlying this belief faded from memory.

By this time, biologists had learned a bit more about heredity. In the nineteenth century, the Austrian monk Gregor Mendel had shown that you could trace the transmission of a trait in plants, like height or seed shape, from one generation to the next. Scientists named whatever it was in a cell that transmitted a trait a *gene*, from the Greek word *genos*, "birth" or "clan." By the late 1920s, biologists were zapping fruit flies with X-rays and tracking the resultant mutations from one generation to the next. Because some traits were always transmitted together, and they were associated with physical changes in chromosomes—threadlike structures in the nucleus—geneticists recognized that genes must be bundled together in chromosomes. They also knew that chromosomes were made of two substances: DNA and protein.

But what *was* a gene? Was it a single molecule? Was it many, perhaps loosely bound together? No one had any idea. Geneticists did agree, however, that genes were likely made of the most impressive molecules in our cells: proteins. Unlike simple DNA, proteins—strings of up to twenty different types of amino acids—came in a dazzling diversity of sizes and shapes. So it seemed obvious that proteins were the brainy ones. Only they had the complexity necessary to transmit heredity.

It was at just the same time that an almost invisible crack appeared in this wall of accepted wisdom. It was introduced not by a geneticist, but by Oswald Avery, a medical researcher studying bacteria. Avery, a friend and colleague of Levene's at the Rockefeller Institute, was recognized as odd, even by his associates, who revered him and called him Fess, short for *professor*. He had begun his career as a physician but had often felt frustrated that he could do nothing to help his patients. The leading cause of death, for instance—pneumonia—with its progression of chills, fever, and hallucinations, killed fifty thousand Americans a year, including Avery's mother. So he turned to research. Although he sparkled as a conversationalist and excelled in public speaking in college, he transformed himself at the Rockefeller into a scientific monk. Avery had a small frame crowned by an oversize balding cranium and haunting bulging eyes (a consequence of his hyperthyroidism). Introverted and intensely private, he lived near the Rockefeller with another bachelor scientist and resented any distraction, even answering his mail, that kept him from thinking about his research. Like Miescher, he loved the chemical hunt. But Avery was disdainful of investigators who launched into experiments willy-nilly. He would sit for days mulling over how to make an experiment more elegant and meaningful before he was ready to pick up a test tube. At his lab bench, he appeared to heighten his senses, and his gaze "focused inwardly as if unconcerned with the surrounding world."

It was Avery who paid more attention than anyone else to a ghoulish discovery. In 1928, Frederick Griffith, a medical officer at the Ministry of Health in London, analyzed the gunk coughed up by pneumonia patients. He was surprised to see that they commonly harbored not one, but two types of pneumonia bacteria. One had an exterior rough coating and was harmless when injected into mice. The other had a smooth coating and was lethal. It seemed unlikely to Griffith that the same person would contract two different strains of pneumonia. His experiments revealed something even more curious. If he killed the lethal bacteria with heat and injected them into mice, the mice remained perfectly healthy. But if he injected dead lethal bacteria and the harmless live bacteria at the same time, the mice would die. And they now harbored lethal bacteria that were alive. The strains seemed to be able to change from one form to the other.

At first, Avery was convinced that Griffith's strange result was caused by contamination and would not let his associates waste their time trying to replicate it. But a researcher in his lab decided to replicate the experiment while Avery was away on vacation. Because they worked on the sixth floor of the Rockefeller Institute for Medical Research, he had only to walk down to the hospital wards to harvest fresh pneumonia. On Avery's return, he was surprised to learn that Griffith was right. Even if the lethal bacteria were heated, ground up, and placed in a test tube with the harmless living strain, the benign bacteria would turn deadly and their descendants remained killers. As Avery put it, something was turning bacterial Dr. Jekylls into Mr. Hydes.

Few others were interested in investigating a strange freak occurrence in lowly bacteria, but Avery couldn't stop thinking about it. What was this mysterious substance—"the transforming principle," he called it—that turned benign bacteria into lethal ones? Was something from the deceased lethal bacteria attaching itself to the benign bacteria and stimulating its enzymes to manufacture a substance that turns the harmless bacteria deadly? Or, he wondered, was something else going on? Was a gene from the dead bacteria picked up and incorporated into the live ones? He alone seemed to have recognized that Griffith's oddball observation about lowly bacteria might illuminate how some molecules in our own cells direct the activities of others.

Avery was determined to find out, but difficulties plagued him. In the early 1930s, his hyperthyroidism flared up, bringing on hand tremors, depression, and frailness. He returned to his usual weight of just over a hundred pounds only after surgery. Meanwhile, his team faced constant "headaches and heartbreaks" as they struggled to obtain reliable results. Sometimes the transforming substance they isolated transformed bacteria, and sometimes it didn't. "Many were the times we were tempted to throw the whole thing out the window," he recalled. "Disappointment is my daily bread," he would often say.

It was not until 1940, when Avery was sixty-two years old, just three years away from mandatory retirement, that he could finally focus all of his time and energy on the "transforming principle." He was rewarded by several breakthroughs. His colleague, Colin MacLeod, developed a reliable method to isolate large quantities of bacteria for their experiments. He brewed pneumonia bacteria in large flasks, and to remove them he adapted a large industrial dairy cream separator as a centrifuge. To keep it from spewing a mist of virulent pneumonia into the air, he enclosed it in a stainless-steel box and designed an apparatus to sterilize its interior with steam, before using a tire wrench to loosen the heavy bolts that sealed it. (Avery always made himself scarce when the box was being opened.)

With plenty of the transforming substance at hand, another colleague, Maclyn McCarty, began putting it through a gauntlet of chemical tests to search for its identity. He discovered that even after he removed all the fats and sugars, it still transformed harmless bacteria into deadly ones. That left only two chemical suspects. The first was proteins. The second, to Avery's surprise, was DNA.

By 1942, they had isolated a white stringy extract that could turn harmless pneumonia into assassins. It was 0.01 percent protein. The other 99.99 percent was a substance they gradually suspected was DNA. As their investigation continued, they treated the extract with enzymes from rabbit bones, swine kidneys, dog intestines, and rabbit, dog, and human blood. Only enzymes known to destroy DNA stopped the substance from working. Every test they could dream up suggested that the transforming substance was DNA.

Avery was taking no chances, though; he was haunted by a ghost of the past. Twenty years earlier, he had announced that a lethal strain of pneumonia could be identified by a protein on its surface. Six years later, he himself discovered that he was wrong; the identifying molecule was a sugar. His public retraction had been met by skepticism and sarcasm. Many years later, he still felt the sting and didn't want to be wrong again. But after consulting with several eminent chemists at Princeton, he could think of no other test to perform. On the trip back to the city, MacLeod asked him, "What else do you want, Fess? What more evidence can we get?" Avery consulted with more chemists at the Rockefeller.

Finally in 1944, he agreed to publish the results of almost fourteen years of research. Jubilantly, he wrote his brother that he had made a momentous discovery, one "that has long been the dream of geneticists." At the end of his lengthy paper, he dropped his bombshell. Despite decades of belief that genes were made of protein, he wrote that the transforming substance might be likened to a gene and was made of DNA. Then, immediately afterward, he hedged with a caveat: it was, of course, possible that his transforming substance might yet be found to be contaminated—in which case, well, never mind.

Avery's paper, tepidly titled "Studies on the Chemical Nature of the Substance Inducing Transformation of Pneumococcal Types: Induction of Transformation by a Deoxyribonucleic Acid Fraction Isolated from Pneumococcus Type III," went over like a lead balloon. It didn't help that Avery's fiercest critic, Alfred Mirsky, one of the world's leading authorities on proteins, worked two floors above him at the Rockefeller. Ignoring Avery's wide variety of tests, Mirsky claimed that no one could purify DNA more than 99 percent, and so if Avery's preparation was contaminated by just a tenth of a percent of protein, that still left millions of proteins that could have caused the transformation. Once again, the "World's Greatest Expert Must Be Correct" bias swayed scientists' thinking. Of course, that's often a reasonable assumption. Moreover, Avery was studying bacteria, which biologists still knew relatively little about. Who knew if their genetics had anything in common with ours? To most geneticists it hardly seemed important whether genes were made of proteins or DNA; neither was well understood. DNA was just "some goddamn other macromolecule," one remembered thinking. To most scientists, the nature of the molecules controlling the operations of our cells remained shrouded in darkness.

Nonetheless, Avery had planted the seeds of a revolution. A small number of scientists did take him seriously, among them the Columbia biochemist Erwin Chargaff. "I saw before me in dark contours,

the beginning of a grammar of biology," he recalled. Chargaff immediately began studying DNA. He decided to test Levene's assumption
that all DNA had the same proportion of bases with a new technique
called paper chromatography, developed just the previous year. It revealed that in an ox's DNA, the ratios of the bases A:G:C:T were
about 30:20:20:30, but in tuberculosis bacteria they were closer to
35:15:15:35, and in humans they were different still. To Chargaff, this
was proof that the bases did not always appear in a fixed repetitive
sequence, as had been assumed. The order of the bases in DNA might
not be "stupid" after all.

Scientists had fallen into not one but two cognitive traps: "Because
It Seems Most Likely, It Must Be True" and the "World's Greatest
Expert Must Be Correct." Levene, in earlier days, did not have the
technology to measure the base pairs precisely, and he himself recognized that. Along the way, however, his speculation that they were
always found in the same proportion and always appeared in the same
simple fixed order turned into accepted wisdom.

Chargaff made one more curious discovery. He found an odd pattern in the ratios of the bases. The base A was always found in the
same proportion as T, and G in the same proportion as C. Chargaff
did not know what to make of it. He would come to regret that he
missed its significance.

The year that Avery's article appeared, a slim book called *What
Is Life?* written by the physicist Erwin Schrödinger also put a few
scientists on DNA's trail. Schrödinger was famed for his thought
experiment—"Schrodinger's cat"—that revealed how bizarrely subatomic particles behave. After helping develop quantum theory, he
looked around for another suitably big problem to solve. He realized that in biology, genes were an almost mystical concept. Scientists talked about genes that could pass on, for instance, eye color, or
height, yet they could tell you nothing about the nature of the gene
associated with a particular trait. Was a gene a molecule? Many molecules working together, who knows how? Schrödinger speculated
that genes are a "code-script" embedded in some kind of biological

molecule (he called it an "aperiodic crystal"). The quest to discover its identity, he proclaimed, was the most pressing scientific problem of the day.

Schrödinger's short, heady book turned the heads of a number of young scientists, inspiring them to switch careers and search for the physical nature of the gene. Among them were an American zoology student named James Watson and two British physicists, Francis Crick and Maurice Wilkins. Unknowingly, they had entered a race for the Nobel Prize.

When he read Schrödinger's book in 1944, Wilkins, a tall, socially reserved physicist with a long face, was a member of a British team contributing to the war effort at Berkeley's Lawrence Livermore Lab. He was working on uranium separation for the atom bomb but was horrified by the threat that nuclear weapons posed to our survival. He decided to switch to the study of life. Back in England, he joined a newly established biophysics laboratory at King's College London. To Wilkins, as to most others, it had seemed obvious that proteins were the molecules that transmitted heredity. Yet when he heard of Avery's work, suddenly DNA also seemed to be a contender. He began developing a new type of optical microscope to seek clues to DNA's structure, but microscopes had limited resolving power. For an even closer view, he turned to the science of X-ray crystallography. Without it, DNA's structure would still be a mystery today.

Just a few decades before Wilkins began his work, British physicists had pioneered this remarkable technique. They placed a crystalline molecule in front of photographic film, barraged it with X-rays, and captured an image of the rays that were diffracted by it, that is, bent as they passed around or through it. Then they applied complex mathematics to the image to reconstruct the crystal's structure. It was like working out the shape of an object from the shadows it cast on a wall—except the target was a million times smaller than anything we could ever see.

In a stroke of luck, at a meeting Wilkins attended in May 1950, a Swiss scientist generously handed out unusually pure samples of

DNA. Back in London, Wilkins and the graduate student Raymond Gosling tried to take an X-ray crystallography image of it. After much experimentation, they were thrilled to find that they had made a photograph much sharper than any that had ever been made before. They joyfully gulped down glasses of sherry to celebrate. Years later, Gosling still savored the feeling.

Even so, Wilkins recognized that they were neophytes at this technique. So when he heard that his department head, John Randall, was hiring Rosalind Franklin, a scientist experienced in X-ray crystallography, Wilkins suggested to Randall that she join his efforts to work on DNA. And that is, unfortunately, where trouble began.

Franklin, thirty, was an accomplished chemist working in Paris who had already made important discoveries about the structure of coal. Randall had initially asked her to study proteins, but before she arrived, he wrote to request that she work on DNA instead—assuring her that she and Raymond Gosling would be the only ones working on the problem. Randall seems to have wanted Wilkins to drop the X-ray work but did not order him to do so, and for reasons known only to him, Randall never told Wilkins about his letter to Franklin. It was at this very same time that Wilkins came to the conclusion that his microscopes would not reveal much more about DNA's structure, so his only hope of making progress lay in redoubling his efforts to use X-ray crystallography. Had they collaborated, they might later have taken a trip to Sweden together. Instead, the stage was set for one of the most famous feuds in science.

Misunderstandings reigned from the start. Wilkins, the assistant director of the lab, was on vacation when Franklin arrived. When he returned, eager to greet his new junior colleague, he met a seasoned scientist with short dark hair and watchful black eyes that projected confidence. Randall had already told Gosling to work for her. She would take over the X-ray crystallography equipment and Wilkins's precious sample of pure DNA. Wilkins still expected that she would join his efforts as his assistant. At a minimum, he expected that they would collaborate and he, a theorist, would help interpret her images.

Franklin wasn't interested. That wasn't her expectation. She had no desire to be anyone's assistant, and Wilkins did not impress her in the least. As she began work, she saw that she knew much more about the tricky techniques needed to take sharp photographs. She was soon making better ones in her basement laboratory. Franklin was puzzled and offended when Wilkins began offering uninvited suggestions about how to interpret her images. Why did he keep trying to move in on her turf?

There was another reason they got on like cats and dogs: they had a classic personality clash. Franklin had always been confident. She expected to do well in everything, and to lead. She was passionate about her science, and she enjoyed blunt direct intellectual sparring. Wilkins, mild mannered and shy, shrank from conflict. He often turned his head away from his listener as he spoke—and that was in *ordinary* conversation. In a disagreement, he might lapse into silence. "She was quite sharp and quick and decisive," a later collaborator, Aaron Klug, said. "That's why she didn't get on with Wilkins. Wilkins was a clever man. Shrewd but slow. She was quick and decisive." In fact, both could be difficult, and their inability to find common ground would cost them dearly.

Meanwhile, many months later, another team grew interested in the same problem. In fact, Wilkins himself unwittingly recruited a gawky American with wide, buggy eyes and a crew cut to join the search for DNA's structure. James Watson, a postdoctorate researcher, was floundering in a position in Denmark that was going nowhere, yet he dreamed of scientific glory. His American professors were among the few who, taking Avery's finding to heart, had taught that genes were made of DNA, not proteins. Watson simply accepted it. As far as he was concerned, the structure of DNA was the most important question in biology, and he wanted to be the one to find it. But he had no idea how to go about it.

Then, at a conference in Italy, he attended a presentation by Wilkins. When Wilkins displayed the sharp X-ray photograph of DNA that Gosling and he had taken, Watson was electrified. Its pattern of lines and dots strongly suggested that DNA had an orderly structure—and that X-ray crystallography might reveal it.

Watson immediately wanted to jump ship to go work with Wilkins at King's College in London, but that wasn't possible, and so he found the next best thing: a postdoc at Cambridge University, an hour-and-a-half train ride away. The researchers there were using X-ray crystallography to search for the structure of proteins. Watson planned to learn this technique and bide his time until he could launch his own hunt for the structure of DNA.

His opportunity came sooner than expected.

On his first day in Cambridge, he met a skinny, dandyish physicist with a loud booming laugh who never stopped talking. It was his new insatiably curious officemate, Francis Crick. Crick had helped design naval mines during the war. A friend of Wilkins, he had also decided that he would prefer to study life than make weapons. Rather than joining Wilkins's lab in London, however, he came to Cambridge to work out the structure of complex proteins, which, he was sure, played the most important role in the transition from dead molecules to life.

Watson and Crick immediately recognized that they were like minds. They made an odd pair. Watson was immature, precocious, just twenty-two years old but already a postdoc, while Crick, married, thirty-five, was still working on his PhD. But Crick's mind was so quick that he scared his colleagues. If he heard them describe a problem, he might go home and solve it before they could do it themselves. Watson was ambitious and equally cocky. At lectures, he would ostentatiously read a newspaper, lowering it only if he heard something he deemed interesting enough to warrant his attention. As Crick later wrote, "A certain youthful arrogance, a ruthlessness, and an impatience with sloppy thinking came naturally to both of us."

Watson and Crick soon agreed that they would much rather hunt for DNA's structure than for proteins'. For one thing, this might be easier to find. Proteins were massive and extremely complex. Crick's advisor, Max Perutz, had been working on the structure of hemoglobin for fifteen years (and he'd work on it for another nine). If genes were proteins, who *knew* how long it would take to make sense of them? DNA might be much simpler, so they decided to give it a shot. And they wouldn't even have to bother with experiments, which was just as well, as they lacked the technical skills, anyway.

They would use other scientists' data instead, and learning a lesson from Linus Pauling, "the lion of chemistry," they would employ a clever shortcut. Pauling was the greatest chemist of his day. At Caltech in the late 1920s, he had used quantum mechanics to discover new rules of atomic bonding, and almost single-handedly transformed chemistry into a more exact science. Now, he had joined the race to find the structure of what he too was sure were the most important molecules of all—proteins. And he was crushing Watson and Crick's Cambridge lab at their own game. Although their department head, Lawrence Bragg, had actually pioneered X-ray crystallography with his father just a few months before, Pauling had beaten their group to a major advance. He found that the structure of a great many proteins contained a three-dimensional spiral called a helix.˙ Pauling made this discovery by inventing a new technique. Instead of simply analyzing X-ray images, he used measurements from them to make scale models of the protein's subunits. Then he played with his models like Tinkertoys, while employing his deep knowledge of atomic bonding to puzzle out a logical structure.

The Cambridge team had been converging on the same solution but had gotten hung up. Small fuzzy dots in their X-ray images seemed to rule out a helix. Meanwhile Pauling had decided that, because the dots didn't match the structure he found, he might as well ignore them. He turned out to be right. The dots were artifacts of the photography process. It was a traumatic defeat for Crick and the entire Cambridge lab.

˙ By the way, Pauling had help. The Black chemist Herman Branson worked out much of the mathematics that supported the model. For whatever reason, Pauling apparently minimized Branson's contribution, although he did give him credit as the paper's third author.

Now, it was obvious to Crick and Watson that they needed to out-Pauling Pauling. They needed to use his modeling technique to hunt for the structure of DNA—before he got around to it himself. The physicist and biologist began work swiftly, untroubled by the fact that neither knew much about chemistry.

They urged Wilkins to start his own model building at King's. But he couldn't without Franklin's cooperation—and she saw no point in such a speculative approach. To her mind, a more patient one that relied entirely on X-ray data would yield more definitive answers.

Watson and Crick began by examining the data already published by others. They quickly realized, however, that they needed more measurements. And these could only be found in one lab—Franklin's. It just so happened that King's College was about to have a departmental colloquium, at which Franklin would summarize her preliminary results. Wilkins obligingly issued them an invitation. So just six weeks after his arrival in Cambridge, Watson took the train to London and slipped into the lecture hall, "like a spy," as he later put it.

Then he raced back to tell Crick what he thought he had learned.

Watson quickly had the lab shop build Tinkertoy models, wire-and-wood stick-and-ball pieces in the shape and relative size of DNA's building blocks. Then, gazing at them on a desk in their brick-lined office, they tried to imagine the unfathomable: what genes, the molecules that direct our cells, look like.

They knew DNA was made of just five elements. It had backbones containing alternating phosphate groups (composed of phosphorus and oxygen) and sugars called deoxyribose (made of carbon, oxygen, and hydrogen). But the key to DNA was that its backbones supported four small bases: adenine, guanine, cytosine, and thymine (made of thirteen to sixteen atoms of nitrogen, carbon, hydrogen, and oxygen). Presumably, if they were right, and DNA carried genes, the order of the bases somehow encoded a vast amount of genetic information.

They were fearful that DNA's structure would turn out to be a complicated nightmare like proteins. It was even depressingly possible that genes were made of some messy combination of *both* DNA and proteins. But if they were lucky, genes were made only of DNA, and its structure would be simple. If it was, they guessed, its likeliest shape was a helix. That seemed to match both the dimensions of the X-ray images and Wilkins's suspicions. They rushed out to buy a copy of a text by Pauling, *The Nature of the Chemical Bond*, and set to work.

Within just two weeks, their Tinkertoy model was complete. Speculating that three backbones might match the patterns in Franklin's images, they planted three helices joined stably in the center. Then they hung the bases awkwardly off the sides, like decorations on a Christmas tree. Nervously, Crick invited Wilkins to take a look.

The next morning, Wilkins, accompanied by Franklin, Gosling, and two other colleagues from King's College, arrived from London on the 10:00 a.m. train. They were in glum spirits, fearing they had been scooped. But as soon as Franklin saw the model, she laughed. It was so wrong, it was inside out. She had even explained at the colloquium why her calculations showed that the bases must fit *between* the helical backbones, not outside them. But this had gone right over Watson's head.

Watson and Crick were humiliated. Worse was to come. Their boss, Bragg, received an irate call from Wilkins and Franklin's department head, Randall. He was incensed by Watson and Crick's exceedingly ungentlemanly behavior. There were only a few biophysics labs in England. It didn't seem right to duplicate efforts. Wilkins and Franklin had begun work on the problem first, so by all rights, it belonged to them. Embarrassed and annoyed, Bragg ordered Watson and Crick to knock it off and return to what they were supposed to be doing.

As a conciliatory gesture, a chastened Watson and Crick sent Wilkins and Franklin their model-making equipment, inviting them to build their own. But Franklin still didn't see the point. She thought "one could build atomic models 'until the cows came home,'" Gosling recalled, "but it would be impossible to say which were nearer to the truth." She still believed that there was only one sensible approach— let the correct structure emerge from the data.

Back at King's College, Franklin and Gosling returned to work, buoyed by a discovery she had made months earlier. Franklin found that a strand of DNA could take two forms. In drier conditions, its diameter grew wider. She called this the A form. But in wet conditions, like those in our cells—it took on a thinner shape called the B form.

This was a major advance.

But she followed it with a decision that slowed her down. Because the image of the A form was more complex and thus contained more data, Franklin felt it would yield more definitive results, so she decided to work out its structure first. She began applying very complex laborious mathematical calculations to interpret the patterns in her X-ray images. And the more she analyzed her data, the more she was convinced that the A form was not a helix.

Meanwhile, Wilkins was growing increasingly frustrated. As far as he could see, he had begun the DNA research at King's and suggested to Randall that she help his efforts, but she had taken his project over and shut him out. Relations were so bad that Randall had to step in to broker a truce. Franklin would analyze the A form (using the pure DNA Wilkins gave her) while Wilkins would work on the B form (with a DNA sample he found elsewhere). By now they barely talked. Wilkins purchased a new, larger camera to study DNA's B form. But he discovered that he could not find another DNA sample as pure as the one he had received from the Swiss biochemist. His agreement with Franklin effectively sidelined him. He occasionally met with Watson and Crick, updated them on her progress, and groused about "Rosy," as they condescendingly called her.

Franklin too had grown increasingly miserable. In Paris, she had felt at home among cultured, intellectually stimulating colleagues. Here, she was offended that women were not allowed to lunch with the men in the senior common room. And some of the staff at King's College were less refined former military men who fostered a teasing atmosphere she detested. In Paris, she had been recognized as an accomplished scientist; here she was an unknown. Above all, Wilkins infuriated her with his continued suggestions that she work with him.

In retrospect, she would have benefited from a collaborator, but it certainly wasn't going to be Wilkins. "Maurice was, unfortunately, acting, let's say, in male-chauvinist fashion," Don Caspar, a close friend of Franklin's, told me. "I'm sure that it always came across as, he wanted Rosalind to help *him*." She was confident his help was not needed, but her situation was so excruciating that she was searching for a position elsewhere.

All the while, in Cambridge, Watson and Crick chafed. As far as they could tell, in over a year, Franklin had made little progress. Then, in January 1953, Watson and Crick saw an advance copy of a paper by Pauling, and at first glance it scared the bejesus out of them.

Pauling had proposed a structure for DNA. At first Watson and Crick were devastated. Then, to their amazement and immense relief, they realized he had been sloppy. He had placed three helices in the center, just as they had in their own failed model. Pauling also made an uncharacteristically elementary mistake in envisioning how some of the molecules bonded. It was obvious that Pauling's DNA molecule would immediately fall apart. Still, Watson was sure that Pauling would quickly learn of his errors and work out the correct structure. Watson saw his chance for glory slipping away. More than Pauling, Crick, Wilkins, or Franklin, Watson was convinced that discovering DNA's structure would be a huge advance. He hoped that it would shed great light on how genes worked. "I was the only person in the world who valued the problem correctly," he recalled later with his usual modesty. Several days after he saw Pauling's paper, he took the train from Cambridge to London to warn Wilkins and Franklin that they had to begin building a model immediately, before Pauling discovered his errors.

Watson, proud of his un-British manners, walked into Franklin's office unannounced, which, he recalled, was simply never done. He showed her Pauling's manuscript and began describing the obviously incorrect triple-helix backbone that Pauling proposed. Franklin was not happy to see Watson barge in, particularly with an important paper she had not seen herself.

However, their famous clash, immortalized in Watson's candid book, *The Double Helix* (consciously written, Watson claimed, from the immature perspective of a twenty-three-year-old), was over more than just turf. Franklin was still not convinced that the A form of DNA was a helix. Watson told her that she was wrong. He placed his trust in Crick, who had been analyzing X-ray crystallography images of helices in proteins. Crick told him that Franklin was putting too much faith in small points of misleading data—the same type of mistake that previously misled his own group at Cambridge. Franklin did not take kindly to Watson's criticism. So, figuring he had nothing to lose, Watson fired back. "Without further hesitation," he wrote in *The Double Helix*, "I implied that she was incompetent in interpreting X-ray pictures. If only she would learn some theory she would understand how her supposed anti-helical features arose from the minor distortions needed to pack regular helices into a crystalline lattice." Then he retreated in the face of Franklin's anger.

Only Wilkins's arrival saved him from further confrontation. "I thought she was going to hit me," Watson told his colleague as they walked down the hall. Then, to demonstrate how much he had also been putting up with, Wilkins opened a drawer to show Watson a remarkable photograph. Franklin had taken it months earlier but had not shared it with him, he complained, and had only allowed him to see it a few days ago.

It was the now-famous image known as Photo 51.

As soon as Watson saw the photograph, his heart began to pound. His throat went dry. Photo 51, made from a sixty-two-hour exposure of the B form of DNA—the form in our cells—was a testament to Franklin's remarkable skill. It was startlingly sharp. What's more, although Watson had long feared that DNA's structure would be maddeningly complex, Franklin's masterful image contained a strikingly clear X pattern. Crick had explained to Watson what the image of a helix should look like and here it was, unmistakable. Watson also extracted a key measurement in the photograph from Wilkins. But the main impact of Photo 51 was much simpler. It lit a fire under

Watson. He was convinced that he and Crick had to return to model building—as fast as possible.

Wilkins has been criticized for showing Franklin's image to Watson without her permission. But the situation was a bit murkier. The image was his, kind of. Franklin had, by now, found a position at another lab in London. In just eight weeks, Wilkins would be in charge of the DNA work at King's; so, in preparation for her departure, she had asked Gosling to give Wilkins the photograph. As soon as she was gone, Wilkins planned to analyze it and begin model making. In the meantime, he thought, what harm would it do, to show it to Watson?

He would soon find out.

Watson raced back to Cambridge and told Bragg, their department head, that Linus Pauling was about to beat him once more. Pauling had already humiliated Bragg twice. Now Bragg saw the threat as a question of national pride; he had no wish to see the British defeated by Americans once again. Instantly, Bragg gave Watson and Crick the green light to begin building a model.

They awkwardly asked Wilkins for permission to infringe on his turf. He was devastated. He couldn't begin his own model building until Franklin left, but he could not think of how to tell them no. He didn't own DNA, he thought. His lab had the field to themselves for so long, it was only fair to give them a shot. In fact, their model building had already begun.

Now, as Watson and Crick looked at their molecular Tinkertoys, they had to hazard new guesses. Crick still thought the backbone had three helices, but the density of the X-ray image suggested to Watson there were only two. He assembled two stick-and-ball helical backbones and placed them in the middle of the model as before. But after a few days, he still could not join them in an arrangement that matched the X-ray measurements. In frustration he decided he may as well try placing the backbones on the outside—where Franklin had told him they belonged.

That same week, manna from heaven fell into Crick's lap. Months earlier, an agency that funded both labs asked Franklin, like other scientists, to summarize her progress. The report was passed on to Crick's supervisor. It contained largely the same data that Franklin had presented a year earlier at the colloquium. But Watson had not taken notes and was so green that much of it went over his head. Now Crick, a brilliant theorist, had more critical data, and he knew how to make sense of it.

Looking at one measurement, Crick deduced something that Franklin had not recognized. By happenstance, he had seen a similar measurement in hemoglobin, so he immediately recognized that there were two helices that were parallel, but they ran in opposite directions. They were like two spiral staircases, with one set of stairs facing up and the other down. They now knew the precise arrangement of the backbones that the bases must fit between.

The pieces representing the bases had still not arrived from the lab's machine shop, so Watson cut some out of stiff cardboard. Then he tried to fit them between the helices, but because the four bases had different shapes, no matter which way he turned, twisted, or tried to pair them, he couldn't see how to fit them in.

Once again, chance intervened. A visiting chemist from Caltech, who was sharing Watson and Crick's office, told Watson that he had placed some hydrogen atoms in the wrong places because the information in his textbook was outdated. Figuring he had nothing to lose, Watson modified his cardboard bases.

The next morning, Saturday, February 23, 1953, he sat down at his desk and began trying to pair the bases so he could fit them between the helices. As before, he tried joining like with like—A with A, T with T, and so on. Nothing worked. Then, as he shuffled them around, he saw that if he matched A with T, it created a pair that was just the same size and shape as C with G. Light bulbs flickered in his head. He realized he could neatly stack as many of these pairs as he liked in very long sequences between the two helices. Moreover, he recognized excitedly, this pairing explained the puzzling fact that Erwin Chargaff had discovered years earlier. A and T and C and G were always found in the exact same proportions because they can bond only in these two combinations.

Crick arrived at about 10:30 in the morning, his customary hour. He looked at the model on the desk, tweaked it, and was ecstatic. It was suddenly clear how DNA could carry genes. DNA's design was unexpectedly simple. And brilliant! It resembled a twisted ladder with two parallel outer rails—the helices—while the rungs of the ladder were long sequences of base pairs containing the genetic code. Watson and Crick marveled at how only five elements had created an amazingly efficient molecule that could preserve and transmit vast amounts of information. And it is remarkable. Just you try fitting as many words as are in a stack of a thousand telephone books into a molecule with building blocks a million times smaller than we can see.

Watson and Crick had long suffered from the gnawing fear that if they ever did find DNA's structure, it might tell them little about how genes actually work. Instead, their model revealed far more than they ever dreamed. Before then, "it was almost impossible," Crick recalled, "to see how a gene might be replicated." Now they immediately saw how a gene could be copied and passed on. Because the two bases on each rung of the ladder are joined in the middle by weak hydrogen bonds, a gene—a section of DNA—could be unzipped. Then the bases attached to one helical strand can be matched with complementary ones (A with T and C with G) to make a mirror-image copy. Furthermore, it was also obvious how a mutation could be created. All it would take was the accidental insertion of the wrong base pair.

For one sweet moment, Watson, Crick, and their officemate were the only ones in the world who knew of the exquisitely elegant design hidden in each of our cells. By lunchtime, Watson and Crick were celebrating with a drink in The Eagle, their favorite pub. They felt like they had found the secret of life (although Crick didn't boast about it publicly, as Watson famously claimed he did in *The Double Helix*).

A week later, after checking their model, they invited Wilkins to take a look. "It seemed that nonliving atoms and chemical bonds had come together to form life itself," he recalled. "I was rather stunned by it all." He felt as if the model had a life of its own. He was over-whelmed by its beauty—and he also felt as if he had been punched in the gut. Just a week before, he had written Crick with the good news that he was about to begin his own model building. His good friends had beaten him to it. Wilkins was momentarily bitter. Franklin was more gracious. She could see how the model agreed with her data, so she recognized how right the structure seemed. "We all stand on each other's shoulders," she said to Gosling. As well, she was on her way to new projects in a new lab.

Just seven weeks later, Watson and Crick's landmark article on the structure of DNA appeared in *Nature*. It was accompanied by papers from Franklin and Gosling and Wilkins, with data and images that supported the double-helix model. But in a move that would hardly be considered ethical today, Watson and Crick acknowledged their colleagues' contributions with just one single line. "We have also been stimulated by a knowledge of the general nature of the unpublished experimental results and ideas of Dr. M. H. F. Wilkins, Dr. R. E. Franklin and their co-workers at King's College, London."

Why were they so stingy? It's likely they were afraid to reveal to Franklin how much of her unpublished data they had used to build their model. She may have died without ever knowing.

Years after, her later collaborator, Aaron Klug, read her lab books and discovered that, in the last month of the race, she came much closer to finding DNA's structure than anyone suspected. She had worked out that it was a double helix and had recognized from Chargaff's observation that the bases A and T and C and G must be in some way "interchangeable." She lost a race that she was not aware she was in until Watson barged into her office. Klug believes that, given another year, she would have found the structure of DNA on her own. But that was not to be. She was on her way to a lab at Birkbeck College in London, where she would make important contributions to finding the structure of viruses.

After their historic discovery, Watson, Crick, and Franklin finally developed respect for one another. Franklin even consulted with

Watson and Crick about her later work. However, in 1962, the Nobel Prize was awarded to Watson, Crick, and Wilkins (who began the research on DNA at King's).

Tragically, Franklin was ineligible. Only the living can receive a Nobel Prize, and four years earlier, at the age of thirty-seven, she died of ovarian cancer, possibly brought on by her exposure to X-rays in the lab. She had grown close to Crick and his wife by then, even staying with them for weeks when she was recovering from her second operation.

Incidentally, Oswald Avery, the first scientist to show that genes were made of DNA, was nominated for the Nobel Prize many times but never won. When he died of liver cancer just two years after Watson and Crick's discovery, DNA's role in heredity was not yet universally accepted. Some scientists believe that Avery deserved two Nobels: one for his work on DNA, the other for his work on pneumonia.

From the time Watson and Crick first took a victory lap and began to show off their model, they remained awestruck by its unexpected and delightful elegance. At one lecture, all a tipsy Watson could manage to say in summation was "It's so beautiful, you see, so beautiful." Yet even in the haze of their euphoria, DNA was silently mocking them. They were all too aware of how little they still knew. When Watson showed the physicist Leo Szilard the model, Szilard immediately asked him, "Can you patent it?" (Szilard himself had many patents, including one on the nuclear chain reaction, which he donated to the British government in 1934.) But Watson knew you can't patent anything without a practical application, and he still had no idea how genes worked. How could insanely long sequences of As, Cs, Ts, and

---

* In the 2000s in the United States, many individual human genes were patented, until, in 2013, the Supreme Court ruled that this would not be allowed. But modified genes are not considered natural and can be patented.

Gs manage to direct the activity of millions of molecules in our cells to create life? How could a code written in bases tell our cells how to break down and rearrange our food to make us? How did that code create a curve in a lip or a bend in a nose—much less the difference between an elephant and a fly?

Crick thought it would take more than half a century to find the answers, but at least he and Watson had an inkling of where to begin. They latched on to an idea that others had previously proposed, that every gene codes for a different type of protein, and each of these proteins plays a unique role in a cell. But that still left them scratching their heads. How could genes make proteins? Genes are sequences of just four types of bases, and they are stuck in the nucleus; while proteins are strings of up to twenty different kinds of amino acids and are found throughout the cell.

For years, scientists trying to make progress tumbled and bush-whacked their way through a twisted thicket of puzzles. One advance came from the recognition that the RNA in our cells might act as an intermediary between genes and proteins. An RNA molecule is a mirror image of a section of DNA. The main differences are that RNA contains a base called uracil wherever DNA has thymine, and RNA is a copy of just one of DNA's strands. By 1961, it was proven that an RNA copy of a gene could escape the nucleus and travel to a recently discovered structure called a ribosome, which used it to manufacture proteins. So far, so good.

But Watson, Crick, and everyone else were still stuck. How could a nonsensical sequence of bases in RNA (such as GAGAUUCAG) tell a ribosome which amino acids to string together to make a protein? If the bases contained a code, what was the key? Since the mid-1950s, many geneticists, physicists, and mathematicians had flailed away as they dreamed up scores of clever mathematical and logical schemes to unlock the code. But they remained baffled. Crick admitted that after they had passed through an initial "vague phase" and then an "optimistic phase," they were now stuck in a "confused phase." His suspicion that it would take at least half a century to understand how DNA worked seemed right on target.

Then in 1961, an unknown researcher, Marshall Nirenberg, at the National Institutes of Health, bested many of the greatest minds in science. Nirenberg, just two years out of graduate school, cut the Gordian knot by putting logic aside and experimenting. Why bother trying to predict the code, he wondered, when you can simply ask ribosomes to tell you? Nirenberg's ingenious idea was to remove ribosomes from a cell, place them in a test tube with a full complement of amino acids, and give the ribosomes artificially made RNA. He hoped that the ribosomes would oblige him by continuing to do what they do in cells naturally: string together amino acids. At about six a.m. one morning, his postdoctoral researcher, Johannes Matthaei, found that when he fed the RNA molecule UUUUUU to ribosomes, they linked two molecules of the amino acid phenylalanine together. Nirenberg and Matthaei had deciphered the first word in the code of life. The code was based on three-letter words; UUU coded for phenylalanine. In the frenzied race that ensued, similar experiments revealed that several triplets code for each of our twenty amino acids. UUU, GUU, and ACG, for instance, all code for phenylalanine. A few triplets, like TAA, send a completely different kind of message. They tell the ribosome: Stop. Enough already. The protein is done.

This code was nothing like the elegant solutions that so many brilliant minds had been seeking for years. No mathematical or logical pattern lay behind it. The code of life was simply a historical accident that worked. It was one of the most successful inventions on Earth. We share that ancient code with every single living creature: with the bacteria *Thiomargarita namibiensis*, vampire squid, pigbutt worms, and extinct Bambiraptors.* Once the code was cracked, genetics and biology would experience an unprecedented information explosion—so

---

* A few rare minor variations of the code exist here and there.

much so that today, as the geneticist Sean Carroll put it to me, keeping up is like drinking from a firehose.

At last, scientists could answer questions that once seemed totally beyond reach. How were atoms that arrived on Earth billions of years ago eventually able to create creatures like us? For that matter, how do we know what to do with the atoms that we ate last night for dinner? Where do the instructions reside that tell us how to turn our nutrients into us?

We know now that the answers lie in our DNA. Over billions of years, countless tiny mutations in DNA created a vast variety of biochemical experiments. The successful ones produced a tremendous diversity of creatures. And the DNA that we ended up with tells our cells how to transform our food into us.

DNA is often called our instruction manual, our blueprint, but it's a very strange one indeed. No offense but, frankly speaking, your DNA is a confusing rat's nest. All of your cells contain the same DNA. It is divided into twenty-three bundles, each a tightly folded chromosome. (With the exception of sperm and eggs, all of your cells have an identical copy of each chromosome.) These are so densely packed that if you straightened out all the chromosomes in just one of your cells and placed them end to end, the code sequence (ATTGACCACAGG . . .) would drone on for a mind-numbing three billion base-pairs and it would stretch out six feet.

If you tried to take a walking tour of the genes in one of your cells, you'd immediately find yourself hopelessly lost. Immense stretches of bases have been called "junk DNA" (a hotly disputed term) because they have no known function. They include relics of viral invaders (known as retroviruses). Some common repetitive sequences are parasitic elements (called jumping genes). And a much smaller number are "ghost genes" that code for extinct ancestral genes disabled by mutations. The amount of useless DNA, however, is fiercely contested: somewhere between 20 and 90 percent, depending on whom you ask.

Scattered within these freeloading stretches are the useful sequences that are our genes. They transmit hereditary information and mastermind our growth, but they also do more. They never stop telling our cells how to use our nutrients to operate and make repairs.

It turns out that DNA's game plan for doing that is spectacularly simple. Your genes have basically just one job—to control the production of new proteins.* In a contest for the Most Important Molecules in Our Cells, proteins easily take second prize after DNA. Although they don't store genetic information or make copies of themselves, as once assumed, they do carry out almost all the other work in your cells. Proteins, like strings of tiny ball magnets, can contort themselves into an amazing variety of convoluted shapes, many of which are huge. Your hemoglobin is built from 574 amino acids—9,272 atoms. Your largest protein, titin, a rubber band–like molecule in your muscles, is made of 34,350 amino acids. That's about 540,000 atoms. We use some proteins, like oxygen-ferrying hemoglobin, as tools. We use others, like the flexible collagen in our skin and bones, as structural elements. But the lion's share of work in our cells is done by the proteins that are enzymes. They are geniuses at promoting chemical reactions. However, their unique shapes allow most of them to accelerate only one or two types of reactions. So your DNA has its work cut out for it. In its lifetime a cell will need so many different types of chemical reactions that its DNA has to call for the production of thousands of different kinds of enzymes.

You may wonder how it is that DNA, which is cloistered in the nucleus, can control an unruly mess in the rest of the cell where countless enzymes and all kinds of other molecules are zipping around willy-nilly. Imagine elementary school teachers stuck in a classroom but

---

* There are many other molecules that also control when genes are expressed to make proteins, but the instructions for producing these molecules are also ultimately coded for by DNA.

trying to control their entire student body running around unattended on the playground outside. Scientists discovered that DNA is able to exert control only because enzymes, like most proteins, live fast and die young. Most degrade after a few hours or days. So it's the highly choreographed sequence in which genes call for (or don't call for) the production of new enzymes that controls most of the action within a cell. It is the changes in the ceaseless pipeline of tens of thousands of copies of enzymes and other proteins that a cell makes every second that determines how it operates, repairs itself, and reproduces.

The principle is simple: DNA tells a cell which proteins to make, and the proteins do most everything else. But the process DNA uses to orchestrate the production of new proteins is absurdly complex. Only about 1 to 2 percent of our DNA's bases code for proteins. Twice as many or more simply help choreograph the devilishly intricate dances that determine when genes turn on and off. Many genes are controlled by other genes located far away, which, in turn, are switched on or off by others elsewhere. One gene may activate a suite of genes, each of which turns on others in a cascade, like a computer program activating subroutines.

How, then, do we grow in the womb from a single-cell embryo to a creature with arms, legs, a heart, and a brain? The secret is in the order and timing in which a cell's genes, and entire suites of genes, are regulated—the patterns in which they are turned on or off. As your embryo grew, each cell created new ones, and you developed three layers, known as the ectoderm, mesoderm, and endoderm. The top one produced much of your skin, brain, and nerves. The middle layer began the development of your heart, blood cells, muscles, bones, and reproductive system. The inner layer produced your lungs, intestines, and liver, among other things. It was the order in which particular genes were turned on and off (a process modified by signals from surrounding cells) that turned one into a cell in your little toe and another into a cell in your upper lip.

Even when you're fully grown, there's no rest for the weary. Your DNA is still in continuous motion. (And that is a lot of DNA. If you laid all the DNA strands in your trillions of cells out in a single line, it would extend twice the diameter of the solar system.) Portions of it are constantly unwinding so genes can be activated. Right now, each of your cells is making RNA copies of thousands of genes.* When you run or lift weights, eat, fall sick, or learn a new language, you are activating genes that churn out new proteins.

That does leave another puzzle. If you simply placed a copy of your DNA and all the nutrients a cell might need in a test tube, nothing would crawl out. At least not in a human lifetime. So how was it that your very first cell, a fertilized egg, was able to begin building you from the instant of fertilization? The answer is that it didn't start entirely from scratch. In addition to DNA, it received enzymes from your mother to encourage reactions. And it also inherited mitochondria and ribosomes from her to make energy and proteins. Like DNA, these too were handed down from one organism to another over billions of years. So, from the second you were conceived, your founding cell contained not just a DNA blueprint, but also the tools it needed to begin turning your food into you.

All of this still raises one more baffling question. Each of your cells is made from a vast number of atoms, about a hundred trillion or so, which once floated in space before reaching the Earth. How does the enormous jumble of atoms that you eat make the leap to life? Does it all come down to DNA, proteins, and ribosomes? Or are other invisible mechanisms also required to breathe life into our cells' once-dead atoms? As recently as the 1920s, this appeared impossible to answer, perhaps forever. But a young open-minded Belgian scientist was determined to try.

---

* The only exceptions are mature red blood cells. After they mature, they jettison their nuclei and DNA because their main job is simply to transport hemoglobin.

13

ELEMENTS AND ALL

What Is Really Inside You?

Man, like other organisms, is so perfectly coordinated
that he may easily forget, whether awake or asleep,
that he is a colony of cells in action, and that it is the cells
which achieve, through him, what he has the illusion
of accomplishing himself.
—*Albert Claude*

You may not realize it, but you are a towering skyscraper, a cooperative apartment building of thirty trillion units, or cells. Stacked end to end, your cells would circle the Earth over four times. They are the smallest units of life; you can't point to anything you take out of a cell and exclaim, "It's alive!" So how do once-inanimate atoms that found their way to Earth ignite life in our cells? Is there more to it than just DNA and enzymes? Although we are conscious of our senses, our thoughts, and our emotions, they reveal as little about what goes on inside us as a street view of the Empire State Building reveals what's happening in its offices and hallways. If you were to zoom in to one of your cells, what would you see? Can we ever know the secrets of how the trillions of atoms inside each of your cells conspire to create life?

In the 1920s, that question was intensely personal for Albert Claude, a dapper Belgian medical student sporting an enthusiastic pompadour. Claude, the son of a baker, had grown up in a small rural hamlet and attended a one-room schoolhouse. Tragically, when he was just three, his mother fell ill from breast cancer. He watched the inexorable painful progression of the disease over four years until she finally succumbed. As Claude grew older, he was seized by the desire to understand the mysterious ailment that had stolen his beloved mother, but his route to a medical degree was as erratic as they come. At age ten, his father pulled him out of school to care for a paralyzed uncle who'd suffered a cerebral hemorrhage. Two years later, Claude began working in a steel factory and learned to be a draftsman. But then came the First World War and the brutal German invasion of Belgium. Risking his life, Claude, by now a teenager, volunteered for a British intelligence service, relaying information about German troop movements. It was this wartime service that unexpectedly allowed him to pursue his dream of curing cancer. For a brief period of time, the Belgian government permitted war veterans without high school diplomas to enroll in college. Claude, a fifth-grade dropout, applied to medical school, despite fearing his classes would all be taught in Latin. He was relieved to find they were not.

In medical school, Claude spent many hours peering at cells through a microscope—hoping to glimpse the mechanisms that created life, and those that fomented cancer and other diseases. But he could see little, no matter how much he squinted his eyes or twisted his focusing dial—just a nucleus, chromosomes, and oval dots called mitochondria whose function was a mystery. There was one other blotch, the Golgi apparatus, first seen by the Italian physician Camillo Golgi. No one could tell if this was an actual structure or just an artifact of the staining process. And that was it. The rest of a cell's interior was a foggy haze. To Claude it appeared as a frustrating "blurred boundary which concealed the mysterious ground substance where the secret mechanisms of cell life might be found."

Even more disheartening, microscopes were no longer of any use; they had reached the theoretical limits of their magnifying power. The interiors of cells and the cause of cancer appeared as mockingly distant as stars and galaxies seemed to astronomers.

That didn't keep biochemists from thinking they had a good idea of what else lay within cells. Their answer was: not much. They were convinced that enzymes, which vastly accelerate chemical reactions, did all the work. They were sure that a cell's interior was simply a "biochemical bog," a thick soup in which enzymes and other molecules collided and reacted. They had fallen under the sway of a cognitive bias we have seen before: "If Our Current Tools Haven't Detected It, It Doesn't Exist."

Claude studied cancer in laboratory animals and earned his MD in 1928. Then he joined a research institute in Berlin whose director was championing the theory that cancer was caused by bacteria. With his strong independent streak, Claude didn't shrink from showing that the director's lab was riddled with contamination that rendered the experiments worthless. He was asked to leave the premises as soon as possible. Fortuitously, he already knew what he wanted to do next. Researchers at the Rockefeller Institute in Manhattan had found an unusual strain of tumorous cells in chickens. Injecting a filtered extract from these cells would transmit cancer, but no one had been able to isolate the cancer-causing substance from the extract. Claude thought that he was the one to do it, and that the best way to get the job was to write directly to its director. Surprisingly, despite his poor English, it worked.

The famed Rockefeller Institute was renowned for its collection of brilliant, strong-minded researchers. Claude arrived in 1929, the same year that scientists discovered that stars are powered by the fusion of hydrogen atoms and that the universe is expanding. He no doubt hoped that cell biology would soon enter an equally revolutionary era.

In a well-equipped lab, Claude plunged into the work of trying to isolate the cancer-transmitting substance found in chicken tumors. To his colleagues, he was likable but a bit odd. He spoke with a strong Belgian accent. He was highly cultured, comfortable conversing about anything from science and music to history and politics. Among his New York friends were many musicians and artists, including the painter Diego Rivera. But at work, a colleague recalled, Claude was like a solitary wild boar. He was short and stocky, asked disarmingly naïve questions, and went his own way. When he had an original idea, he would isolate himself to push it as far as he could on his own, usually working late into the night (which, not surprisingly, doomed his first marriage).

Yet despite Claude's best efforts, after three years he had made only slight progress. He must have felt a chilly wind; the director of the Institute wanted to replace him with an actual chemist. But Claude's lab director praised his ingenuity in developing new techniques and argued for keeping him on. It was a full two years later, in 1935, that a restless Claude heard that researchers in England were making progress in isolating the cancerous agent with a high-speed centrifuge.

A benchtop device the size of a large cooking pot that whirled solutions around like a carnival ride may not sound high-tech. For cell biologists, however, it was revolutionary. Spinning separates a solution into layers by weight, which is why farmers used early centrifuges to make cream rise to the top of milk. By Claude's day, engineers had put a tremendous amount of work into making ultracentrifuges whirl exceedingly fast—over 10,000 revolutions per minute, creating a force of about 17,000 g.

Claude began by gently grinding chicken tumor cells with a mortar and pestle and adding saline. He found that spinning the solution separated it into layers. Then Claude had a brainstorm. He centrifuged his layers again. That produced even more layers. And centrifuging these produced yet more. To Claude, this suggested that each layer contained large molecules of different characteristic weights. His technique finally allowed him to isolate much higher concentrations of the cancer-causing substance. Moreover, he determined that it contained RNA. Years later, he would even argue that this RNA belonged to a virus, thereby lending support to the suggestion that some viruses can cause cancer.

Claude's greatest breakthrough was different. He realized that his technique might also isolate very large molecules in normal cells, not just in cancerous ones. He was thrilled. These might be structures in our cells too small to be seen with a microscope. Studying them might divulge new clues about how cells operate, perhaps even about how things go wrong. Immediately Claude changed gears. He put his cancer work on hold, convinced that we would have to understand how normal cells work before we could hope to understand cancer.* He was determined to go inside the cell. If a microscope couldn't help him, he would take a hammer to cells, as he later put it, smash them, and investigate what was inside.

Many were scornful when Claude began grinding and centrifuging tissue from chickens, mice, and other laboratory animals. "When he started tearing cells apart, taking pieces out and examining them, everybody who called himself a decent . . . cell biologist was at him," his colleague Keith Porter recalled. "What was the good of doing that, breaking up that gorgeous structure?" He was mocked for making a "cellular mayonnaise." Some colleagues saw it as a betrayal. His first crime was breaking beautiful cells apart. His second was to pretend that anything he then isolated remained intact and had not also been destroyed.

But Claude was keenly aware that scientists' understanding was limited by the acuity of their instruments. He knew that, sometimes, great advances arrive only with an "accident of technical progress," the

---

* His timing was fortuitous. Biologists' efforts to find the root cause of cancer had hit a wall. They lacked the right tools, and their knowledge of the cell was rudimentary. Oswald Avery, Claude's colleague at the Rockefeller, had not yet discovered that genes were made of DNA. Over the next several decades, some researchers would continue trying to show that viruses or bacteria were the primary causes of cancer. Only in the 1970s would scientists recognize that cancer is triggered by genetic mutations. The relatively few viruses and fewer bacteria that induce cancer are far outnumbered by other sources of cancerous mutations.

introduction of a new tool. And as one historian observed, Claude was a master in taking advantage of them.

Peering at his mayonnaise layers under the microscope, Claude saw barely visible dots that he variously called "granules" or "particulates." He was sure these were structures in the cell whose existence no one had suspected. The layers also contained distinct enzymes. Presumably, they offered clues to what the structures did. He even suspected that his particulates were the chemical factories of the cell. Yet he despaired of ever seeing them, and many colleagues continued to doubt that his tiny dots were meaningful at all.

Exploiting a strange fact about our universe, a German electrical engineer in Berlin came to Claude's rescue. Over a decade earlier, Ernst Ruska had realized that, theoretically, one could use an electromagnet to focus a fine beam of electrons, just as a lens focuses light. If this could be done precisely enough, the deflected electron waves should create an image on a screen. If you are not confused by this last sentence, you should be. To physicists' bewilderment, quantum theory had just revealed that, though we may think of electrons as particles, they also behave like waves. That is a baffling paradox that we still don't understand. But for Ruska, this paradox was a blessing. Because an electron's wavelength is a thousand times smaller than that of visible light, the resolution of an "electron microscope" could, at least in principle, be a thousand times greater than a regular microscope's.

Producing an image from a stream of electrons sounded like a pipe dream to most experts. Ruska ignored them. In 1933, he built a prototype of an electron microscope that few thought would ever work. He continued developing it, even after a journal rejected a paper he submitted with accompanying images because his device "would serve no useful purpose." Eventually, though, the Siemens company decided that perhaps there was something to his idea and backed him. In the United States, the RCA Corporation developed one of its own.

By the summer of 1943, wartime New York City had only one electron microscope. A paint and chemical company was using it to develop new products. Its research director, a man by the name of Albert Gessler, was also interested in cancer; it had killed one of his close friends. One day as he was paging through *Science*, he read an article by Claude. It suggested that a closer look at the particulates in normal cells and how they replicated might yield clues to the cause of cancer. Gessler invited Claude to work with the company's electron microscopist—after hours. Claude eagerly accepted.

The technical challenges were daunting. They had to learn how to coat a cell with chemicals to create contrast and how to keep the vacuum in the microscope's chamber from destroying it. Luckily, Claude's associate, Keith Porter, was extraordinarily skilled at culturing cells. He developed an ingenious technique to grow chicken cells that were flat and thin enough for the electron beam to penetrate. After a year, in 1944, the scientists finally produced the first electron microscope image of a single cell. Suddenly, they were looking at an entirely new landscape. "It was wonderful," Porter later recalled. "Believe me, we had never seen anything like it. Men have visited the moon . . . but we were the first . . . to see particles, to see structures that the light microscope had not been able to resolve." The Dark Ages of cell biology were about to end.

Now there was no longer any reason to debate whether the atoms of life build other kinds of structures in our cells. Further refinements of electron microscopy techniques brought them into startlingly sharp focus. Brilliantly exploiting Claude's techniques, a "small band of explorers" at the Rockefeller, most notably Keith Porter, Christian de Duve, and George Palade, were shocked by how many new structures they found. There was the ribosome (which Francis Crick and others would discover uses RNA to make proteins). Many of the other "particles" that Claude had isolated in his centrifuge turned out to be large membrane-bound structures that we now call organelles. Best of all, they discovered that they could determine their functions by combining Claude's techniques. An electron microscope could reveal the organelle's structure, while a centrifuge allowed them to identify its enzymes. They discovered that, among their other jobs, the organelle called the endoplasmic reticulum and its neighbor, the Golgi apparatus, chemically modify new proteins, package them in vesicles, and ship them to other locations in the cell. Another structure called the lysosome is the cell's garbage disposal. Cell biologists were ecstatic. In a boon to medicine, researchers would even discover that disorders in organelles cause disease. For instance, faulty ribosomes cause Diamond-Blackfan anemia. A mitochondrial disorder causes a form of epilepsy called MERRF, and the lack of an enzyme in lysosomes causes Tay-Sachs disease.

Claude himself played a direct role in investigating only a few organelles. In 1949, he was tempted away from the Rockefeller by an offer to direct a cancer research center in his native Belgium. His genius was apparently less in using his techniques to plumb the secrets of cells than in creating new tools for doing so—but this he had done in spades. He would win the Nobel Prize, along with de Duve and Palade, who used Claude's innovations to make many groundbreaking discoveries.

By the early 1950s, it was clear that, in addition to the nucleus with DNA and the ribosomes that make proteins, organelles with distinct enzymes also carry out specialized tasks. Nonetheless, much of the old view of the cell still prevailed. When researchers looked outside of the organelles, even with an electron microscope, they could still make out only a hazy soup. "They thought that organelles were like solid islands in a sea of soluble molecules that were just bumping around into each other and diffusing," the cell biologist Franklin Harold recalled. The rest of the cell's interior appeared as murky as ever. Presumably, it was dominated by enzymes. Few thought there was much else there.

One of the greatest challenges to that view would come from a "crackpot" biochemist who would spark the discovery of radically new mechanisms that no one imagined could exist in cells. Peter Mitchell may not be known throughout the world. Among biochemists, however, he is widely admired, even revered. From his earliest days at the University of Cambridge, he was a hard figure to forget. He arrived as a student in late 1939 driving a stylish Morgan automobile, later replaced by a used Rolls-Royce, thanks to the largesse of an uncle, the owner of a construction company. "Many of his friends remember Mitchell," his biographers wrote, "for his flamboyant dress: a burgundy purple jacket, shirt open sometimes as far as the waist, long hair almost to his shoulders." They thought he resembled Beethoven, whose bust Mitchell kept in his quarters. Mitchell always had a fondness for unconventional and philosophical thinking. His unusual notions about how the atoms in our cells create life would ignite one of the bitterest wars in all of science, one that would be waged for almost two decades.

In 1954, Mitchell left Cambridge, where he received his doctorate and held a research position, to direct a biochemical unit at the University of Edinburgh. There he began grappling with a fundamental question at the root of the hostilities: How do our cells generate energy? Mitchell thought everyone else was going at it all wrong.

Most scientists didn't see a problem. It hardly seemed like it should be difficult to answer. They had already pieced together almost all of the puzzle.

Using a centrifuge and an electron microscope, Claude had determined that mitochondria burn sugar and oxygen to liberate energy. He coined an expression that has echoed through classrooms ever since: mitochondria are the "power plants" of the cell.

That left another puzzle. How does a cell distribute this energy to its far-flung molecules and organelles? There are, after all, no power cables in our cells fanning out from the mitochondria.

Biochemists had solved that part of the puzzle too. They had discovered a small molecule of only forty-seven atoms called ATP (adenosine triphosphate) and had been astonished to realize that it was nothing less than a tiny portable battery pack. ATP contains phosphorus, once called the "devil's element" because it glows and has an unsettling tendency to spontaneously combust. ATP is so crucial because snapping off its last phosphate liberates a bonanza of energy.

Scientists recognized that ATP performs the final step of the process that allows us to draw energy from the Sun. Plants first capture that energy during photosynthesis and pack it away in sugar. Once we eat that sugar (or fats made from it), the mitochondria in our cells release the energy. And ATP transports that energy from the Sun to all the molecules in our cells that need it. It was obvious to biochemists that we must churn out ATP molecules in huge numbers. We now know that a typical human cell consumes an astounding ten to one hundred million ATPs every second. The fact that all life uses ATP (or a similar molecule) to distribute energy suggests that these tiny power packs are as old as life itself. Researchers also knew that, once an ATP releases energy by losing a phosphate, the rest of the molecule circulates back to a mother ship—a mitochondrion—where an enzyme adds a new phosphate to recharge the tiny battery pack.

There remained just one small puzzle. How did this enzyme, ATP synthase, make ATP? It should have been a piece of cake. Biochemists knew how to determine the chemical paths of enzymes. But for the life of them, the "mitochondrists" could not explain how the enzyme worked.

As the news spread, more labs joined the hunt—chemists and physicists, dozens of labs, hundreds of researchers. Big labs and big scientists competed. Explaining the enzyme's operation became a burning issue. To their continued dismay, however, "only shadows of moving parts were seen," recalled the biochemist Efraim Racker, who spent over twenty years working on the problem. Occasionally a group would triumphantly announce that they had solved it. Soon after, their claim of victory would fall apart, leaving everyone more baffled than ever. At one conference, Racker declared that "anyone who was not thoroughly confused, simply did not understand the situation." They had no clue they were on a wild-goose chase, which would last for decades.

By now, it was clear to the philosophically minded Mitchell that they were wasting their time on a fool's errand. Mitchell, who preferred to think problems through for himself, proposed an alternative so novel, it appeared to verge on kookdom. He was intrigued by the observation of the pre-Socratic philosopher Heraclitus that you can't step into the same river twice. Mitchell began thinking of a cell in a similar way—as a structure that remains constant, like a river or candle flame, although its atoms are continuously being replaced. He grew interested in how membranes regulate the endless stream of molecules that flow in and out of them. Ultimately, he became convinced that membranes do much more than act as simple gatekeepers.

Mitchell reexamined a mitochondrion's architecture. Electron microscopy showed that it was oval and bounded by a membrane. Within lay another closed membrane that created an inner chamber. His heresy was to suggest that the structure of this inner membrane was essential to making ATP. If so, all the researchers trying to explain how a traditional enzyme created ATP were looking for a ghost that didn't exist. He argued that a mitochondrion used the energy freed from sugar to transport immense numbers of positively charged hydrogen ions out of its inner membrane. Then it used the resulting difference in electric charge between the two sides of the membrane to make ATP.

In other words, Mitchell proposed—and this was weird—that a mitochondrion used its membrane to create an electric current that powers the cell. It made this current not from negatively charged electrons, like the ones flowing through our electric wires and appliances, but from positively charged protons—hydrogen nuclei. And it used the current to power a mechanism in the membrane that generated ATP.

In 1961, his idea simply seemed bizarre. It was nothing like any of the ways of making molecules that were well understood. And it was only a prediction. He had no experimental evidence to back up his claim. "I remember thinking to myself that I would bet anything that [it] didn't work that way," wrote Leslie Orgel. "These formulations sounded like the pronouncements of a court jester or of a prophet of doom," the biochemist Racker recalled. Mitchell found virtually no supporters. In a knee-jerk reaction, many scientists were swayed once more by the "Too Weird to Be True" bias. They dismissed a theory because it seemed unlikely, went against everything they'd been taught, and was ungainly to boot.

Mitchell would not yield, but while he may have been a genius, he hardly made his case easier. He presented his theory in obscure terms of his own making, not those familiar to mitochondrists. Nor would many of the mechanisms he proposed to explain how his proton current made ATP turn out to be right.

After a few years at the University of Edinburgh, Mitchell suffered from stomach ulcers so dire that he was forced to resign. To recuperate, he moved to a decrepit country manor in rural Cornwall. Yet he refused to give up. After two years, he had renovated the elegant building and turned one of its wings into a private laboratory. He staffed it with Jennifer Moyle, an able collaborator who had worked with him in Cambridge and Edinburgh. In their lab overlooking wooded countryside, they conducted experiments to bolster his theory. Revenue from his prize dairy cows helped support their work.

For a decade, Mitchell made little headway with his colleagues. One recalled a conference at which the future Nobel Prize winner's words "went into one of my ears and out the other, leaving me feeling annoyed they allowed such a ridiculous and incompetent speaker in." One opponent grew so incensed when speaking to Mitchell that he hopped on one foot in anger.

On a map of the world in his home, Mitchell marked the locations of his critics with red pins. If a skeptic began to waiver, Mitchell replaced the red pin on his "persuasion monitor" with a white one. If they fully converted, Mitchell changed it to green.

Eventually, experiments by Mitchell and Moyle, and by many others who were attempting to disprove him, slowly turned the tide. At last, biochemists had to concede that Mitchell was right. To make our ATP, mitochondria use hydrogen ions—positively charged protons—to create a difference in electric charge between the two sides of their inner membranes, and that charge is staggeringly strong—almost 100 million volts per foot—as powerful as lightning.

One of Mitchell's long-standing opponents, Paul Boyer at UCLA, made a remarkable discovery that helped make Mitchell's theory more comprehensible. In the early 1960s, Boyer became convinced that the enzyme that makes ATP could not be an ordinary one. He first thought it must act in concert with neighboring proteins that changed their shapes to facilitate the process. By the 1970s, however, he had a clearer vision. He believed that the enzyme that produced ATP was itself extraordinary. It was complex, with many interlinked moving parts that made it, in effect, a tiny molecular machine. Boyer came around to believing that Mitchell's current of hydrogen atoms must power his invisible machine.

He discovered that the structure of this machine, known as ATP synthase, *is* extraordinary. Here was a mechanism that Albert Claude had not imagined. It looks like the kind of waterwheel-powered contraption that Leonardo da Vinci might have designed, except that it's turned by a buildup of protons behind a membrane, not by water behind a dam. The device, a tiny rotary motor, spans the inner membrane, and has a shaft that can rotate about three hundred times a second. Sir John Walker, the X-ray crystallographer, who later won the Nobel Prize for working out its precise molecular structure, describes it as having a bearing, pistons or valves, a crank or cam, and a spring or flywheel. It is powered by a proton current that spins the rotor and its attached mechanisms. One grabs a phosphate and an adenosine *di*phosphate molecule. A second part of the machine combines the two to make adenosine *tri*phosphate—ATP. Then a third gives the newly made ATP a strong kick, to launch it back into the cell.

In 1975, Boyer sent Mitchell an olive branch. He offered to write a review article with Mitchell and some other former opponents to acknowledge that the war was over. Three years later, Mitchell, wearing an earring, flew to Stockholm to pick up the Nobel Prize in Chemistry—for his "bioimagination," as one scientist put it. Mitchell used the prize money to bail himself out of the debt he had incurred funding his research and farm. Years later, Boyer also received a Nobel. They had discovered a mechanism in our cells that was nothing like the old picture of enzymes simply accelerating reactions. Instead, mitochondria use the energy in sugar to create networks of electric current. That proton electricity powers exquisite molecular machines, and as they crank away, they recharge a constant stream of tiny batteries that fuel everything that we do.

Remarkably, as we saw in chapter 7 on the origin of life, Mike Russell and William Martin believe that a similar current of protons helped kick-start the very first life. They see ancient traces of Mitchell's electric currents in deep-sea hydrothermal vents. There, hydrogen ions bubble up through mineral deposits full of tiny chambers with membrane-thin walls—much like cells. The researchers suggest that different concentrations of positively charged hydrogen protons built up on opposite sides of these walls, creating electric currents. As that proton current flowed, they believe it powered the formation of more complex organic compounds—and ultimately life.

Whether or not this is so, Mitchell and Boyer's mechanism does explain how mitochondria supercharged evolution. Recall, as Lynn Margulis argued, that microbes dominated our planet until one type became especially efficient at producing energy. When one of these was engulfed by another cell, its descendants were domesticated. They became mitochondria, and the rest is history. Our history. On average, each of our cells contains a thousand to ten thousand mitochondria. Mitochondria take up about 35 percent of a heart muscle cell's volume. They allow one of our cells to make tens of thousands of times more energy than a bacterium. With this supercaffeinated boost, our DNA can direct ribosomes to churn out many more proteins and enzymes and turn our cells into hives of activity. Every second in your body thousands of trillions of former bacteria are pumping protons across membranes, producing electricity for rotary motors that make ATP. You breathe in about two-thirds of a pint of oxygen a minute to keep those motors running, enabling them to generate as much energy as a 100-watt light bulb.

To impart life to our cells, our molecules build even more than DNA, RNA, ribosomes, enzymes, and organelles; they also make tiny machines and produce electric currents from protons. In fact, our bodies also rely on a different kind of current. It was discovered by the physician Luigi Galvani in 1780 when he observed that an electric spark made the leg of a dead frog twitch. By the late 1950s, biologists had determined that we generate this electricity in our nerves with the help of another kind of invisible mechanism—a pump that moves charged atoms. These sodium-potassium pumps are in all of your cells, and you spend about a third of your energy just to keep those sodium-potassium pumps running. But you won't begrudge them the outsize share of ATP they consume once you realize that, without them you couldn't think, or even send a message from your brain to your feet saying "Run!"

Those critical pumps keep more charged sodium molecules outside our cells and more potassium inside. This helps maintain the cell's internal pressure, keeping it from bursting and preserving its balance of other chemicals.

Scientists discovered that our nerves also use these pumps for something else. They send electrical messages, but not with electrons or protons. Researchers began learning how this works in 1939 when they found that, for defense, armorless giant squid rely mainly on fast escapes made possible by giant neurons. These nerves were so wide, investigators could actually thread a wire through one to measure the electrical potential inside and out. They learned that the nerve's electrical current is created by molecules of negatively charged sodium and positively charged potassium. Stranger yet, these charged molecules, called ions, don't speed down the nerve from one end to the other. Instead, like football fans, they do the wave. Picture a long row of tiny channels in a cell membrane that either allows sodium to enter or potassium to leave. As each channel opens, more than a million sodium ions a second rush in or potassium ions rush out. These trigger adjoining channels to open, just as the movement of football fans' arms triggers the motion of their neighbors'. As the sodium and potassium ions race in and out of the membrane, they propagate a wave of electrical charge that travels along the nerve.

What enables this zany scheme to work is that a resting nerve has many more sodium ions outside than inside, and more potassium ions inside than out. And what maintains that difference are these little machines that span the cell's outer membrane. With an energy kick from ATP, each pump expels three sodium ions. Then the pump changes shape and allows just two potassium in. The difference in charge created by perhaps a million sodium-potassium pumps in one nerve enables messages to zip along at up to 350 feet a second. Other cells have found even more uses for these tiny machines. Your muscles, including your heart, use them to help contract.

You may not know it, but your life depends on a cool quadrillion or so minuscule sodium-potassium pumps (over a thousand million million of them). There is no way that you could live, much less think or escape predators, without those little machines. That, by the way, is why sodium chloride, or rather salt, tastes so good. Although the plants we eat have a lot of potassium, they don't have much sodium. You need a bit less than a teaspoon of salt a day to maintain your internal electrical charges. So, although hunter-gatherers got their salt from meat, agriculturalists had to supplement their diet. The salt in your saltshaker makes it possible for you to cross your fingers, touch your ear, think, and talk.

Elegant ATP synthases and sodium-potassium pumps are not the end of it. Since the 1960s and '70s, scientists have found a remarkable menagerie of other ATP-fueled molecular machines churning away in our cells. These are tiny, just wheelbarrow-size compared to large factory-size organelles. Little two-footed walking machines called kinesins race around the cell, each step powered by an ATP. They travel on roads made of proteins arranged in microtubules that are constantly being laid down wherever mitochondria or other organelles are needed, and pulled up where they're not. Kinesins spend their days hauling protein-packed vesicles from one side of a cell to the other. In your muscles, little ATP-powered motors called myosins change shape as they slide along ratcheted tracks, lengthening and contracting your muscle filaments. "The reason animals become rigid when they die," the biophysicist Holly Goodson told me with relish, "is that their myosin gets stuck in a strongly bound state like glue." In other words, the muscles' tiny motors seize up. Meanwhile in all of your cells, newly made proteins enter little machines called chaperonins that close and help fold a protein into its proper shape before releasing it. All the while, like woodchippers, proteasomes are chopping up unneeded proteins bound for the scrap heap.

"Basically, the interior of a cell is like a construction site, a place full of machinery doing things, moving things about. There is still a certain amount of liquid, but with every year there are new discoveries, so the amount of room left in between for soup is diminishing," Franklin Harold said, amazed at how many discoveries of new molecular machines he has seen in his lifetime.

All of these mechanisms—factory-like organelles, electric currents, and tiny molecular machines—go a long way toward explaining how trillions upon trillions of atoms create life in our cells. But even they are not the whole story.

First, there is the remarkable fact that simple molecular attraction and repulsion allows some structures to assemble themselves. As we saw, Alec Bangham first spotted this in the early 1960s while testing his new electron microscope. He was startled to see that as fat molecules pivoted to hide their water-hating ends, they spontaneously formed membranes. In the same way, some regions of proteins love water while others want nothing to do with it. These attractions and repulsions help proteins create stable origami-like shapes.

Here's a clue to another reason our molecules are able to create life. Your cells contain a staggering number of structures, including DNA with about three billion base pairs, yet when you grow, heal, or replace a doddering cell, a cell can make a brand-new one in about only twenty-four hours. It would take me more than fifteen years to type out three billion letters, without taking the slightest break. How can our cells work so much faster?

In 1827, the Scottish botanist Robert Brown found a hint in his microscope. He could not figure out why pollen grains that he had placed in water would not settle down and stop colliding. At first, he wondered if they were alive. He changed his mind when he saw dust from rocks behaving just the same way. Why were they moving? In 1905, Albert Einstein proved that the cause of their erratic drunken behavior was that the invisible water molecules surrounding them were constantly randomly colliding. (In fact, it was this paper by Einstein that helped convince many physicists that atoms actually exist.) But what Brown and Einstein didn't know is that those unremitting collisions help impart life to our cells. That's because on the nano level—the level of atoms and molecules—the world is a blur. Heat makes the molecules in our cells vibrate and collide at random in what the physicist Peter Hoffmann calls a molecular storm. And it's way more intense than a category-five hurricane. The average molecule in your cells withstands countless collisions every millisecond.

You might expect that these unending bumper-car collisions would wreak havoc, but surprisingly, our cells don't seem to mind. Instead, they harness this incessant movement to help create life. The ceaseless buffeting helps push molecules in and out of cells, helps proteins change shape, and helps enzymes get around. Collisions make a globular protein, for instance, rotate over two million times a second. On average, every molecule looking to react collides with every protein in the cell about once every second. Those ceaseless impacts send small water molecules zigzagging around your cells at the astonishing speed of over 1,000 miles an hour (although they only go about four billionths of an inch before they smack into another molecule). Somewhat larger molecules, like glucose, bounce around at about 260 miles an hour. Even huge proteins cruise around at 20 miles per hour. The fastest sprinter in the world can race at only 27 miles an hour: most of the activity in your cells is insanely faster.

But since our cells are filled with tiny molecular machines, shouldn't they suffer the same fate as our cars and dishwashers? Shouldn't they constantly break down? The biophysicist Dan Kirschner told me that just thinking about everything that could go wrong in cells used to keep him awake at night. He was learning about cell development in a graduate school course just as his wife was about to have a baby. He was so overwhelmed by the many opportunities for mistakes that he feared his daughter would be born with a neck like a giraffe. She wasn't. Our cells have come up with a number of clever strategies to avoid living short lives. The first is that their machinery is astonishingly reliable. Ribosomes, for instance, insert the wrong amino acid into a protein on the order of once every ten thousand times. The machines that copy our DNA make a mistake only about one in a million to ten million or so.

Nonetheless, nothing is perfect. Sometimes, mistakes happen. Battering collisions, UV light, and dangerous molecules like free radicals also cause damage. Ingeniously, our cells have several ways to meet these threats. For one, they are full of clever repair mechanisms—machines whose jobs are to go on patrol to look for mistakes and fix them. Our cells have error-checking molecular machines and auto-correcting feedback loops that ensure remarkable fidelity.

A 1954 newspaper story in the *Atlanta Constitution* suggests a second strategy our cells have adopted to stay alive. "Bored with yourself? Tired of the same old frame and face? Take another look then. In a manner of speaking you're constantly being reborn. Mankind, like the automobile industry, goes in for a radical chassis change each year." The science behind this odd claim was the work of an inventive nuclear physicist named Paul Aebersold. Aebersold began his career at the cyclotron in Berkeley's Radiation Lab (where Martin Kamen of carbon-14 fame was developing new radioactive isotopes). Later, at the Atomic Energy Commission, Aebersold oversaw the development of isotopes for medical uses. At some point, he realized he could use his isotopes to find out how often we replace the atoms in our bodies. All he had to do was irradiate a substance like table salt, ask an extremely accommodating subject to swallow it, and trace the salt's path with a radiation-tracking device like a Geiger counter. You can follow radioactive atoms in quantities as small as "a billion billionth of an ounce," Aebersold proudly told a television interviewer.

He found that we swap out half of our carbon atoms every one to two months, and we replace a full 98 percent of all our atoms every year.

Wait, what? Is that even possible? Apparently it is. Over half of you is water, and we know that we constantly replace that. Another large percentage of you is protein, and as you may recall, most proteins degrade within hours or days. We even disassemble and replace our ribosomes and large organelles such as mitochondria, which are made primarily of protein.

Aebersold had discovered another strategy that enables our cells to live so long. They are constantly replacing their seemingly permanent structures and old battered molecular machines with new ones. The only ones they don't replace are our massive chromosomes. Instead, we have machines that swarm along them looking for problems and fixing them.

What if the damage to a cell is too great to repair? We have a fallback plan for that too. We simply destroy the entire cell, chop it up into recyclable units and make a fresh one. On average, you replace most of your cells every ten years, which amounts to about 330 billion cells a day. Those that work in the harshest conditions are retired most frequently. The damage to many cells in your intestines, which are exposed to harsh acids, is so predictable that they commit planned suicide and are replaced every two to four days. You replace your skin cells, which endure scrapes and UV light, every month or so. Your red blood cells, which take a beating as they careen through your bloodstream, are replaced every 120 days. That means you have to make almost three and a half million new red blood cells every second. Other cells, like those in our bones, are taken out of commission less often, only about once every ten years.

So, in addition to using reliable machines, our cells have a three-pronged motto to stay alive: ceaselessly check for errors, constantly repair, and continually replace. Your body is like a major New York highway—always open, and always under repair.

That all sounds well and good. By now you may be thinking,

Shouldn't I be immortal? If only. Although we constantly rebuild and replace our cells, when the wrong kinds of genetic mutations occur, a cell can go rogue—particularly in genes that tell the cell when to divide, or how to repair damaged DNA. Then the cell breaks its covenant with its brethren and selfishly reproduces unchecked at the expense of everyone else. It's called cancer, and it can come along at any time.

To put a check on this process, we have evolved a way to reduce its likelihood. Unfortunately, it also keeps us from immortality. The problem for large animals like us is that because we have so many more cells than, say, a mouse, we have greater odds that one of our many cells will turn cancerous. To deal with that, we have a strategy. Before we are born, our fetal cells, like those of other animals, produce an enzyme called telomerase. It protects the ends of replicating DNA from damage by preventing them from shortening. Once we are born, however, our cells stop producing the enzyme. Without it, our cells can divide only a limited number of times before the tails of their DNA become so frayed that they can no longer replicate. The good news is that, even if these cells acquire new mutations, they can no longer multiply, so they can't create cancer. The bad news is that the cells can begin to deteriorate. And if they are stem cells, which create replacements for some of our cells, they're no longer able to do their jobs.

There are other reasons you can't live forever. Even if they don't cause cancer, debilitating mutations still pile up, including in energy-generating mitochondria and stem cells.

There is another, perhaps even more fundamental reason we can never be immortal. There are some cells that we can't replace at all. It would be welcome reassurance to know that if you suffered a stroke or heart attack, you could grow new brain or heart cells to repair the damage, just as you do if you skin a knee, but you can't, with some exceptions—for a good reason. Your brain is a web of about eighty-six billion neurons, each of which may form ten thousand connections with neighbors. Those intricate links encode the memories and experiences that created your identity. They made you you. If you were to try to replace them, you would lose your hard-won knowledge of the world. You would lose yourself. You grow few new brain cells after infancy (with the possible exception of some in the part of the brain called the hippocampus). Likewise, your beating heart has largely lost its ability to make new cells. No one knows exactly why. Perhaps because in order to pump so powerfully, highly specialized heart muscle cells had to give up genetic pathways that once allowed them to regenerate. Adults replace about 1 percent of their heart muscle cells a year. This means, the biologist Nick Lane points out, "the things that make us human—our brains and our hearts—are also what make us age and die. Some of our cells are irreplaceable." That is why, he observes in *The Vital Question*, "I doubt we will ever find a way of living much beyond 120 merely by fine-tuning our physiology." In the end, there is no avoiding the fact that you will become obsolete.

Despite this ultimate limitation, our bodies' strategies of repair, rebuild, and replace give us a remarkably good run.

At long last we've arrived at the heart of the matter. Now we can begin to understand how intrepid atoms, which journeyed here from the Big Bang and the stars, conspire to create life within our cells. We've found the "secret mechanisms" that Albert Claude vainly searched for in cell biology's Dark Ages. Mind you, even now, most of us have a hard time appreciating a cell's sophistication, because we are used to seeing it represented by a simple childlike drawing, but there is nothing simple about it. If you zoomed down to enter one, you would find yourself in a massive jam-packed city of dizzying complexity.

In the nucleus, thousands of genes in our DNA are being copied every second. RNA clones made from genes direct ribosomes to produce a pipeline of proteins and enzymes that tell the cell how to operate, maintain itself, and reproduce. But there is so much more.

Power stations—mitochondria—generate electric currents of hydrogen protons that course through minuscule rotary motors that produce hundreds of millions to a billion ATPs a second. The cell is full of other large organelles—specialized factories and manufacturing centers, storage depots, and garbage dumps—and it is spanned by networks of shipping highways full of cargo.

Meanwhile, many other kinds of molecular machines are whirring, pumping, ratcheting, and walking. Chemical attractions and repulsions help some structures assemble themselves. Cleverly, the cell takes advantage of the fact that its molecules are crammed together so tightly that the physicist Peter Hoffmann likens their density to a parking lot with a foot or less between cars. The random energetic jostling of those closely packed molecules accelerates movements and interactions. Finally, the cell has strategies to stay young. It constantly replaces old, damaged machines and structures with new ones, just as your body constantly replaces entire cells.

In fact, our cells are complex enough that they don't really need the rest of our bodies. If you take a human cell and give it all the nutrients it dreams of, it will do just fine, at least for a while. Some cancerous cells, like the HeLa cells named after the long-deceased cancer sufferer Henrietta Lacks, who passed away in 1951, are even immortal. These abnormal cells can keep reproducing forever and are still used for many kinds of biological research.

No scientist will pretend that we entirely understand how all those atoms and molecules add up to living cells. The sheer scale of everything happening in a single cell is hard to comprehend. "When someone starts thinking about how many molecules are in the so-called simplest living system," George Cody of the Carnegie Institution for Science told me, "how many molecules are interacting with each other simultaneously at any point—it would become almost beyond our comprehension to see all of it operating at once." The lines on flow charts that scientists make to represent all of the chemical reactions going on are so densely packed they are hard to read. Cody suspects that some breakthroughs in understanding how cells work will arise not from chemistry, but from theoretical physics. Holly Goodson expects them to come from computer modeling. The cell biologist

Franklin Harold thinks that the ways in which the staggering numbers of molecules within cells find their proper locations is still one of life's greatest secrets. To the plant physiologist Tony Trewavas, the complexity of even a single-cell organism demands what Aretha Franklin asked for—respect. "All unicellular organisms are able to do things in ways we don't know or understand yet," he said. "We're still finding out bits and pieces. Whether we will get a full understanding, I have my doubts. I may be wrong, but I think there are probably things going on there that we can't appreciate. We need to be a little more respectful of the organisms that we see around us."

And so beneath their foggy haze, the interior of our cells, as Albert Claude discovered, are not simply enzyme soups. They are cities, huge metropolises, as much solid matter as liquid, full of not just reacting molecules, but of machines, coordinated movement, feedback loops, self-correcting processes propelled, in part, by the constant motion of their molecular citizens, as though everyone in New York City was out on the street at the same time, colliding and elbowing one another as they raced through its streets.

Needless to say, when our atoms first arrived on Earth, they had no idea they would find their way into an enormous revolving cycle of self-sustaining life, much less in the dazzling mechanisms in our cells that create our thoughts, desires, plans, and actions.

Thus, at last, we arrive at one of the profoundest questions of all, as much philosophical as scientific. Our cells are packed with all kinds of extraordinary machinery, but what about our selves? We now know what's gotten into us, but what, ultimately, are we?

Conclusion

WHAT A LONG STRANGE
TRIP IT'S BEEN

Science, truly understood, is not the death, but the birth
of mystery, awe, and reverence.
Frederick G. Donnan

I began the idle brainstorming that led to this book by wondering: What are we actually made of? And where did it come from? As I delved further, I began to wonder how what I was learning adds up. What did all those particles that sprang from the Big Bang create? Physically and philosophically, what the heck are we? What does science tell us about what we are and our ultimate origins?

That's not easy to answer, even on a material level. As we've seen, there are so many different levels on which we can contemplate our existence. Ask scientists what we are ultimately made of, and you will receive many different answers.

In one sense, we *are* utterly extraordinary biological machines whose details are so staggeringly complex they are hard to wrap your head around. A typical cell within you is made of a galaxy of atoms—a hundred trillion or so of them. A stack of that many dollar bills would reach to the Moon and back over twenty-five times. Every second within each of your cells, many hundreds of millions of molecules are shooting in and out of membranes. Thousands of genes are being locked and unlocked. Millions of ribosomes and organelles are working. Electric currents are surging. Many hundreds of thousands if not millions of motors are turning and pumps churning. That is just one cell. You are composed of about a hundred times more cells than there are stars in the Milky Way.

In another sense, as the biochemist Peter Mitchell observed, you are more like a flame whose atoms are constantly being replaced. Although we may die, our atoms don't. They revolve through life, soil, oceans, and sky in a chemical merry-go-round. "I don't think of us as something that's made," the geologist Mike Russell said. "I think of us as processors." The cell biologist Franklin Harold agrees. He sees our cells as organized patterns, systems that must remain in constant motion, like bicycles, which stay upright only so long as their wheels continue to spin.

At another basic level, we are simply temporary gatherings of elements forged by the Big Bang and stars. All told, you're made of about 60 of the 132 or so elements in the periodic table.

A physicist would say there is an even more fundamental you. Your seventy octillion atoms are made of electrons, protons, and neutrons, and those protons and neutrons are built from smaller particles, quarks and gluons. All of which makes you simply a very large collection of overachieving subatomic particles who in combination vastly exceed their seeming potential.

If that is not strange enough, on an even deeper physical level, most quantum physicists say that your elementary particles are localized excitations of energy fields (called quantum fields) that permeate space, so the smallest bits of you are simultaneously particles and waves. That makes all of the universe, including you, an interwoven web of rippling energy fields. You may not believe you've attained enlightenment; nevertheless, in some sense, you are already at one with the universe.

If you're not dizzy yet, consider this. About 99 percent of the volume of your atoms is simply empty space between elementary particles. Yet, on closer examination, even that emptiness is not truly nothing. It contains energy fields from which particles of matter and antimatter continuously bubble up and annihilate each other.

Somehow, all those minuscule fields, waves, particles, and atoms add up to you and me . . . with all our foibles.

Over 150 years after the biochemistry pioneer Justus Liebig brashly extended the study of chemistry to animals, we no longer need to resort to an inexplicable "vital force" to explain the workings of the human body. Yet many deep mysteries remain. The greatest challenge of all is to explain how our quintessential human qualities—our consciousness, spirituality, language, and thought—emerge from our atoms and cells. Are our sensations, like our delight at smelling a rose or the awe we feel looking at the Grand Canyon, solely a creation of chemical reactions, nanomachines, and the known forces of physics? Will science ever offer an answer?

On this, I have come to understand, researchers widely disagree. There is plenty of room to debate whether we are the sum of our atoms or if there is more to us than that.

Many scientists, including Murray Gell-Mann, the brilliant discoverer of the quark, are confident that the most fundamental particles and physical forces that we have discovered are all that there is. These, ultimately, can explain everything. In Gell-Mann's wide rearview mirror, the universe began in a state of "wound up" energy. As it expanded, the overall disorder in the universe increased. Yet small pockets of greater order also formed, and that is where we find life.

A few scientists, like the renowned Oxford physicist Roger Penrose, are hopeful that still-to-be-discovered quantum mechanical interactions in our brains will ultimately explain our conscious experience.

Others think that we have yet to discover new kinds of physical phenomena that will reveal the roots of consciousness.

The origin-of-life researcher and chemist Günter Wächtershäuser is firmly of the opinion that life began on the ocean floor. Yet this former patent lawyer also believes that, alongside the material world, there must be a world of the mind. He is confident that our ideas, culture, and the mental inventions that make us human exist independently of the physical domain.

The Nobel Prize winner Charles Townes, the first scientist to discover organic molecules in space, suspected that once these molecules reached Earth, they may have helped create life. Yet he also firmly believed that God created our universe with laws of physics so finely tuned that our appearance was inevitable. The priest and cosmologist Georges Lemaître, who first conceived of the Big Bang, might have said something similar. He believed one should look to science to learn about nature, and the Bible to learn about salvation.

In some ways, both science and religion answer similar fundamental human needs. "The real object of science is to make the world intelligible," the cell biologist Franklin Harold said to me. We are meaning makers, born with a craving to explain the world's mysteries and understand our place in it.

Of course, science will never tell us everything. There is so much we don't know about the cosmos. Science has not yet provided us with a full explanation of how our molecules ultimately animate our consciousness, nor can it explain why our universe exists in the first place.

But it is able to answer another big question.

Science *can* tell us how we got here—at least over the last 13.8 billion years. From that point on, our atoms all shared the same remarkable journey—an awe-inspiring odyssey that began with a Big Bang.* The

---

* Today, what scientists mean by "the Big Bang" is a matter of scientific debate. At a minimum, it means that all of the matter in the universe was tucked into an unbelievably tiny and unimaginably hot volume of space and time, and that since then the universe has been growing larger and cooling down. Whether the universe began as a "singularity," a point so small that space and time did not exist, as Einstein's equations suggest, is not known. To avoid that apparent impossibility, physicists have proposed many theories for what might have existed before the Big Bang. We might live in a universe that continuously expands, contracts, and then expands again in a Big Bounce. Or we may live in one of a vast multitude of universes that somehow spawn new ones. There are many theories, but conclusive evidence is lacking.

stuff that makes up every living person and all of life once came from the same tiny point in time and space. Our atoms were born as smaller particles that attraction forged into hydrogen. Gravity sucked them into huge stars, and the reactions between their subatomic particles worked out in just the right way that they were able to create carbon, oxygen, and the other elements of life. After sailing through cold dark space, our molecules were trapped in the monstrously large dust cloud that created our solar system. Those that collided in our vicinity produced a rocky, watery Earth rich in the elements we are made of. And once they met here in close quarters, sticky building blocks capable of forming long chains of fantastic variety spawned new types of molecules, self-perpetuating cells, and then, remarkably, us.

Incidentally, many scientists suspect that the appearance of life on our planet was inevitable. That is one of many reasons for us to remain humble: we may not be so special. Our universe may contain life elsewhere, and some of our cosmic brothers and sisters could be more intelligent than we are.

Here on Earth, we are part of a majestic, inventive tree of life that, once established, stubbornly hung on and has never stopped changing. Our atoms' journeys were paved by an unbroken chain of organisms that stretches back at least 3.8 billion years. Learning their story has left me with a greater appreciation of the debt we owe our earliest bacterial kin, who developed the template for life. They pioneered many of our basic tools, handy things like RNA, DNA, ATP, ribosomes, and sodium-potassium pumps. Photosynthesizing bacteria oxygenated our atmosphere, making it possible for plants to pluck molecules from the air and rock and create the sugars, proteins, fats, vitamins, and minerals that we eat at dinnertime in order to build ourselves.

It is good to remember that life has fundamentally altered our planet. Without it, our atmosphere would have virtually no oxygen, and it would have more carbon dioxide. Photosynthesis pulled some CO_2 from the air that was then buried in fossil fuels and stored in soil and plants. By releasing some of that insulating gas back into the atmosphere, the Earth will warm, as it has in response to rising CO_2 levels many times before.

The fact that all life sprang from microbes means that we have a deep biological bond with all other living things on Earth. We came from single-cell organisms, and even the smallest cell, I've come to see, is an incredible creature that deserves our utmost respect. In one sense, we share so much with our microbial ancestors, our greatest aunts and uncles, that we are just variations on a theme. We are, as Lynn Margulis liked to point out, simply massive overgrown colonies of microbes.

Yet we are also more than that. Within us, atoms from the stars create hundreds of specialized kinds of cells that cooperate and talk to each other in ways no bacteria do, such as discussing spirituality and the nature of the universe.

It is a remarkable fact about our universe that our atoms, which once came chaotically crashing down from space, can now look back in time to reconstruct their own journeys. In other words, self-replicating humans, the product of chemical and biological evolution, can now "see" the world, probe it, and in a stunning recursiveness, study the origin and voyages of the very molecules of which we are made. In Carl Sagan's words, "We are a way for the cosmos to know itself." How is it possible that we have been able to learn so much, to peer back so far, even to the beginning of time?

That understanding is a testament to the obsessiveness and doggedness of countless researchers driven by a deep thirst for knowledge for its own sake. Some researchers who made radical discoveries were introverted. Many came from humble beginnings. They included high school dropouts who read on their own and willed their way into college. The sheer satisfaction that comes from understanding our world, the thrill that arises from being the first to make a new discovery, has been powerful motivation throughout human history. Sir John Walker, who worked out the intricate molecular structure of ATP synthase, sums it up with Winston Churchill's observation, "Success consists of going from failure to failure without loss of enthusiasm."

Scientists have been helped immeasurably by the profound fact that the structure of our universe accords with patterns first discovered by mathematicians and physicists scribbling equations on paper. Einstein's general theory of relativity helped reveal the Big Bang, which was later confirmed by observations. Quantum mechanics and obscure geometry helped Murray Gell-Mann predict the existence of quarks and gluons. The structure of our universe resonates in many deep ways with mathematics that humans derived in their minds.

We've also been able to peer so far back through the veils of time because our senses are so acute, so remarkably fine-tuned. Our eyes are capable of detecting just a few photons of light, particles ten million times smaller than a speck of dust. On a very clear night, our naked eyes can see Andromeda, the galaxy nearest to our Milky Way. Those photons began sailing toward us from that distant galaxy 2.5 million years ago. Our senses and the instruments that extend them can detect atoms, particles, and waves ridiculously smaller than we are. They've enabled us to learn an amazing amount from visible light, X-rays, microwaves, and other wavelengths of electromagnetic radiation—even from tracks of invisible subatomic particles.

Mind you, we could never have reconstructed our ancient journeys without the brilliant and formidable reasoning power of our minds: our ability to detect patterns, to apply logic to what we see, and to follow the evidence wherever it leads.

That's not to say that we are perfect in this respect. As we've seen, scientists are human. They have ambitions and egos. Rivalries and personal interests cloud their vision—just like the rest of us.

Of course, we also fall prey to cognitive biases; none of us escapes them. As the economist Pierre Cremieux pointed out to me, cowbirds lay their eggs in other birds' nests because they know that the nest owners have a cognitive bias: they assume that any egg lying in their nest is their own. Similarly, we don't go to bed every night wondering if the Sun will come up in the morning. If we constantly questioned everything and assumed nothing, we would be paralyzed. For scientists, like the rest of us, cognitive biases provide useful shortcuts. You would have a hard time being a productive scientist if you didn't build on earlier work, if you were unwilling to assume that experts in your field are often correct.

But biases have a cost. If you don't recheck your assumptions, they can lead you to miss the possibility that "simple" DNA might not be so simple after all. And beyond the cognitive biases that we all succumb to, we've seen in this story that when scientists have been confronted by radical scientific breakthroughs (or conspicuous clues to them), a particular set of biases have repeatedly led entire communities astray. These are the thinking traps, such as the "Too Weird to Be True" bias and the "As an Expert, I've Lost Sight of How Much Is Still Unknown" bias.

It's not surprising that many researchers who made radical discoveries were fiercely independent spirits, willing to go their own ways despite ridicule and scorn.

Scientists have been most successful not just when they have come up with testable hypotheses, but also when they were able to distinguish between what is known, what seems likely, and what—however far-fetched and absurd—might nonetheless still be possible. The history of science is littered with elder statesmen's grand pronouncements of certainties that would soon be overturned. Cosmologists proclaimed that of course the universe is static and always existed. Physicists were sure that nothing could be smaller than the fundamental trinity of electrons, protons, and neutrons. The most respected astronomers agreed that stars and planets had the same composition and that organic molecules could never survive in outer space. Biologists were sure that life could not exist in the deepest ocean depths. The greatest paleontologists agreed we would never find fossils older than 500 million years. Plants could never communicate, chemistry could never explain life, DNA could not transmit heredity, and so on.

The interplay between cognitive biases and our willingness to challenge them is at the core of science. It makes the bumpy progress of science possible. Ironically, the knee-jerk impulse to reject new ideas that often retards science is also a strength. Scientists are in the business of thinking outside the box, of generating multitudes of theories, most of which will be proven wrong. Progress is made only because investigators ruthlessly examine new theories and force their proponents to demonstrate that they've ruled out every possible source of error. Without a necessary combination of imagination, fierce skepticism, and intellectual honesty, researchers could never separate the scientific wheat from the chaff. They would make little headway. In the larger scheme of things, those who were right owe a great debt to those who were wrong.

We began this book as the Catholic priest Georges Lemaître, in black suit and clerical collar, presented the kernel of his theory, later to be known as the Big Bang, to a large and largely bewildered audience. Despite Lemaître's mystifying address, the attendees had much to celebrate that day. In 1931 they were attending the centenary anniversary of the British Association for the Advancement of Science. In the previous one hundred years, biologists had discovered the immense power of evolution. Geologists had revealed vast ancient transformations of our planet. Physicists had detected electrons and protons. Chemists were learning the intricacies of how atoms bond. And cosmologists had discovered the inconceivable vastness of our universe, and even, as Lemaître himself was astonished to discover, the bizarre fact that it is expanding.

In celebration of that momentous centenary, Ernest Barnes, the bishop of Birmingham, delivered a commemorative sermon to a large audience at Liverpool Cathedral. "The most astonishing fact in the whole picture of . . . scientific progress," Barnes said, "is man himself. The modern astronomer thinks in terms of thousands of millions of years and lives perchance for three score years and ten. His mind embraces the whole of space: In thought he travels distances measured by many thousands of millions of millions of miles—and seven feet suffice for his resting place."

Barnes that day captured a feeling that I have come to share. Writing this book has been a continual source of wonder, stupefaction, exhilaration, and gratitude. Even if we were somehow to learn that the physical forces we have discovered in the universe are all that is needed to explain how we arrived here, that doesn't mean we must feel gloom, or despair at life's meaninglessness. There will always be a million reasons to feel awe and gratitude at the nature of our universe.

As I was finishing this book, my mother died, at the age of ninety-three. We buried her ashes in a beautiful cemetery in Cambridge, Massachusetts. Just before, I held in my hands the small carved wooden box that received her ashes. It was smaller than a shoebox. It represented only a few percent of the atoms that made her so lovable. Nonetheless, I felt grateful to them. Those atoms I held in my hands had begun their journeys in the Big Bang. They underwent many tribulations to bring life to her. They will remain in the ground for years, and then find release, perhaps to become part of other life, whether a blade of grass or a bird. But where is she? She is at least in my heart. And my mind. My relationship with her will never die, so long as I continue to breathe. The complexity of the universe that can create a person like her—a wonderful passionate advocate for and lover of other people—is a thing of beauty to be treasured. As I write now, I feel grief. But as my family and friends celebrated her life, I also celebrate the wider world we live in.

To retrace the journeys of our atoms is to appreciate the world anew.

ACKNOWLEDGMENTS

When I began this book, I wondered how I could tell such a sprawling story. Only the generous help of an astounding number of scientists, friends, family, and other accomplices made it possible.

First, I'm deeply indebted to my agent, Suzanne Gluck at William Morris Endeavor, who saw the promise in this book from the very start. I'm more grateful to her than I can say. Noah Eaker, my wonderful editor, was equally enthusiastic and brilliantly contributed so much, not least by asking his "English-major questions." Others, including the eagle-eyed copy editor Gary Stimeling, editorial assistant Edie Astley, and Andrea Blatt at WME provided great assistance in ushering the book into print. Thanks, as well, to Meghan Hauser for her many suggestions, and to Toby Lester, who offered sage advice and much needed encouragement throughout.

I'm so appreciative of Doron Weber and the Alfred P. Sloan Foundation, who provided a generous grant for the research and writing of this book.

Following the twists and turns in scientific theorizing in so many subjects is, to put it mildly, challenging. I would have been lost without the many scientists and historians who so graciously offered insights. Among them: František Baluška, Janet Braam, Ted Bergin, Lawrence Brody, Don Caspar, Thomas Cech, James Collins, Kent Condie, Dale Cruikshank, Brian Fields, Simon Gilroy, Owen Gingerich, Alfred Goldberg, Stjepko Golubic, Douglas Green, Linda Hirst, Nicholas Hud, Joe Kirschvink, Keith Kvenvolden, Jack Lissauer, Nick Lane, Avi Loeb, Steven Long, Tim Lyons, Simone Marchi, Jim Mauseth, Jay Melosh, Carol Moberg, Alessandro Morbidelli, Wayne Nicholson, Rob Phillips, Jonathan Rosner, Dave Rubie, Mike Russell, Kim Sharp, Fred Spiegal, Paul Steinhardt, Tony Trewavas,

John Valley, Elizabeth Van Volkenburgh, Günter Wächtershäuser, and Jack Welch.

I'm also indebted to the scientists who not only spoke with me but also kindly took time away from their busy lives to comment on portions of the manuscript. Thank you to: John Archibald, Alisa Bokulich, Peter Bokulich, David Catling, Frank Close, George Cody, Gerald Combs, Jr., Don Davis, David Devorkin, Holly Goodson, Govindjee, Franklin Harold, Dave Jewitt, Paul Kenrick, Andy Knoll, Simon Mitton, Hans-Jörg Rheinberger, William Schopf, Jack Schultz, Thomas Sharkey, Ruth Lewin Sime, Martha Stampfer, Christopher T. Walsh, and Martin Wuhr. My deep appreciation, as well, to Conel Alexander, Lynn DiBenedetto, Daniel Kirschner, and Anna Sajina. You were wonderfully helpful and saved me from more inaccuracies than I care to contemplate. Whatever factual errors remain are, of course, mine.

To my friends who kept me on track and offered unsparingly candid comments: you made the writing of this book so much more fun! Huge thanks to Larry Braman, Isabel Bradburn, Anne Braude, Steve Collier, Pierre Cremieux, Polly Farnham, Marc Freedman, Jennifer Gilbert, Alex Hoffinger, John Jelesko, and Tse Wei Lim. I owe an extra special debt to Georgann Kane, Megan McCarthy, and Carol Thomson, who were not only brilliant readers but whose unflagging support and elegant suggestions made this book better on so many different levels.

I'm especially grateful to my parents, Lore and Dave, who inspired my love of science from an early age, beginning with the gift of a chemistry kit when I was seven. Dave, a wonderful reader, also chased down facts. My children, Eli and Zoe, were more than patient with me as I rambled on about the book at every opportunity. They contributed much assistance and insight. And finally, I could never have written this without the encouragement and love of my wonderful wife, Ariadne, who happens to be one of my most astute readers. I owe her, and her seven billion quintillion atoms, many enormous debts for which I will be eternally grateful.

NOTES

Epigraph

vii   "We are an example": Quoted in Cott, "The Cosmos: An Interview with Carl Sagan."

Introduction: $1942.29 in the Bank

xii   thirty trillion cells: Sender, Fuchs, and Milo, "Revised Estimates for the Number of Human and Bacteria Cells in the Body," 9.

xii   over a hundred trillion atoms: Milo and Phillips, *Cell Biology by the Numbers*, 68.

xii   a billion times more atoms than all the grains of sand: Blatner, *Spectrums*, 20.

xii   a cool $1,942.29: Calculations of the value of the elements in the human body vary wildly. Here is how $1,942.29 was calculated. The mass of each element in the body comes from *Nature's Building Blocks: An A–Z Guide to the Elements*, by John Emsley. The cost of each element is from the website Chemicool.com. Of course, the actual costs will vary—depending, for instance, on whether you plan to make water from scratch or buy it from the supermarket.

Chapter 1: Happy Birthday to Everyone: The Priest Who Discovered the Beginning of Time

3   "All great truths": Shaw, *Annajanska, the Bolshevik Empress*, 139.

3   Central Hall, Westminster: The *Times*, "The British Association: Evolution of the Universe."

4   "primeval atom": Lemaître, "Contributions to a British Association Discussion on the Evolution of the Universe," 706.

4   "who are immeasurably beyond": Barnes, "Contributions to a British Association Discussion on the Evolution of the Universe," 722.

4   cycling trip: Mitton, "The Expanding Universe of Georges Lemaître," 28.

4   outdated single-loading rifles: Mitton, "Georges Lemaître and the Foundations of Big Bang Cosmology," 4.

5   "The madness of it": Deprit, "Monsignor Georges Lemaître," 365.

5   He lacked, it seemed: Deprit, "Monsignor," 366.

5   he somehow found the concentration: Lambert, *The Atom of the Universe: The Life and Work of Georges Lemaître*, 56–57.

5   What was the universe: Lambert, "Georges Lemaître: The Priest Who Invented the Big Bang," 11.

5   "two ways of arriving": Aikman, "Lemaitre Follows Two Paths to Truth."

5   Amis de Jésus: Lambert, "Georges Lemaître," 16.

6   "wonderfully quick and clear-sighted": Kragh, "'The Wildest Speculation of All': Lemaître and the Primeval-Atom Universe," 24.

6   "'island universes' similar to our own": *New York Times*, "Finds Spiral Nebulae Are Stellar Systems: Dr. Hubbell [*sic*] Confirms View That They Are 'Island Universes' Similar to Our Own."

6   the latest measurements taken: Mitton, "The Expanding Universe," 29–30.

7   the space between them was actually expanding: Although the empty space between galaxies is expanding, the galaxies themselves and the matter within them are not. That is because the gravitational attraction between the clusters of matter in galaxies is much stronger than the force that is pulling space, and them, apart.

7   in a little-known Belgian periodical: Kragh, "'The Wildest Speculation,'" 34.

7   As he strolled through: Lambert, "Einstein and Lemaître: Two Friends, Two Cosmologies."

7   He hated it: Frenkel and Grib, "Einstein, Friedmann, Lemaître," 13.

8   "Your calculations": Deprit, "Monsignor," 370.

8   began speaking with Piccard: Lemaître, "My Encounters with A. Einstein."

8   embarrassed to find: Farrell, *The Day without Yesterday: Lemaître, Einstein, and the Birth of Modern Cosmology*, 97.

9   "disintegration": Lemaître, "Contributions," 706.

9   "The evolution of the universe": Lemaître, *The Primeval Atom: An Essay on Cosmogony*, 78.

9   "Bart, I've had a funny idea": DeVorkin, AIP oral history interview with Bart Bok.

9   "Out of a single bursting atom": Menzel, "Blast of Giant Atom Created Our Universe."

9   "an example of speculation run mad": Kragh, "'The Wildest Speculation,'" 35–36.

10   in 1978: Godart, "The Scientific Work of Georges Lemaître," 395.

10   "Physics provides a veil": Quoted in Lambert, "Georges Lemaître," 16.

10   "There is no conflict": Aikman, "Lemaitre Follows."

10   "biggest blunder": O'Raifeartaigh and Mitton, "Interrogating the Legend of Einstein's 'Biggest Blunder.'" While some historians have questioned whether Einstein actually said this, there are those who believe that he did.

10   "This is the most beautiful": Aikman, "Lemaitre Follows."

11   "*the* Catastrophe to begin": Lemaître, *The Primeval Atom*, vi.

11   "The hypothesis that all matter": Cooper, *Origins of the Universe*.

11   "the Big Bang man": Lambert, "Georges Lemaître," 17.

12   "It is the same way that": Author interview with Avi Loeb, Harvard Smithsonian, August 2018.

13   gravitational echoes of the Big Bang: Webb, "Listening for Gravitational Waves from the Birth of the Universe."

Chapter 2: "That's Funny": What the Eye Can Never See

15   But many physicists were dubious: Rhodes, *The Making of the Atomic Bomb*, 30–31.

16   "Atoms and molecules . . . from their very nature": Blackmore, *Ernst Mach*, 321.

17   a hundred thousand times smaller: Close, *Particle Physics: A Very Short Introduction*, 14.

17   thick metal boxes: De Angelis, "Atmospheric Ionization and Cosmic Rays," 3.

18   deep into caves: Gbur, "Paris: City of Lights and Cosmic Rays."

18   Enlisting the help: Bertolotti, *Celestial Messengers: Cosmic Rays: The Story of a Scientific Adventure*, 36.

18   a twelve-story orange-and-black: Kraus, "A Strange Radiation from Above," 20.

18   squeezed himself into: Part of Hess's account is translated into English in Steinmaurer, "Erinnerungen an V. F. Hess, Den Entdecker der Kosmischen Strahlung, und an Die ersten Jahre des Betriebes des Hafelekar-Labors."

19   "an inner joy is felt": "The *Zenith* Tragedy"; and Oliveira, "Martyrs Made in the Sky."

19   Wasn't it more likely: Ziegler, "Technology and the Process of Scientific Discovery," 950.

20   especially fierce opponent: Walter, "From the Discovery of Radioactivity to the First Accelerator Experiments," 28.

20   until Hess bitterly objected: De Maria, Ianniello, and Russo, "The Discovery of Cosmic Rays," 178.

21   "the most original and wonderful instrument": Quoted in *Nobel Lectures Physics: Including Presentation Speeches and Laureates' Biographies, 1922–1941*, 215.

21   Fearing others would think him crazy: Pais, *Inward Bound: Of Matter and Forces in the Physical World*, 38.

21   when cathode rays: Two years later, J. J. Thomson would discover that the cathode "rays" inside the glass tube were actually streams of electrons.

21   "Nearly every professor": Pais, *Inward Bound*, 39.

21   He was startled to see: Crowther, *Scientific Types*, 38.

21   Wilson was ecstatic: BBC Interview with Wilson in transcript of the BBC documentary "Wilson of the Cloud Chamber."

22   "little wisps and threads": *Nobel Lectures Physics*, 216.

22   insisted that Anderson: Anderson, *The Discovery*, 25–26.

22   the tracks must instead be from positively charged protons: Anderson, *The Discovery*, 29–30.

22   None of the famous gods: Hanson, "Discovering the Positron (I)," 199.

23   almost four thousand positrons a day: Sundermier, "The Particle Physics of You."

23   "Who ordered that?": Close, Marten, and Sutton, *The Particle Odyssey: A Journey to the Heart of Matter*, 69.

24   radium-fortified soap: Rentetzi, *Trafficking Materials and Gendered Experimental Practices*, 2; and Miklós, "Seriously Scary Radioactive Products from the 20th Century."

24   Could she use a photographic plate to detect: Sime, "Marietta Blau: Pioneer of Photographic Nuclear Emulsions and Particle Physics," 7.

25   That was impossible: Rentetzi, AIP oral history interview with Leopold Halpern.

25   generously offered help: Rentetzi, AIP oral history interview with Leopold Halpern.

25   an early member: Galison, "Marietta Blau: Between Nazis and Nuclei," 44.

25   Wambacher began an affair: Sime, "Marietta Blau," 14.

26   up to twelve smaller particles: Rosner and Strohmaier, *Marietta Blau, Stars of Disintegration*, 159.

26   taking her newest plates: Rentetzi, "Blau, Marietta," 301.

26   seized her photographic plates: Rentetzi, AIP oral history interview with Leopold Halpern.

27   she was embittered: Rentetzi, AIP oral history interview with Leopold Halpern.

28   130,000 miles in just four-fifths of a second: Plumb, "Brookhaven Cosmotron Achieves the Miracle of Changing Energy Back into Matter."

28   less than a billionth of a second: Close, Marten, and Sutton, *The Particle Odyssey*, 13.

28   Joy turned to bafflement: Riordan, *The Hunting of the Quark: A True Story of Modern Physics*, 69.

28   "If I could remember the names": Quoted in Riordan, *The Hunting*, 69.

29   At age three: Johnson, *Strange Beauty: Murray Gell-Mann and the Revolution in Twentieth-Century Physics*, 35.

29   "on first acquaintance": Glashow, "Book Review of *Strange Beauty: Murray Gell-Mann and the Revolution in Twentieth-Century Physics*," 582.

30   Perhaps it would finally lead: Bernstein, *A Palette of Particles*, 95.

30   Gell-Mann was apprehensive: Johnson, *Strange Beauty*, 194.

30   Collegially, they shared: Johnson, *Strange Beauty*, 208.

30   Long Island housewives: Crease and Mann, *The Second Creation: Makers of the Revolution in Twentieth-Century Physics*, 275.

30   But in the 97,025th: Johnson, *Strange Beauty*, 217.

31   "That would be a funny quirk": Riordan, *The Hunting*, 101.

31   *What the hell, why not?*: Crease and Mann, *The Second Creation*, 281.

31   a useful mathematical fiction: Johnson, *Strange Beauty*, 283–84.

32   more open to publishing "crazy" ideas: Crease and Mann, *The Second Creation*, 284.

32   the barriers: Charitos, "Interview with George Zweig."

32   labeled him a charlatan: Zweig, "Origin of the Quark Model," 36.

32   a million times smaller than a grain of sand: Butterworth, "How Big Is a Quark?"

32   "they [had] begun opening": Sullivan, "Subatomic Tests Suggest a New Layer of Matter."

33   a billion billion billion times: Chu, "Physicists Calculate Proton's Pressure Distribution for First Time."

33   You could fit all of humanity: Sundermier, "The Particle Physics of You."

35   a trillion quadrillion H-bombs: Cottrell, *Matter: A Very Short Introduction*, 127.

Chapter 3: The Best Man at Harvard: The Woman Who Changed How We See the Stars

36   "There are three stages": Hoyle, *Home Is Where the Wind Blows: Chapters from a Cosmologist's Life*, 154.

36   "I saw an abyss opening": Payne-Gaposchkin, *Cecilia Payne-Gaposchkin: An Autobiography and Other Recollections*, 124.

37   She prayed for high marks: Payne-Gaposchkin, *Cecilia Payne-Gaposchkin*, 97.

37   "prostituting her gifts": Payne-Gaposchkin, *Cecilia Payne-Gaposchkin*, 98.

37   expected to become a botanist: Payne-Gaposchkin, *Cecilia Payne-Gaposchkin*, 102.

37   "like a nervous breakdown": Payne-Gaposchkin, *Cecilia Payne-Gaposchkin*, 117–18.

38   she was eager to tackle: Gingerich, AIP oral history interview with Cecilia Payne-Gaposchkin.

40   thousands of individual stars: Moore, *What Stars Are Made Of: The Life of Cecilia Payne-Gaposchkin*, 172.

40   "As you look at it": Author interview with Owen Gingerich, Harvard University, February 2018.

40   "utter bewilderment": Payne-Gaposchkin, *Cecilia Payne-Gaposchkin*, 163.

40   "Miss Payne? You're very brave": Payne-Gaposchkin, *Cecilia Payne-Gaposchkin*, 165.

41   "clearly impossible": Payne-Gaposchkin, *Cecilia Payne-Gaposchkin*, 19.

41   There were strong reasons: Gingerich, "The Most Brilliant Ph.D. Thesis Ever Written in Astronomy," 11.

41   "His word could": Payne-Gaposchkin, *Cecilia Payne-Gaposchkin*, 201.

41   "almost certainly not real": Payne-Gaposchkin, *Cecilia Payne-Gaposchkin*, 5.

41   told the writer Donovan Moore: Moore, *What Stars Are Made Of*, 183.

41   Russell himself: DeVorkin, *Henry Norris Russell: Dean of American Astronomers*, 213–16; and Gingerich, "The Most Brilliant Ph.D. Thesis Ever Written in Astronomy," 13–14.

41   "the best man at Harvard": Payne-Gaposchkin, *Cecilia Payne-Gaposchkin*, 184.

41   would not be listed: Payne-Gaposchkin, *Cecilia Payne-Gaposchkin*, 26.

42   offended by the "stupidity": Hoyle, *The Small World of Fred Hoyle: An Autobiography*, 72.

42   When he was not "ill": Hoyle, *The Small World*, 64.

42   "one of the most innovative": Couper and Henbest, *The History of Astronomy*, 217.

42   "the most creative and original": Martin Rees quoted in Livio, *Brilliant Blunders: From Darwin to Einstein—Colossal Mistakes by Great Scientists That Changed Our Understanding of Life and the Universe*, 219.

42   "in less time than it takes": Livio, *Brilliant Blunders*, 180.

43   nowhere near hot enough: Hoyle, *Home Is Where*, 150.

43   top-secret meeting: Mitton, *Fred Hoyle: A Life in Science*, 99.

44   A nighttime curfew: Mitton, *Fred Hoyle*, 104–5.

44   vastly more heat: Gregory, *Fred Hoyle's Universe*, 31.

44   trying to glean: Hoyle, *Home Is Where*, 229.

44   When a star ran out of fuel: Hoyle, *Home Is Where*, 230.

45   actually be hot enough: Mitton, *Fred Hoyle*, 200.

46   almost 23 percent: Emsley, *Nature's Building Blocks: An A–Z Guide to the Elements*, 111.

46   "Here was this funny little man": Weiner, AIP oral history interview with William Fowler.

46   like a prisoner in a dock: Hoyle, *Home Is Where*, 265.

46   After several months: In *Home Is Where the Wind Blows*, Hoyle writes that the wait was ten days, but according to his biographer, Simon Mitton, Hoyle heard the results several months later.

48   less than 1 percent: Emsley, *Nature's Building Blocks*, 112.

48   heavier than iron on Earth: Uranium, with the atomic number 92, is the heaviest element that exists naturally on Earth.

49   as a hundred billion suns: Gribbin and Gribbin, *Stardust: Supernovae and Life—the Cosmic Connection*, 156.

49   newly released data: Burbidge, "Sir Fred Hoyle 24 June 1915: 20 August 2001," 225.

49   Hoyle's team found evidence: Hoyle, *Home Is Where*, 296–97.

50   27 million degrees: "The Sun," NASA, https://www.nasa.gov/sun.

50   clouds of viruses and bacteria: Horgan, "Remembering Big Bang Basher Fred Hoyle."

Chapter 4: Catastrophes to Be Thankful For: How to Make a World from Gravity and Dust

52   "My own suspicion": Haldane, *Possible Worlds*, 286.

53   "innocent entertainment": Wetherill, "The Formation of the Earth from Planetesimals," 174.

54   someone with the technical skill: Burns, Lissauer, and Makalkin, "Victor Sergeyevich Safronov (1917–1999)."

55   Soviet colleagues were skeptical: E-mail to author from Andrei Makalkin, Institute of Earth Physics of the Russian Academy of Sciences, May 2018.

55   He presented a copy: Author interview with that former graduate student: astronomer Dale Cruikshank, NASA Ames Research Center, May 2018.

55   a groundbreaking program: Wetherill, "Contemplation of Things Past," 17.

56   runaway effect: Wetherill, "Contemplation," 19.

57   Venus spins backward: Hazen, *The Story of Earth: The First 4.5 Billion Years, from Stardust to Living Planet*, 45.

57   violently assaulted: Fisher, "Birth of the Moon," 63.

58   "His contributions are of overwhelming proportion": Wetherill, "Contemplation," 18.

58   "first scientist": Gribbin, *The Scientists*, 68.

58   far-off magnetic mountains: Hockey et al., "Gilbert, William."

61   "As far as I'm concerned": Cooper, "Letter from the Space Center," 50. Cooper tells this story beautifully in a series of articles in the *New Yorker* and his book *Apollo on the Moon*.

61   King had helped persuade NASA: Compton, *Where No Man Has Gone Before*, 52.

61      $25 billion: Wilford, "Moon Rocks Go to Houston; Studies to Begin Today: Lunar Rocks and Soil Are Flown to Houston Lab."

61   scientists were debating whether the massive craters: Corfield, "One Giant Leap," 50.

61   A lunar lander would be swallowed: Powell, "To a Rocky Moon," 200.

62   the same alarms: Eyles, "Tales from the Lunar Module Guidance Computer."

62   about to set them down: Wagener, *One Giant Leap*, 182.

62   Armstrong's pulse doubled: Portree, "The *Eagle* Has Crashed (1966)."

63   watched impatiently: King, *Moon Trip: A Personal Account of the Apollo Program and Its Science*, 92.

63      $24 billion: Wilford, "Moon Rocks."

63   police closed the road: Wilford, "Moon Rocks."

63   "a radical group of hippies": King, *Moon Trip*, 101.

64   exposing algae: West, "Moon Rocks Go to Experts on Friday."

64   with gas masks: Weaver, "What the Moon Rocks Tell Us."

65   "I mean, big deal": Author interview with Bill Schopf, UCLA, July 2019.

65   as some scientists expected: Cooper, *Apollo on the Moon*, 96–99.

65   tunnels and two-thirds of a mile of wire: Marvin, "Gerald J. Wasserburg," 186.

65   the Lunatic Asylum: Hammond, *A Passion to Know: 20 Profiles in Science*, 52–53.

66   to a Pasadena bar: Wolchover, "Geological Explorers Discover a Passage to Earth's Dark Age."

66   "It must in any event": Tera, Papanastassiou, and Wasserburg, "A Lunar Cataclysm at ~3.95 AE and the Structure of the Lunar Crust," 725.

67   a slight chance that Mercury: Laskar and Gastineau, "Existence of Collisional Trajectories of Mercury, Mars and Venus with the Earth."

68   7,000 degrees Fahrenheit: Interview of Peter Schultz of Brown University in the 2005 documentary "The Violent Past" from *Miracle Planet*.

Chapter 5: Dirty Snowballs and Space Rocks: The Biggest Flood of All Time

73   "If there is magic": Eiseley, *The Immense Journey*, 15.

74   "How inappropriate": Lovelock, "Hands Up for the Gaia Hypothesis," 102.

74   more than a hundred species: LaCapra, "Bird, Plane, Bacteria?"

74   aqueous to the core: If you placed your blood vessels—your interior rivers and streams that carry water to your cells—end to end, they would extend fifty thousand miles, about two times the circumference of the globe. Sender, Fuchs, and Milo, "Revised Estimates for the Number of Human and Bacteria Cells in the Body," 7.

74   about 75 percent: Krulwich, "Born Wet, Human Babies Are 75 Percent Water: Then Comes the Drying."

74   same as a banana: USDA FoodData Central website.

74   60 percent: Aitkenhead, Smith, and Rowbotham, *Textbook of Anaesthesia*, 417.

75   eleven cups of water: Emsley, *Nature's Building Blocks: An A–Z Guide to the Elements*, 228.

75   every 1.5 *trillionth* of a second: Hoffmann, *Life's Ratchet: How Molecular Machines Extract Order from Chaos*, 116.

75   350 feet per second: Ashcroft, *The Spark of Life: Electricity in the Human Body*, 56.

75   If you ever feel foggy: Adan, "Cognitive Performance and Dehydration," 73.

76   penned a book: Von Braun, Whipple, and Ley, *Conquest of the Moon*.

76   "For a number of years": DeVorkin, AIP oral history interview with Fred Whipple.

76   better academic opportunities: Marsden, "Fred Lawrence Whipple (1906–2004)," 1452.

76   It was undemanding enough: Whipple, "Of Comets and Meteors," 728.

76   as deadly boring as any: Marvin, "Fred L. Whipple," A199.

77   just weeks after: Marsden, "Fred Lawrence Whipple (1906–2004)," 1452.

77   "orbit computing business": DeVorkin, AIP oral history interview with Fred Whipple.

77   to check the accuracy: Hughes, "Fred L. Whipple 1906–2004," 6.35.

77   bagged over thirty: Levy, *David Levy's Guide to Observing and Discovering Comets*, 26.

77   orbited the Sun over a thousand times: DeVorkin, AIP oral history interview with Fred Whipple.

78   a half hour to an hour: Whipple, "Of Comets and Meteors," 728.

78   "what's happening to comets!": DeVorkin, AIP oral history interview with Fred Whipple.

79   Whipple's theory only survived: Calder, *Giotto to the Comets*, 38.

79   flying at 41,000 feet: Cowan, "Scientists Uncover First Direct Evidence of Water in Halley's Comet: New Way to Study Comets Will Help Yield Clues to Solar System's Origin."

79   "Well Fred": Levy, *The Quest for Comets*, 70.

79   "kamikaze mission": Quoted in Markham, "European Spacecraft Grazes Comet."

79   over 40 miles a second: Calder, *Giotto*, 107.

80   62,000 miles away: Calder, *Giotto*, 110.

80   set the half-ton machine wobbling: Calder, *Giotto*, 112.

80   80 percent of the gas: Calder, *Giotto*, 130.

81   "The usual thing you get is": Author interview with Dave Jewitt, UCLA, January 2018.

82   "It can't possibly be real": Couper and Henbest, *The History of Astronomy*, 196.

82   comets from the Kuiper Belt: Harder, "Water for the Rock," 184.

85   Their simulations appeared to reveal: Morbidelli et al., "Source Regions and Timescales for the Delivery of Water to the Earth."

86   an immaculately clean desk: Righter et al., "Michael J. Drake (1946–2011)."

86   dust surrounded by water vapor: Drake interview in the National Geographic Channel documentary "Birth of the Oceans."

87   perhaps several times as much: Jewitt and Young, "Oceans from the Skies," 39; and author conversation with David Rubie, Universitaet Bayreuth, February 2021.

88   Rain poured down for thousands: Kunzig, *Mapping the Deep: The Extraordinary Story of Ocean Science*, 17–18.

88   "Nothing is without controversy": Author interview with John Valley, University of Wisconsin–Madison, June 2018.

89   30 to 40 percent of all the gold: Hart, *Gold*, 12.

90   decided to ask Wilde: Valley, "A Cool Early Earth?" 63.

90   Others soon confirmed: At UCLA, Stephen Mojzsis, Mark Harrison, and Robert Pidgeon made a similar finding at roughly the same time.

Chapter 6: The Most Famous Experiment: The Search for the Origin of the Molecules of Life

92   "They are good company": Wald, Nobel Banquet Speech, Nobel Prize in Physiology or Medicine 1967.

93   "the world of the living": Oparin, *The Origin of Life*. An English-language translation by Ann Synge of Oparin's original paper appears in the appendix of Bernal, *The Origin of Life*, 206–7.

93   delighted in the fantastic variety: Mikhailov, *Put' k istinye*, 9–10.

93   single "scientific" worldview: Lazcano, "Alexandr I. Oparin and the Origin of Life," 215.

94   Another 1 percent is ions: Cooper and Hausman, *The Cell*, 44.

94   70 percent amino acids: Woodard and White, "The Composition of Body Tissues," 1214.

95   "In living Nature": Quoted in Hunter, *Vital Forces*, 56.

95   "Dead matter cannot become": Kelvin, *Popular Lectures and Addresses: Geology and General Physics*, II:198.

95   "It is mere rubbish": Quoted in Peretó, Bada, and Lazcano, "Charles Darwin and the Origin of Life," 396.

96   "Who knows," Helmholtz argued, "whether": Helmholtz, *Science and Culture: Popular and Philosophical Essays*, 275.

96   in contrast to: Kursanov, "Sketches to a Portrait of A. I. Oparin," 4.

96   "missing its very first chapter": Schopf, *Cradle of Life: The Discovery of Earth's Earliest Fossils*, 112.

98   it was photosynthesizing algae: Schopf, *Cradle of Life*, 120–21.

98   "wild speculation": Graham, *Science, Philosophy, and Human Behavior in the Soviet Union*, 73.

99   neighboring vacation dachas: Schopf, *Cradle of Life*, 123.

99   "imprisoned in Siberia?": Quoted in Graham, *Science in Russia and the Soviet Union*, 276.

100   "All the scientists I know": Quoted in Shindell, *The Life and Science of Harold C. Urey*, 114.

100   someone should try testing: Miller, "The First Laboratory Synthesis of Organic Compounds Under Primitive Earth Conditions," 230.

100   "The first thing he tried": Henahan, "From Primordial Soup to the Prebiotic Beach: An Interview with the Exobiology Pioneer Dr. Stanley L. Miller."

101   "dungeon": Davidson, *Carl Sagan: A Life*, 23.

101   Urey gave a tour: Sagan, *Conversations with Carl Sagan*, 30.

102   "It looks like fly shit": Bada and Lazcano, "Biographical Memoirs: Stanley L. Miller: 1930–2007," 18.

102   "three feet off the floor": Wade, "Stanley Miller, Who Examined Origins of Life, Dies at 77."

102   at least eight more: Wills and Bada, *The Spark of Life: Darwin and the Primeval Soup*, 49.

103   just the kind that Oparin predicted: Mesler and Cleaves II, *A Brief History of Creation*, 178.

103   "if I'd submitted it": Henahan, "From Primordial Soup to the Prebiotic Beach."

103   "They didn't take it seriously": Sagan, *Conversations with Carl Sagan*, 30.

103   Even Oparin did not believe: Lazcano and Bada, "Stanley L. Miller (1930–2007)," 374.

104   even a high school student: Henahan, "From Primordial Soup to the Prebiotic Beach."

104   "If God did not": Mesler and Cleaves II, *A Brief History*, 173.

104   "The road ahead is hard": Oparin, *The Origin of Life*, 252.

104   *not* full of hydrogen, methane, and ammonia: Radetsky, "How Did Life Start?" 78.

105   primarily nitrogen, carbon dioxide, and water vapor: Zahnle, Schaefer, and Fegley, "Earth's Earliest Atmospheres," 2.

105   hundreds of thousands of enzymes: Author interview with Laura Lindsey-Boltz, University of North Carolina, October 2021.

107   Townes had published: Townes, "Microwave and Radio-Frequency Resonance Lines of Interest to Radio Astronomy."

107   One graduate student: Townes, "The Discovery of Interstellar Water Vapor and Ammonia at the Hat Creek Radio Observatory," 82.

107   "When he came": Author interview with Jack Welch, University of California, Berkeley, June 2018.

108   "You know it's not going to work": Townes, *How the Laser Happened: Adventures of a Scientist*, 65.

108   "I got the feeling": Townes, "The Discovery," 82.

110   when hydrogen cyanide combines: Patel et al., "Common Origins of RNA, Protein and Lipid Precursors in a Cyanosulfidic Protometabolism."

110   "We heard this *ba-boom*": Interview in video, Jess and Kendrew, "Murchison Meteorite Continues to Dazzle Scientists."

110   punched through the metal roof: Meteoritical Society, "Murchison."

110   methylated spirits: Deamer, *First Life: Discovering the Connections between Stars, Cells, and How Life Began*, 53.

111   couldn't rule contamination out: Sullivan, *We Are Not Alone: The Search for Intelligent Life on Other Worlds*, 114.

111   New York City ragweed: Sullivan, *We Are Not Alone*, 123–24.

112   they found two more: Schopf, *Major Events in the History of Life*, 17.

112   the very same ones: Miller, "The First Laboratory Synthesis of Organic Compounds under Primitive Earth Conditions," 240.

114   forty thousand tons: Brownlee, "Cosmic Dust: Building Blocks of Planets Falling from the Sky," 166.

114   ten to a thousand times the mass: Segré and Lancet, ""Theoretical and Computational Approaches to the Study of the Origin of Life," 94–95.

114   fragments might have recombined: Barras, "Formation of Life's Building Blocks Recreated in Lab."

Chapter 7: The Greatest Mystery: The Puzzling Origin of the First Cells

116   "Life is a cosmic imperative": de Duve, "The Beginnings of Life on Earth," 437.

117   failed his qualifying exams: Heap and Gregoriadis, "Alec Douglas Bangham, 10 November 1921–9 March 2010," 28.

117   "renege": Bangham, "Surrogate Cells or Trojan Horses: The Discovery of Liposomes," 1081.

118   "Membranes came first": Deamer, "From 'Banghasomes' to Liposomes: A Memoir of Alec Bangham, 1921–2010," 1309.

119   Their lives are "Greek tragedies": Robert Singer quoted in Albert Einstein College of Medicine press release, "Built-In 'Self-Destruct Timer' Causes Ultimate Death of Messenger RNA in Cells."

119   once every million to billion years: Milo and Phillips, *Cell Biology by the Numbers*, 215–16.

120   it was easy for him: Echols, *Operators and Promoters: The Story of Molecular Biology and Its Creators*, 215.

121   "more and more desperate": Gitschier, "Meeting a Fork in the Road: An Interview with Tom Cech," 0624.

121   "by desperation to the opposite hypothesis": Cech interview in Howard Hughes Medical Institute video, *The Discovery of Ribozymes*.

121   "I didn't even know": Quoted in Dick and Strick, *The Living Universe: NASA and the Development of Astrobiology*, 128.

121   "never thought much about it": Author interview with Thomas Cech, University of Colorado Boulder, September 2021.

121   "Unknown to us": Cech interview in HHMI video, *The Discovery of Ribozymes*.

123   would support the theory of plate tectonics: Kaharl, *Water Baby: The Story of Alvin*, 168–69.

123   tossed overboard after a shipboard feast: Crane, *Sea Legs: Tales of a Woman Oceanographer*, 112–13.

124   "Debra, isn't the deep ocean": Kaharl, *Water Baby*, 173.

124   He was gazing at clams: Kaharl, *Water Baby*, 173.

125   Russian vodka they'd purchased: Ballard, *The Eternal Darkness*, 171.

125   "RETURN TO PORT": Kusek, "Through the Porthole 30 Years Ago," 141.

125   "We all started jumping up and down": Kaharl, *Water Baby*, 175.

126   that lived at high temperatures: Wade, "Meet Luca, the Ancestor of All Living Things."

126   about a year later: Hazen, *Genesis: The Scientific Quest for Life's Origin*, 98–99.

127   "head off": Hazen, *Genesis*, 109.

127   "The vents would": Miller and Bada, "Submarine Hot Springs and the Origin of Life," 610.

128   "Ideas fly around": Author interview with Günter Wächtershäuser, December 2018.

128   the supposedly essential: Wächtershäuser, "The Origin of Life and Its Methodological Challenge," 488.

130   "The prebiotic broth theory": Wächtershäuser, "Before Enzymes and Templates: Theory of Surface Metabolism," 453.

130   "The vent hypothesis is a real loser": Radetsky, "How Did Life Start?" 82.

130   "not relevant to the question": Lucentini, "Darkness Before the Dawn—of Biology," 29.

130   "paper chemistry": Bada interview in BBC *Horizon* documentary, "Life Is Impossible."

130   "As far as I'm concerned": Hagmann, "Between a Rock and a Hard Place."

130   "runaway enthusiasm": Monroe, "2 Dispute Popular Theory on Life Origin."

131   When we spoke: Author interview with Mike Russell, December 2018.

131   origin of life much easier to envision: Lane, *Life Ascending*, 19–23.

132   ten million to one hundred million of them: Flamholz, Phillips, and Milo, "The Quantified Cell," 3498.

132   once provided the energy: Lane, *The Vital Question: Why Is Life the Way It Is?* 117–19.

134   John Sutherland has found: Wade, "Making Sense of the Chemistry That Led to Life on Earth."

134   "We have got to be open": Author interview with George Cody, Carnegie Institution for Science, June 2018.

134   "if we need a location": Author interview with Jay Melosh, Purdue University, May 2018.

135   higher than 104 degrees: California Institute of Technology press release, "Caltech Geologists Find New Evidence That Martian Meteorite Could Have Harbored Life"; and Weiss et al., "A Low Temperature Transfer of ALH84001 from Mars to Earth."

135   the vacuum of space is not a deal breaker: Nicholson et al., "Resistance of Bacillus Endospores to Extreme Terrestrial and Extraterrestrial Environments."

135   have survived a 553-day joyride: Amos, "Beer Microbes Live 553 Days Outside ISS."

135   life certainly existed by 3.5 billion years ago: Knoll, *A Brief History of Earth: Four Billion Years in Eight Chapters*, 81–83.

136   Kirschvink has complex additional reasons: See Kirschvink and Weiss, "Mars, Panspermia, and the Origin of Life: Where Did It All Begin?" Additionally, Kirschvink and the biochemist Steve Benner argue that it is difficult to make RNA without the stabilization provided by the chemical borate; and while borate is rare on Earth, it is plentiful on Mars.

Chapter 8: Light Assembly Required: Discovering Photosynthesis

141   "Food is simply sunlight": Kellogg, *The New Dietetics: What to Eat and How*, 29.

142   to bolster the fortunes of his father: Beale and Beale, *Echoes of Ingen Housz: The Long Lost Story of the Genius Who Rescued the Habsburgs from Smallpox and Became the Father of Photosynthesis*, 29.

143   She was desperate to save: Van Klooster, "Jan Ingenhousz," 353.

143   clergymen railed against the thought: Magiels, *From Sunlight to Insight*, 87.

143   "I feared I should remain": Quoted in Beale and Beale, *Echoes*, 322.

143   some London doctors did: Beaudreau and Finger, "Medical Electricity and Madness in the 18th Century," 338.

144   attempts by Swiss chemist Carl Scheele to replicate: Beale and Beale, *Echoes*, 270–71.

145   "secret operations of plants": Quoted in Beale and Beale, *Echoes,* 279.

146   "When two dogs fight for a bone": Quoted in Beale and Beale, *Echoes*, 323.

146   Nonetheless, Priestley promised: Magiels, "Dr. Jan IngenHousz, or Why Don't We Know Who Discovered Photosynthesis?" 14.

146   he found no acknowledgment: Magiels, *From Sunlight*, 109.

146   "a sultan who did not tolerate": Quoted in Magiels, *From Sunlight*, 109.

146   "If you have realy publish'd this doctrine before me": Quoted in Magiels, *From Sunlight*, 238–39.

146   in the appendix: Gest, "A 'Misplaced Chapter' in the History of Photosynthesis Research: The Second Publication (1796) on Plant Processes by Dr. Jan Ingen-Housz, MD, Discoverer of Photosynthesis," 65.

147   "seeking truth, and knowledge": Debus, *Chemistry and Medical Debate: Van Helmont to Boerhaave*, 33.

147   studied alchemy and magic: Hedesan, "The Influence of Louvain Teaching on Jan Baptist Van Helmont's Adoption of Paracelsianism and Alchemy," 240.

148   Nor did he endear himself: Rosenfeld, "The Last Alchemist—the First Biochemist: J. B. van Helmont (1577–1644)," 1756.

148   "Put a pair of sweaty underwear": Quoted in Cockell, *The Equations of Life: How Physics Shapes Evolution*, 240.

148   "monstrous pamphlet": Quoted in Pagel, *Joan Baptista van Helmont*, 12.

148   soil had lost just 2 ounces: Pagel, *Joan Baptista van Helmont*, 53.

148   not from water: Ingenhousz, *An Essay on the Food of Plants and the Renovation of Soils*, 2.

150   "Be a chemist and make millions": Kamen, *Radiant Science, Dark Politics: A Memoir of the Nuclear Age*, 21.

150   on his fourth version: Yarris, "Ernest Lawrence's Cyclotron: Invention for the Ages."

151   coached by Jack Dempsey: Johnston, *A Bridge Not Attacked: Chemical Warfare Civilian Research During World War II*, 90.

151   "outspoken, abrasive": Kamen, "Onward into a Fabulous Half-Century," 139.

152   "During a recital of these troubles": Kamen, *Radiant Science*, 84.

153   shouldn't take more than a few months: Kamen, "A Cupful of Luck, a Pinch of Sagacity," 6.

153   proton-neutron pairs: Larson, interview with Martin Kamen, Pioneers in Science and Technology Series, Center for Oak Ridge Oral History, 11.

153   "three mad men hopping about": Kamen, *Radiant Science*, 86.

154   Robert Oppenheimer told him: Kamen, "Early History of Carbon-14," 586.

155   began firing alpha particles: Kamen, "Early History," 588.

157   50 percent of its mass is carbon, and 44 percent is oxygen: Petterson, "The Chemical Composition of Wood," 58.

157   About 83 percent: Russell and Williams, *The Nutrition and Health Dictionary*, 137.

158   fell asleep at the wheel: Kamen, *Radiant Science*, 165.

158   perhaps he was too impatient: Benson, "Following the Path of Carbon in Photosynthesis," 35.

159   might leak atom bomb secrets: Larson, interview with Martin Kamen.

159   which were both trailing him: Kelly, "John Earl Haynes's Interview."

160   he chose to sit at the physicists' table: Calvin, *Following the Trail of Light: A Scientific Odyssey*, 51.

160   "Time to quit": Hargittai and Hargittai, *Candid Science V*, 386.

160   unlimited artificial food: Alsop, "Political Impact Is Seen in New Atomic Experiments."

160   solve the world's energy problem: Hargittai and Hargittai, *Candid Science V*, 388.

161   Benson realized: Buchanan and Wong, "A Conversation with Andrew Benson: Reflections on the Discovery of the Calvin–Benson Cycle," 210.

161   "What's new?": Buchanan and Wong, "A Conversation," 213.

161   "He would come tearing into the lab": Moses and Moses, "Interview with Rod Quayle," 6.

162   "He could make interpretations": Moses and Moses, "Interview with Al Bassham," 14.

162   jumped to his feet: Benson, "Following," 809.

163   while Benson didn't bother telling him: Sharkey, "Discovery of the Canonical Calvin-Benson Cycle," 242.

163   "Time to go": Buchanan and Wong, "A Conversation," 213.

164   with a kick of energy from another light beam: Research into the "light reactions" has also been the subject of tremendous amount of research. Govindjee, Shevela, and Björn, "Evolution of the Z-Scheme of Photosynthesis."

164   to simulate the process in a computer: Author interview with Stephen Long, University of Illinois Urbana-Champaign, November 2021.

164   a hundred times more slowly: Falkowski, *Life's Engines: How Microbes Made Earth Habitable*, 99.

164   "Rubisco is a silly enzyme": Author interview with Govindjee, University of Illinois Urbana-Champaign, May 2019.

165   700 million tons: Bar-On and Milo, "The Global Mass and Average Rate of Rubisco," 4738.

165   artificial photosynthetic device: Calvin, "Photosynthesis as a Resource for Energy and Materials," 277.

165   Researchers are still pursuing: Bourzac, "To Feed the World, Improve Photosynthesis."

165   "region of transformation of cosmic energy": Vernadsky, *The Biosphere*, 47.

Chapter 9: Lucky Breaks: From Ocean Scum to Green Planet

167   "Today photosynthesis runs our planet": Author interview with Stjepko Golubic, July 2019.

168   as extreme as a nuclear holocaust: Margulis and Sagan, *Microcosmos: Four Billion Years of Evolution from Our Microbial Ancestors*, 109.

168   No one had found any evidence: The geologist John Dawson thought he had found an older fossil called *Eozoön*, but his claim did not hold up. Schopf, *Cradle of Life: The Discovery of Earth's Earliest Fossils*, 19–21.

168   "driving frozen mist": Walcott, "Pre-Carboniferous Strata in the Grand Canyon of the Colorado, Arizona," 438.

168   food to last three months: Walcott, "Report of Mr. Charles D. Walcott, July 2," 160.

168   "So much snow": Schuchert, "Charles Doolittle Walcott, (1850–1927)," 279.

169   "rocks-rocks-rocks": Yochelson, *Charles Doolittle Walcott, Paleontologist*, 145.

169   were forced to pile ice: Walcott, "Report of Mr. Charles D. Walcott, July 2," 47.

169   created by some kind of life: Walcott, *Pre-Cambrian Fossiliferous Formations*, 234.

170   Other paleontologists also found unusual patterns: Schopf, *Life in Deep Time: Darwin's "Missing" Fossil Record*, 49.

170   one long-disputed fossil: Schopf, *Cradle of Life*, 19–21. The fossil was called *Eozoön*.

170   As the paleobiologist William Schopf put it: Schopf, *Cradle of Life*, 31.

170   calcium-rich mud: Seward, *Plant Life through the Ages: A Geological and Botanical Retrospect*, 87.

170   we could never expect creatures as small as bacteria: Seward, *Plant Life*, 92.

170   many scientists used the term: Author interview with Stjepko Golubic, July 2019.

171   it dawned on them: Although earlier researchers working elsewhere, particularly in the Bahamas, had previously made the connection between cyanobacteria and ancient stromatolites, their claims were not widely accepted. The "living" stromatolites they pointed to looked very different from ancient *Cryptozoön*. In contrast, Logan's stromatolites had an obvious resemblance to the ancient fossils. Hoffman, "Recent and Ancient Algal Stromatolites," 180–81.

171   The mats trapped sediments: Prothero, *The Story of Life in 25 Fossils: Tales of Intrepid Fossil Hunters and the Wonders of Evolution*, 11.

172   They were microbial Bolsheviks: Falkowski, *Life's Engines: How Microbes Made Earth Habitable*, 72.

172   desk and chair on four-inch risers: Author interview with William Schopf, UCLA, July 2019.

172   bantam-weight boxing champion: Crowell, "Preston Cloud," 45.

173   They kept it a secret: Author interview with William Schopf, UCLA, July 2019.

174   "Many kinds of microbes were immediately wiped out": Margulis and Sagan, *Microcosmos*, 108.

177   Budyko had even created a model: Walker, *Snowball Earth: The Story of the Great Global Catastrophe That Spawned Life as We Know It*, 113.

179   could have possibly formed: Walker, *Snowball Earth*, 122–28.

180   "we would never have come out of the snowball": Kirschvink believes that if the Earth had been slightly more distant from the Sun, the temperatures at Earth's poles would have been so frigid that the insulating carbon dioxide gas vented by volcanoes would have frozen when it reached the poles. The planet would have been too cold to ever escape Snowball Earth. This suggests to Kirschvink that there may be many other Earthlike planets where life evolved and then completely froze over.

181   quick-tempered: *The Telegraph*, "Lynn Margulis."

181   "Lynn was good as a needler": Author interview with Fred Spiegel, University of Arkansas, March 2019.

182   "She liked to start trouble": Dorion Sagan interview in *Symbiotic Earth*.

182   without bothering to tell her parents: Margulis, "Mixing It Up," 103–4.

182   "big shot": Quoted in Goldscheider, "Evolution Revolution," 46.

182   statement by one of her professors: Quammen, *The Tangled Tree: A Radical New History of Life*, 120.

182   Despite her thesis advisor's skepticism: Quammen, *The Tangled Tree*, 120.

182   like looking for Father Christmas: Poundstone, *Carl Sagan: A Life in the Cosmos*, 63.

183   as Margulis sat reading: Otis, *Rethinking Thought: Inside the Minds of Creative Scientists and Artists*, 36.

183   hit her like lightning: Otis, *Rethinking Thought*, 19.

183   "never changed a diaper in his life": Quoted in Davidson, *Carl Sagan: A Life*, 112.

184   "a torture chamber": Quoted in Poundstone, *Carl Sagan: A Life in the Cosmos*, 47.

184   two scientists in Sweden: Sagan, *Lynn Margulis: The Life and Legacy of a Scientific Rebel*, 59.

185   "Your research is crap": *The Telegraph*, "Lynn Margulis."

185   "it avoids the difficult thought": Sapp, *Evolution by Association*, 185.

186   "the greatest chemical inventors": Margulis and Sagan, *What Is Life?* 52.

187   "It may come as a blow": Quoted in Goldscheider, "Evolution Revolution," 44.

187   "As her career progressed": Author interview with John Archibald, Dalhousie University, March 2019.

187   by 1.7 billion years ago, if not earlier: Knoll, *A Brief History of Earth: Four Billion Years in Eight Chapters*, 108–11.

188   may have had three thousand of them: Author interview with Nick Lane, University College London, September 2019.

188   without a lot of overhead: Lane and Martin argue that there is a net energy savings. A mitochondrion living inside another cell no longer has to duplicate some jobs like building a cell wall. Overall, a mitochondrion and its host have to perform less work than if they lived separately.

188   about a quadrillion: Lane, "Why Is Life the Way It Is?" 23.

188   It will look like microorganisms: Lane, "Why Is Life the Way It Is?" 27; and Catling et al., "Why O_2 Is Required by Complex Life on Habitable Planets and the Concept of Planetary 'Oxygenation Time.'"

188   by about 1.25 billion years ago: Gibson et al., "Precise Age of *Bangiomorpha pubescens* Dates the Origin of Eukaryotic Photosynthesis." The oldest fossils found so far date to 1.047 billion years, but molecular clock evidence suggests that their ancestors appeared at least 1.25 billion years ago.

189   large fast-moving animals in the oceans didn't show up: There were strange, slow-moving animals during the earlier Ediacaran period that began about 635 million years ago.

189   less than 1 percent oxygen: Falkowski, *Life's Engines*, 130.

190   until 800 million years ago: Reinhard et al., "Evolution of the Global Phosphorus Cycle," 386.

192   30 percent of our protein is collagen: Milo and Phillips, *Cell Biology by the Numbers*, 111.

193   to a staggering 30 to 35 percent: Falkowski, *Life's Engines*, 141.

193   thirty-six thousand gallons: Kahn, "How Much Oxygen Does a Person Consume in a Day?"

Chapter 10: Planting the Seeds: How Greenery and Its Allies Made Us Possible

195   "Shall I not have intelligence": Thoreau, *Walden*, 130.

195   from fiercely defending: Zimmermann, "Nachrufe: Simon Schwendener," 59.

196   a tenth of 1 percent: Bar-On, Phillips, and Milo, "The Biomass Distribution on Earth."

196   kept him from marrying: Honegger, "Simon Schwendener (1829–1919) and the Dual Hypothesis of Lichens," 312.

196   "master is a fungus": Plitt, "A Short History of Lichenology," 89.

197   "Destructiveness is a character of fungi": Ralfs, "The Lichens of West Cornwall," 211.

197   "an assertion either of pure fantasy": Plitt, "A Short History," 82.

197   "Romance of Lichenology": James Crombie, quoted in Smith, *Lichens*, xxv.

197   "met with the ridicule it deserved": Step, *Plant-Life*, 149.

197   still dismissed Schwendener's claim: Schmidt, "Essai d'une biologie de l'holophyte des Lichens," 7.

199   preventing the ends of the roots: Ryan, *Darwin's Blind Spot*, 22.

199   on trees both young and old: Frank, "On the Nutritional Dependence of Certain Trees on Root Symbiosis with Belowground Fungi (an English Translation of A. B. Frank's Classic Paper of 1885)," 271.

199   Frank coined the word: A year after Frank coined *symbiotismus*, the botanist Anton de Bary introduced the term *symbiosis*, meaning "the living together of unlike organisms."

199   "wet nurse": Frank, "On the Nutritional Dependence," 274.

199   "calculated to try our patience": Ryan, *Darwin's Blind Spot*, 49.

199   structures that look just like mycorrhizal fungi: Beerling, *Making Eden*, 125–26.

200   "revolutionary announcement": "Hermann Hellriegel," 11.

200   "Their children suffered": Aulie, "Boussingault and the Nitrogen Cycle," doctoral thesis, 39.

201   "we passed from class to class": Mccosh, *Boussingault*, 4.

202   In one impressive trial: Aulie, "Boussingault and the Nitrogen Cycle," 448.

202   increased its nitrogen content by a third: Aulie, "Boussingault and the Nitrogen Cycle," 447.

203   if something in the soil was helping plants: Nutman, "Centenary Lecture," 72.

203   Cries of "bravo!": Finlay, "Science, Promotion, and Scandal," 209.

203   "highly gifted": MacFarlane, "The Transmutation of Nitrogen," 49.

204   almost 50 percent smaller: Erisman et al., "How a Century of Ammonia Synthesis Changed the World," 637.

205   Lignin is the second most abundant: Walker, *Plants: A Very Short Introduction*, 30.

205   fourteen billion of them: Datta et al., "Root Hairs," 1.

205   "They suck up all the nutrients": Author interview with Simon Gilroy, University of Wisconsin–Madison, November 2021.

206   more than 1,150 prairie plants: Tobey, *Saving the Prairies: The Life Cycle of the Founding School of American Plant Ecology, 1895–1955*, 192–93.

206   burrowed thirty-one feet down: Wilson, *Roots: Miracles Below*, 84.

207   "Why do plants make cocaine?": Author interview with Tony Trewavas, University of Edinburgh, September 2019.

207   at least 100,000 genes: Wade, "Number of Human Genes Is Put at 140,000, a Significant Gain."

207   About a third of your genes: Author interview with the scientist who made this finding: Lawrence Brody, National Institutes of Health, September 2021.

209   "Well, *we* could actually": Author interview with Jack Schultz, University of Toledo, September 2019.

210   "It seemed too woo-woo": Author interview with Elizabeth Van Volkenburgh, University of Washington, September 2019.

213   "What long-term scientific benefits": Alpi et al., "Plant Neurobiology: No Brain, No Gain?" 136.

214   plants have over fifteen senses: Mancuso and Viola, *Brilliant Green: The Surprising History and Science of Plant Intelligence*, 77.

214   They detect neighboring plants with photoreceptors: Trewavas, "Mindless Mastery," 841.

214   "If you grow plants": Author interview with Janet Braam, Rice University, September 2019.

215   can end up high in a neighboring spruce: Yong, "Trees Have Their Own Internet."

215   which should receive: Trewavas, "The Foundations of Plant Intelligence," 11.

215   "explosive growth": Trewavas, "Mindless Mastery," 841.

216   "purpose driven": Trewavas and Baluška, "The Ubiquity of Consciousness," 1225.

216   "we should be aware": Baluška and Mancuso, "Deep Evolutionary Origins of Neurobiology," 63.

217   Simply covering our skin with chloroplasts: Milo and Phillips, *Cell Biology by the Numbers*, 169.

Chapter 11: So Much Depends on So Little: What Do You Need to Eat to Survive?

221   "Imagine all the food": Tegmark, "Solid. Liquid. Consciousness."

221   "fiery and impetuous": Thorpe, *Essays in Historical Chemistry*, 316.

222   "built up new kingdoms": Hofmann, *The Life-Work of Liebig*, 17.

222   *schafskopf*: Brock, *Justus Von Liebig: The Chemical Gatekeeper*, 6.

223   Gay-Lussac insisted that they dance: Brock, *Justus Von Liebig*, 32.

223   "provincial backwater": Brock, *Justus Von Liebig*, 38.

223   "rules useful for making soda and soap": Turner, "Justus Liebig versus Prussian Chemistry," 131.

223   "The consciousness dawned on me": Liebig, "Justus Von Liebig: An Autobiographical Sketch," 661.

223   fume hoods: Morris, *The Matter Factory: A History of the Chemistry Laboratory*, 93.

224   "Storming and raging": Mulder, *Liebig's Question to Mulder Tested by Morality and Science*, 6.

224   "has arisen out of a complete ignorance": Phillips, "Liebig and Kolbe, Critical Editors," 91.

225   "In living Nature": Hunter, *Vital Forces*, 56.

225   "the principles of chemistry and vitality": Klickstein, "Charles Caldwell and the Controversy in America over Liebig's 'Animal Chemistry,'" 141.

226   feces of boa constrictors: Brucer, "Nuclear Medicine Begins with a Boa Constrictor," 280.

226   about six tablespoons a day: We produce roughly eight cups of gastric juice that is about 5 percent hydrochloric acid.

226 God would not have put them there: Carpenter, *Protein and Energy: A Study of Changing Ideas in Nutrition*, 59.

226 "Vegetables produce in their organism": Liebig, *Animal Chemistry: Or Organic Chemistry in Its Application to Physiology and Pathology*, 48.

226 failed to find carbohydrates or fats: Carpenter, *Protein and Energy*, 48.

227 "According to Liebig": Thoreau, *Walden*, 11.

227 they needed to drink beer: Bissonnette, *It's All about Nutrition*, 45.

227 "experienced the highest admiration": Liebig, *Animal Chemistry*, vi.

227 "filled me with admiration": Bence-Jones, *Henry Bence-Jones, M.D., F.R.S. 1813–1873: Autobiography with Elucidations at Later Dates*, 16.

227 "living scientific pioneer": Morris, *The Matter Factory*, 30.

228 it occurred to the Swiss scientists: Carpenter, Harper, and Olson, "Experiments That Changed Nutritional Thinking," 1120S–1121S.

228 faithfully collected their urine: Carpenter, Harper, and Olson, "Experiments," 1021.

228 that turned out to be equally damaging: Carpenter, *Protein and Energy*, 71–72.

229 with convoluted arguments: Carpenter, "A Short History of Nutritional Science: Part 1 (1785–1885)," 642.

230 "the most perfect substitute": Apple, "Science Gendered: Nutrition in the United States 1840–1940," 133.

230 babies raised solely on his formula did not thrive: Carpenter, *Protein and Energy*, 74.

230 scurvy killed about two million sailors: Carpenter, *The History of Scurvy and Vitamin C*, 253.

231 he needed thirty-two wagons: Bown, *Scurvy: How a Surgeon, a Mariner, and a Gentlemen Solved the Greatest Medical Mystery of the Age of Sail*, 68.

231 about 400 of his 1,900 men: Frankenburg, *Vitamin Discoveries and Disasters*, 72.

231 captains made mad dashes from port to port: Bown, *Scurvy*, 75.

231 recommended lemon juice daily: Roddis, *James Lind, Founder of Nautical Medicine*, 55.

232 Over time, unfortunately, the knowledge: Bown, *Scurvy*, 74.

232 there were even "anti-fruiters": Harvie, *Limeys*, 56.

232 Lind had seen relatively little scurvy: Lind, *A Treatise on the Scurvy, in Three Parts: Containing an Inquiry into the Nature, Causes, and Cure of That Disease, Together with a Critical and Chronological View of What Has Been Published on the Subject*, 72.

232 "They had been afflicted by scurvy": Lind, *A Treatise*, 62–63.

232 only sluggish and lazy sailors succumbed: Gratzer, *Terrors of the Table*, 17.

232 it simply seemed more expedient: Harvie, *Limeys*, 18.

234 "If there was ever a researcher": Frankenburg, *Vitamin*, 78.

234 "Dr. Lind reckons the want": Meiklejohn, "The Curious Obscurity of Dr. James Lind," 307.

234   Another 133,708 expired: Bown, *Scurvy*, 26.

235   afflicted 7 percent: Braddon, *The Cause and Prevention of Beri-Beri*, 248.

236   at the elegant Café Bauer: Beek, *Dutch Pioneers of Science*, 138.

236   "legs and feet perfectly numbed": Carpenter, *Beriberi, White Rice, and Vitamin B: A Disease, a Cause, and a Cure*, 27.

236   tantamount to a death sentence: Eijkman, "Christiaan Eijkman Nobel Lecture, 1929."

237   the physicians recommended sterilizing: Carpenter, *Beriberi*, 35.

237   10 miles an hour: "Tracing the Lost Railway Lines of Indonesia."

237   seemed more appetizing: Carpenter, *Beriberi*, 41.

237   they were cheaper to keep: Carpenter, *Beriberi*, 198.

238   "his successor refused to allow": Eijkman, "Christiaan Eijkman Nobel Lecture, 1929."

238   "chance favors only": Houston, *A Treasury of the World's Great Speeches*, 470.

238   a flurry of experiments: Carpenter, *Beriberi*, 40–41.

239   when bacteria in our stomachs feed on white rice: Carpenter, *Beriberi*, 45.

239   "as eating fish had to do with leprosy": Vedder, *Beriberi*, 160.

239   prompted a British physician: Gratzer, *Terrors of the Table*, 141–42.

240   "So much careful scientific work": Hopkins, *Newer Aspects of the Nutrition Problem*, 15.

240   Working alone at "full blast": Maltz, "Casimer Funk, Nonconformist Nomenclature, and Networks Surrounding the Discovery of Vitamins," 1016.

241   still questioned the validity of his "cure": Maltz, "Casimer Funk," 1016.

242   "a vitamin is a substance": Quoted in Gratzer, *Terrors of the Table*, 162.

243   "Scientists Find Indication": *New York Times*, "Scientists Find Indication of a Vitamin Which Prevents Softening of the Brain."

243   and prevent cancer: *St. Louis Post-Dispatch*, "Is Vitamine Starvation the True Cause of Cancer?"

243   vitamin-deficient troops: Price, *Vitamania: How Vitamins Revolutionized the Way We Think about Food*, 75–78.

243   "You're in the Army, too!": Quoted in Bobrow-Strain, *White Bread: A Social History of the Store-Bought Loaf*, 119.

244   "Vitamins are another name": BBC radio, "Enzymes," *In Our Time*.

244   around 60 million years ago: Zimmer, "Vitamins' Old, Old Edge."

245   Harold White suspects: Zimmer, "Vitamins' Old, Old Edge."

246   nylon, acetone, formaldehyde, and coal tar: Price, *Vitamania*, 17.

247   "the most expensive urine": Author interview with Gerald Combs Jr., Tufts University November 2019.

247   Beginning in the 1930s: Carpenter, "A Short History of Nutritional Science: Part 3 (1912–1944)," 3030.

248   arsenic: Collins, *Molecular, Genetic, and Nutritional Aspects of Major and Trace Minerals*, 528.

248   "Anything that's in the soil": Author interview with James F. Collins, University of Florida, February 2020.

249   mineral and vitamin deficiencies: Lieberman, *The Story of the Human Body: Evolution, Health, and Disease*, 191.

249   a handful from bacteria: Some bacteria in our guts make vitamins for us, including B vitamins and vitamin K.

Chapter 12: Hidden in Plain Sight: The Discovery of Your Master Blueprint

250   "Exploratory research": Horgan, "Francis H. C. Crick: The Mephistopheles of Neurobiology," 33.

251   thirty-some years after: Miescher came close to making this prediction in 1892.

251   among the simplest cells of all: Dahm, "Discovering DNA," 576.

251   "cloudy, thick, slimy mass": Olby, "Cell Chemistry in Miescher's Day," 379.

251   something never done before: Dahm, "The First Discovery of DNA," 321.

252   On his wedding day: Meuron-Landolt, "Johannes Friedrich Miescher: sa personnalité et l'importance de son œuvre," 20.

253   "If one . . . wants to assume": Dahm, "Friedrich Miescher and the Discovery of DNA," 282.

253   in a remarkable letter to his uncle: Lamm, Harman, and Veigl, "Before Watson and Crick in 1953 Came Friedrich Miescher in 1869," 294–95.

253   Overwork weakened his immune system: Dahm, "The First," 327.

253   it was nuclein, not protein: Mirsky, "The Discovery of DNA," 86–88.

255   killed fifty thousand Americans: Perutz, "Co-Chairman's Remarks: Before the Double Helix," 10.

255   would sit for days mulling: MacLeod, "Obituary Notice, Oswald Theodore Avery, 1877–1955," 544.

255   "focused inwardly as if unconcerned": Dubos, "Oswald Theodore Avery, 1877–1955," 35.

256   would not let his associates: Williams, *Unravelling the Double Helix: The Lost Heroes of DNA*, 148–49.

256   while Avery was away on vacation: Dubos, "Rene Dubos's Memories of Working in Oswald Avery's Laboratory."

256   Dr. Jekylls into Mr. Hydes: Dubos, *The Professor, the Institute, and DNA*, 116.

256   something from the deceased lethal bacteria: McCarty, *The Transforming Principle: Discovering That Genes Are Made of DNA*, 92.

256   just over a hundred pounds: McCarty, *The Transforming Principle*, 87.

256   "headaches and heartbreaks": In a letter to his brother Roy: Dubos, *The Professor*, 217.

256   "Disappointment is my daily bread": Dubos, *The Professor*, 139.

257   treated the extract with enzymes: Letter from Avery to his brother, in Dubos, *The Professor*, 219.

257   skepticism and sarcasm: Dubos, *The Professor*, 106.

257   "What else do you want, Fess?": McCarty, *The Transforming Principle*, 163.

258   "has long been the dream of geneticists": Dubos, *The Professor*, 245.

258   just a tenth of a percent of protein: McCarty, *The Transforming Principle*, 173.

258   "some goddamn other macromolecule": Judson, *The Eighth Day of Creation: Makers of the Revolution in Biology*, 60.

258   "I saw before me": Chargaff, *Heraclitean Fire: Sketches from a Life Before Nature*, 83.

259   in an ox's DNA, the ratios of the bases: Williams, *Unravelling*, 246.

261   he wrote to request: Wilkins, *Maurice Wilkins: The Third Man of the Double Helix: An Autobiography*, 143–50.

261   It was at this very same time: Wilkins, *Maurice Wilkins*, 129.

262   she knew much more about the tricky techniques: Maddox, *Rosalind Franklin: The Dark Lady of DNA*, 144–45.

262   Why did he keep trying to move in on her turf?: Maddox, *Rosalind Franklin*, 153–55.

262   "She was quite sharp and quick and decisive": Cold Spring Harbor Laboratory, "Aaron Klug on Rosalind Franklin."

263   "A certain youthful arrogance": Crick, *What Mad Pursuit*, 64.

265   she saw no point: Maddox, *Rosalind Franklin*, 161.

265   "like a spy": Watson interview in PBS documentary, Babcock and Eriksson, *DNA: The Secret of Life*.

266   "'until the cows come home'": Quoted in Watson, Gann, and Witkowski, *The Annotated and Illustrated Double Helix*, 91.

268   "in male-chauvinist fashion": Author interview with Don Caspar, May 2020.

268   "I was the only person in the world": Web of Stories interview with Watson, "Complementarity and My Place in History."

269   a sixty-two-hour exposure: Williams, *Unravelling*, 327.

270   she had asked Gosling: Wilkins, *Maurice Wilkins*, 198.

270   the density of the X-ray image suggested: Watson and Berry, *DNA: The Secret of Life*, 51.

271   he had seen a similar measurement: Olby, *The Path to the Double Helix*, 403.

272   "it was almost impossible": Web of Stories interview with Crick, "Molecular Biology in the Late 1940s."

272   although Crick didn't boast about it publicly: Markel, *The Secret of Life*, 12.

272   "It seemed that nonliving atoms": Wilkins, *Maurice Wilkins*, 212.

273   "We all stand on each other's shoulders": "Due Credit," 270.

273   must be in some way "interchangeable": Maddox, *Rosalind Franklin*, 202.

274   "It's so beautiful, you see": Crick, *What Mad Pursuit*, 79.

274   "Can you patent it?": Watson and Berry, *DNA*, 58.

275   in a "confused phase": Crick, "Biochemical Activities of Nucleic Acids: The Present Position of the Coding Problem," 35.

279   Most degrade after a few hours or days: Milo and Phillips, *Cell Biology by the Numbers*, 248.

279   tens of thousands of copies: A cell contains about ten billion proteins, according to *Cell Biology by the Numbers*, and the average half-life of a protein is seven hours. That means every seven hours you replace half of your ten billion proteins or over thirty-nine thousand a second.

279   when genes turn on and off: The base sequences controlling when genes are expressed are known as transcription factor binding sites, activators, promoters, enhancers, repressors, silencers, and control elements.

Chapter 13: Elements and All: What Is Really Inside You?

281   "Man, like other organisms": Claude, "The Coming of Age of the Cell," 434.

281   thirty trillion units, or cells: Sender, Fuchs, and Milo, "Revised Estimates for the Number of Human and Bacteria Cells in the Body," 9.

282   he was seized by the desire: Brachet, "Notice sur Albert Claude," 95.

282   Risking his life: Gompel, *Le destin extraordinaire d'Albert Claude (1898–1983)*, 26.

282   despite fearing his classes would all be taught in Latin: de Duve and Palade, "Obituary: Albert Claude, 1899–1983," 588.

282   "blurred boundary which concealed": Claude, "The Coming," 433.

283   as mockingly distant as stars: Claude, "The Coming," 433.

283   "biochemical bog": Moberg, *Entering an Unseen World: A Founding Laboratory and Origins of Modern Cell Biology, 1910–1974*, 137.

283   leave the premises as soon as possible: Brachet, "Notice," 100.

284   like a solitary wild boar: Brachet, "Notice," 118.

284   wanted to replace him with an actual chemist: Moberg, *Entering*, 23.

284   about 17,000 g: Claude, "Fractionation of Chicken Tumor Extracts by High Speed Centrifugation," 743.

284   with a mortar and pestle: de Duve and Beaufay, "A Short History of Tissue Fractionation," 24.

284   he determined that it contained RNA: de Duve and Palade, "Obituary," 588.

285   take a hammer to cells: *Interview with Albert Claude*, Rockefeller Institute Archive Center, RAC FA1444 (Box 1, Folder 5).

285   "When he started tearing cells apart": Moberg, *Entering*, 38.

285   "cellular mayonnaise": Rheinberger, "Claude, Albert," 146.

285   Some colleagues saw it as a betrayal: Brachet, "Notice," 108.

285   "accident of technical progress": Claude, "Albert Claude, 1948," 121.

286   master in taking advantage of them: Rheinberger, "Claude, Albert," 146.

286   chemical factories: Moberg, *Entering*, 76.

286   "would serve no useful purpose": Hawkes, "Ernst Ruska," 84.

287   it had killed one of his close friends: Moberg, *Entering*, 55.

287   "It was wonderful": Moberg, *Entering*, 60.

288 His genius was apparently less in using his techniques: Palade, "Albert Claude and the Beginnings of Biological Electron Microscopy," 15–17.

289 "Many of his friends remember Mitchell": Prebble and Weber, *Wandering in the Gardens of the Mind*, 15.

289 the "power plants": Claude, "The Coming," 434.

290 ten to one hundred million ATPs: Flamholz, Phillips, and Milo, "The Quantified Cell," 3499.

290 Big labs and big scientists competed: Gilbert and Mulkay, *Opening Pandora's Box*, 26. This entire book examines how scientists discussed and reacted to Mitchell's theory.

290 became a burning issue: Harold, *To Make the World Intelligible*, 121.

290 "only shadows of moving parts": Racker, "Reconstitution, Mechanism of Action and Control of Ion Pumps," 787.

290 "anyone who was not thoroughly confused": Racker, "Reconstitution," 787.

291 Heraclitus: Prebble, "The Philosophical Origins of Mitchell's Chemiosmotic Concepts," 443.

291 He had no experimental evidence: Prebble, "Peter Mitchell and the Ox Phos Wars," 209.

291 "I remember thinking to myself": Orgel, "Are You Serious, Dr. Mitchell?" 17.

291 "These formulations sounded like": Racker, "Reconstitution," 787.

292 He presented his theory in obscure terms: Harold, *To Make the World Intelligible*, 49.

292 Revenue from his prize dairy cows: Lane, *Power, Sex, Suicide*, 102.

292 "went into one of my ears": Govindjee and Krogmann, "A List of Personal Perspectives with Selected Quotations, along with Lists of Tributes, Historical Notes, Nobel and Kettering Awards Related to Photosynthesis," 16.

292 hopped on one foot in anger: Prebble, "Peter Mitchell and the Ox Phos Wars," 210.

292 Mitchell marked the locations: Saier, "Peter Mitchell and the Life Force," chapter 8, page 10 of 14.

292 almost 100 million volts per foot: Lane, *The Vital Question: Why Is Life the Way It Is?* 73.

293 three hundred times a second: Milo and Phillips, *Cell Biology by the Numbers*, 357.

293 describes it as: Walker, *Fuel of Life*.

293 for his "bioimagination": Roskoski, "Wandering in the Gardens of the Mind," 64–65.

293 Mitchell used the prize money: Saier, "Peter Mitchell and the Life Force," chapter 9, page 2 of 8.

294 Mike Russell and William Martin believe: Lane, *Life Ascending*, 32–33.

294 a thousand to ten thousand mitochondria: Milo and Phillips, *Cell Biology*, 34.

294 35 percent of a heart muscle cell's volume: Hom and Sheu, "Morphological Dynamics of Mitochondria: A Special Emphasis on Cardiac Muscle Cells," 7.

294   tens of thousands of times more energy: Author interview with Nick Lane, University College London, December 2021.

294   two-thirds of a pint of oxygen: Flamholz, Phillips, and Milo, "The Quantified Cell," 3499.

295   about a third of your energy: Hoffmann, *Life's Ratchet: How Molecular Machines Extract Order from Chaos*, 212.

295   more than a million sodium ions a second: Ashcroft, *The Spark of Life: Electricity in the Human Body*, 42.

296   a million sodium-potassium pumps: Stevens, "The Neuron," 57.

296   350 feet a second: Ashcroft, *The Spark of Life*, 56.

296   a cool quadrillion or so minuscule sodium-potassium pumps: Each of our hundred billion nerve cells has about a million sodium-potassium pumps. Each of our several million cardiac muscle cells contains a few million pumps. These alone add up to quadrillions of sodium-potassium pumps. Our other cells possess them in lesser numbers.

296   hunter-gatherers got their salt from meat: Lieberman, *The Story of the Human Body: Evolution, Health, and Disease*, 283.

298   a molecular storm: Hoffmann, *Life's Ratchet*, 72.

298   two million times a second: E-mail to author from Kim Sharp, University of Pennsylvania.

298   collides with every protein: Milo and Phillips, *Cell Biology*, 220.

298   four billionths of an inch: E-mail from Kim Sharp, University of Pennsylvania.

298   20 miles per hour: Bray, *Cell Movements*, 4.

299   once every ten thousand times: Lane, *The Vital*, 12.

299   one in a million to ten million or so: Estimates of incorporating the wrong base in DNA vary from one in a million to one in ten million. Repair mechanisms that immediately follow along decrease the error rate to perhaps one in ten billion.

299   "Bored with yourself?": *Atlanta Constitution*, "Each of Us Is Charged with Busy Little Atoms."

299   Aebersold proudly told: "Paul C. Aebersold Interview," *Longines Chronoscope*.

300   98 percent of all our atoms every year: Stager, *Your Atomic Self*, 213.

300   every ten years: Kirsty Spalding and Jonas Frisén were the first to recognize this. Wade, "Your Body Is Younger Than You Think." See also Milo and Phillips, *Cell Biology by the Numbers*, 279. While a few types of cells are not replaced at all, you replace the vast preponderance within ten years.

300   330 billion cells a day: Sender and Milo, "The Distribution of Cellular Turnover in the Human Body," 45.

300   replaced every two to four days: Milo and Phillips, *Cell Biology*, 279.

300   replaced every 120 days: Milo and Phillips, *Cell Biology*, 279.

300   three and a half million new red blood cells every second: Sender and Milo, "The Distribution," 45.

300   once every ten years: Milo and Phillips, *Cell Biology*, 279.

301   eighty-six billion neurons: Herculano-Houzel, "The Human Brain in Numbers," 7.

302   about 1 percent: You replace heart cells at a rate of about 1 percent a year until you are about fifty, at which point the rate declines. Wade, "Heart Muscle Renewed over Lifetime, Study Finds."

302   "I doubt we will ever find a way of living much beyond 120": Lane, *The Vital*, 278.

303   hundreds of millions to a billion ATPs a second: Milo and Phillips, *Cell Biology*, 201. Milo and Phillips estimate that a mammalian cell with a volume of 3,000 μm^3 consumes on the order of one billion ATPs a second.

303   a parking lot with a foot or less: Hoffmann, *Life's Ratchet*, 107.

Conclusion: What a Long Strange Trip It's Been

305   "Science, truly understood": Donnan, "The Mystery of Life," 514.

306   than there are stars in the Milky Way: The number of cells in the human body is on the order of thirty trillion. Sender, Fuchs, and Milo, "Revised Estimates for the Number of Human and Bacteria Cells in the Body." The number of stars in the Milky Way is estimated to be one hundred billion to four hundred billion.

307   In Gell-Mann's wide rearview mirror: Horgan, "From My Archives: Quark Inventor Murray Gell-Mann Doubts Science Will Discover 'Something Else.'"

310   "We are a way for the cosmos to know itself": Carl Sagan in the television series *Cosmos*.

BIBLIOGRAPHY

Adan, Ana. "Cognitive Performance and Dehydration." *Journal of the American College of Nutrition* 31, no. 2 (April 1, 2012).

Aikman, Duncan. "Lemaitre Follows Two Paths to Truth." *New York Times*, February 19, 1933.

Aitkenhead, Alan R., Graham Smith, and David J. Rowbotham. *Textbook of Anaesthesia*, 5th ed. London: Elsevier, 2007.

Albert Einstein College of Medicine. "Built-In 'Self-Destruct Timer' Causes Ultimate Death of Messenger RNA in Cells." Press release, December 22, 2011.

Alpi, Amedeo, Nikolaus Amrhein, et al. "Plant Neurobiology: No Brain, No Gain?" *Trends in Plant Science* 12, no. 4 (April 2007).

Alsop, Stewart. "Political Impact Is Seen in New Atomic Experiments." *Toledo Blade*, January 6, 1949.

Amos, Jonathan. "Beer Microbes Live 553 Days Outside ISS." BBC News, August 23, 2010, https://www.bbc.com/news/science-environment-11039206.

Anderson, Carl D., and Richard J. Weiss. *The Discovery of Anti-Matter: The Autobiography of Carl David Anderson, the Youngest Man to Win the Nobel Prize.* Singapore: World Scientific, 1999.

Apple, Rima. "Science Gendered: Nutrition in the United States 1840–1940," in *The Science and Culture of Nutrition, 1840–1940*, ed. Harmke Kamminga and Andrew Cunningham. Amsterdam: Rodopi, 1995.

Ashcroft, Frances. *The Spark of Life: Electricity in the Human Body*. New York: Norton, 2012.

Atlanta Constitution, "Each of Us Is Charged with Busy Little Atoms, November 8, 1954.

Aulie, Richard P. "Boussingault and the Nitrogen Cycle." Doctoral thesis, Yale University, 1969.

———. "Boussingault and the Nitrogen Cycle." *Proceedings of the American Philosophical Society* 114, no. 6 (December 18, 1970).

Babcock, Viki, and Magdalena Eriksson, writers; Ian Duncan and David Glover, directors. *DNA: The Secret of Life*, episode 1. Arlington, VA: Public Broadcasting Service, 2003.

Bada, Jeffrey, and Antonio Lazcano. "Biographical Memoirs: Stanley L. Miller: 1930–2007." National Academy of Sciences, 2012, http://www.nasonline.org/publications/biographical-memoirs/memoir-pdfs/miller-stanley.pdf.

Ballard, Robert D. *The Eternal Darkness: A Personal History of Deep-Sea Exploration*. Princeton, NJ: Princeton University Press, 2000.

Baluška, František, and Stefano Mancuso. "Deep Evolutionary Origins of Neurobiology: Turning the Essence of 'Neural' Upside-Down." *Communicative & Integrative Biology* 2, no. 1 (December 1, 2009).

Bangham, Alec D. "Surrogate Cells or Trojan Horses: The Discovery of Liposomes." *BioEssays* 17, no. 12 (1995).

Barnes, E. W. "Contributions to a British Association Discussion on the Evolution of the Universe." *Nature*, no. 128 (October 24, 1931).

Bar-On, Yinon M., and Ron Milo. "The Global Mass and Average Rate of Rubisco." *Proceedings of the National Academy of Sciences of the United States of America* 116, no. 10 (March 5, 2019).

Bar-On, Yinon M., Rob Phillips, and Ron Milo. "The Biomass Distribution on Earth." *Proceedings of the National Academy of Sciences* 115, no. 25 (June 19, 2018).

Barras, Colin. "Formation of Life's Building Blocks Recreated in Lab." *New Scientist*, no. 2999 (December 13, 2014).

BBC documentary transcript. "Wilson of the Cloud Chamber," 1959.

BBC *Horizon* documentary. "Life Is Impossible," 1993.

BBC radio. "Enzymes." *In Our Time*, June 1, 2017.

Beale, Norman, and Elaine Beale. *Echoes of Ingen Housz: The Long Lost Story of the Genius Who Rescued the Habsburgs from Smallpox and Became the Father of Photosynthesis.* Gloucester, UK: Hobnob Press, 2011.

Beaudreau, Sherry Ann, and Stanley Finger. "Medical Electricity and Madness in the 18th Century: The Legacies of Benjamin Franklin and Jan Ingenhousz." *Perspectives in Biology and Medicine* 49, no. 3 (July 27, 2006).

Beek, Leo. *Dutch Pioneers of Science.* Assen, Netherlands: Van Gorcum, 1985.

Beerling, David. *Making Eden: How Plants Transformed a Barren Planet.* Oxford, UK: Oxford University Press, 2019.

Bence-Jones, Henry. *Henry Bence-Jones, M.D., F.R.S. 1813–1873: Autobiography with Elucidations at Later Dates.* London: Crusha & Son, 1929.

Benson, Andrew A. "Following the Path of Carbon in Photosynthesis: A Personal Story." *Photosynthesis Research* 73, (July 1, 2002).

Bernal, J. D. *The Origin of Life.* London: Weidenfeld & Nicolson, 1967.

Bernstein, Jeremy. *A Palette of Particles.* Cambridge, MA: Harvard University Press, 2013.

Bertolotti, Mario. *Celestial Messengers: Cosmic Rays: The Story of a Scientific Adventure.* Berlin: Springer, 2013.

Bissonnette, David. *It's All about Nutrition: Saving the Health of Americans.* Lanham, MD: University Press of America, 2014.

Blackmore, John T. *Ernst Mach: His Life, Work, and Influence.* Berkeley: University of California Press, 1972.

Blatner, David. *Spectrums: Our Mind-Boggling Universe from Infinitesimal to Infinity.* London: Bloomsbury, 2013.

Bobrow-Strain, Aaron. *White Bread: A Social History of the Store-Bought Loaf.* Boston: Beacon Press, 2012.

Bourzac, Katherine. "To Feed the World, Improve Photosynthesis." *MIT Technology Review* 120, no. 5 (September 2017).

Bown, Stephen R. *Scurvy: How a Surgeon, a Mariner, and a Gentleman Solved the Greatest Medical Mystery of the Age of Sail.* New York: St. Martin's Press, 2003.

Brachet, Jean. "Notice sur Albert Claude." *Annuaire de l'Académie royale de Belgique,* 1988.

Braddon, William Leonard. *The Cause and Prevention of Beri-Beri.* London: Rebman Limited, 1907.

Bray, Dennis. *Cell Movements: From Molecules to Motility.* New York: Garland Science, 2001.

Brock, William H. *Justus Von Liebig: The Chemical Gatekeeper.* Cambridge, UK: Cambridge University Press, 2002.

Brownlee, Donald E. "Cosmic Dust: Building Blocks of Planets Falling from the Sky." *Elements* 12, no. 3 (June 1, 2016).

Brucer, Marshall. "Nuclear Medicine Begins with a Boa Constrictor." *Journal of Nuclear Medicine Technology* 24, no. 4 (1996).

Buchanan, Bob B., and Joshua H. Wong. "A Conversation with Andrew Benson: Reflections on the Discovery of the Calvin–Benson Cycle." *Photosynthesis Research* 114, no. 3 (March 1, 2013).

Burbidge, Geoffrey. "Sir Fred Hoyle 24 June 1915–20 August 2001." *Biographical Memoirs of Fellows of the Royal Society* 49 (2003).

Burns, Joseph A., Jack J. Lissauer, and Andrei Makalkin. "Victor Sergeyevich Safronov (1917–1999)." *Icarus* 145, no. 1 (May 1, 2000).

Butterworth, Jon. "How Big Is a Quark?" *The Guardian,* April 7, 2016, https://www.theguardian.com/science/life-and-physics/2016/apr/07/how-big-is-a-quark.

Calder, Nigel. *Giotto to the Comets.* London: Presswork, 1992.

California Institute of Technology. "Caltech Geologists Find New Evidence That Martian Meteorite Could Have Harbored Life," press release, March 13, 1997, https://www2.jpl.nasa.gov/snc/news8.html.

Calvin, Melvin. *Following the Trail of Light: A Scientific Odyssey.* Washington, DC: American Chemical Society, 1992.

———. "Photosynthesis as a Resource for Energy and Materials: The Natural Photosynthetic Quantum-Capturing Mechanism of Some Plants May Provide a Design for a Synthetic System That Will Serve as a Renewable Resource for Material and Fuel." *American Scientist* 64, no. 3 (1976).

Carpenter, Kenneth J. *Beriberi, White Rice, and Vitamin B: A Disease, a Cause, and a Cure.* Berkeley: University of California Press, 2000.

———. *The History of Scurvy and Vitamin C.* Cambridge, UK: Cambridge University Press, 1988.

———. *Protein and Energy: A Study of Changing Ideas in Nutrition.* Cambridge, UK: Cambridge University Press, 1994.

———. "A Short History of Nutritional Science: Part 1 (1785–1885)." *Journal of Nutrition* 133, no. 3 (March 2003).

————."A Short History of Nutritional Science: Part 3 (1785–1885)." *Journal of Nutrition* 133, no. 10 (October 2003).

Carpenter, Kenneth J., Alfred E. Harper, and Robert E. Olson. "Experiments That Changed Nutritional Thinking." *Journal of Nutrition* 127, no. 5 (May 1997).

Catling, David C., Christopher R. Glein, et al. "Why O_2 Is Required by Complex Life on Habitable Planets and the Concept of Planetary 'Oxygenation Time.'" *Astrobiology* 5, no. 3 (June 2005).

Chargaff, Erwin. *Heraclitean Fire: Sketches from a Life before Nature.* New York: Rockefeller University Press, 1978.

Charitos, Panos. "Interview with George Zweig." *CERN EP News*, December 13, 2013, https://ep-news.web.cern.ch/content/interview-george-zweig.

Chu, Jennifer. "Physicists Calculate Proton's Pressure Distribution for First Time." *MIT News*, February 22, 2019, https://news.mit.edu/2019/physicists-calculate -proton-pressure-distribution-0222.

Claude, Albert. "Albert Claude, 1948." Harvey Society Lectures, Rockefeller University, January 1, 1950.

————. "The Coming of Age of the Cell." *Science* 189, no. 4201 (August 8, 1975).

————. "Fractionation of Chicken Tumor Extracts by High Speed Centrifugation." *American Journal of Cancer* 30, no. 4 (August 1, 1937).

Close, Frank. *Particle Physics: A Very Short Introduction.* Oxford, UK: Oxford University Press, 2004.

Close, Frank, Michael Marten, and Christine Sutton. *The Particle Odyssey: A Journey to the Heart of Matter.* Oxford, UK: Oxford University Press, 2004.

Cockell, Charles S. *The Equations of Life: How Physics Shapes Evolution.* New York: Basic Books, 2018.

Cold Spring Harbor Laboratory, Oral History Collection. "Aaron Klug on Rosalind Franklin," June 17, 2005, http://library.cshl.edu/oralhistory/interview/scientific -experience/women-science/aaron-rosalind-franklin/.

Collins, James F. *Molecular, Genetic, and Nutritional Aspects of Major and Trace Minerals.* San Diego: Academic Press, 2016.

Compton, William. *Where No Man Has Gone Before: A History of Apollo Lunar Exploration Missions.* Washington, DC: NASA, 1988.

Cooper, Geoffrey M., and Robert E. Hausman. *The Cell: A Molecular Approach.* Sunderland, MA: Sinauer Associates, 2013.

Cooper, Henry S. F. *Apollo on the Moon.* New York: Dial Press, 1969.

————. "Letter from the Space Center." *New Yorker*, July 25, 1969.

Cooper, Keith. *Origins of the Universe: The Cosmic Microwave Background and the Search for Quantum Gravity.* London: Icon Books, 2020.

Corfield, Richard. "One Giant Leap." *Chemistry World*, August 2009.

Cott, Jonathan. "The Cosmos: An Interview with Carl Sagan." *Rolling Stone*, December 25, 1980.

Cottrell, Geoff. *Matter: A Very Short Introduction.* Oxford, UK: Oxford University Press, 2019.

Couper, Heather, and Nigel Henbest. *The History of Astronomy*. Richmond Hill, Ontario: Firefly Books, 2007.

Cowan, Robert. "Scientists Uncover First Direct Evidence of Water in Halley's Comet: New Way to Study Comets Will Help Yield Clues to Solar System's Origin." *Christian Science Monitor*, January 13, 1986.

Crane, Kathleen. *Sea Legs: Tales of a Woman Oceanographer*. Boulder, CO: Westview Press, 2003.

Crease, Robert P., and Charles C. Mann. *The Second Creation: Makers of the Revolution in Twentieth-Century Physics*. New Brunswick, NJ: Rutgers University Press, 1996.

Crick, Francis. "Biochemical Activities of Nucleic Acids: The Present Position of the Coding Problem." *Brookhaven Symposia in Biology* 12 (1959).

———. *What Mad Pursuit: A Personal View of Scientific Discovery*. New York: Basic Books, 1988.

Crowell, John. "Preston Cloud," in *National Academy of Sciences: Biographical Memoirs*, vol. 67. Washington, DC: National Academy Press, 1995.

Crowther, James. *Scientific Types*. Chester Springs, PA: Dufour, 1970.

Dahm, Ralf. "Discovering DNA: Friedrich Miescher and the Early Years of Nucleic Acid Research." *Human Genetics* 122, no. 6 (January 2008).

———. "The First Discovery of DNA: Few Remember the Man Who Discovered the 'Molecule of Life' Three-Quarters of a Century before Watson and Crick Revealed Its Structure." *American Scientist* 96, no. 4 (2008).

———. "Friedrich Miescher and the Discovery of DNA." *Developmental Biology* 278, no. 2 (February 15, 2005).

Datta, Sourav, Chul Min Kim, et al. "Root Hairs: Development, Growth and Evolution at the Plant-Soil Interface." *Plant and Soil* 346, no. 1 (September 1, 2011).

Davidson, Keay. *Carl Sagan: A Life*. New York: Wiley, 1999.

Deamer, David. *First Life: Discovering the Connections between Stars, Cells, and How Life Began*. Berkeley: University of California Press, 2012.

Deamer, David W. "From 'Banghasomes' to Liposomes: A Memoir of Alec Bangham, 1921–2010." *FASEB Journal* 24, no. 5 (May 2010).

de Angelis, Alessandro. "Atmospheric Ionization and Cosmic Rays: Studies and Measurements before 1912." *Astroparticle Physics* 53 (January 2014).

Debus, Allen G. *Chemistry and Medical Debate: Van Helmont to Boerhaave*. Canton, MA: Science History, 2001.

de Duve, Christian. "The Beginnings of Life on Earth." *American Scientist* 83, no. 5 (1995).

de Duve, Christian, and Henri Beaufay. "A Short History of Tissue Fractionation." *Journal of Cell Biology* 91, no. 3 (December 1, 1981).

de Duve, Christian, and George E. Palade. "Obituary: Albert Claude, 1899–1983." *Nature* 304, no. 5927 (August 18, 1983).

Deprit, Andre. "Monsignor Georges Lemaître," in *The Big Bang and Georges Lemaître:*

Proceedings of the Symposium, Louvain-La-Neuve, Belgium, October 10–13, 1983, ed. A. Berger. Dordrecht, Netherlands: D. Reidel, 1984.

de Maria, M., M. G. Ianniello, and A. Russo. "The Discovery of Cosmic Rays: Rivalries and Controversies between Europe and the United States." *Historical Studies in the Physical and Biological Sciences* 22, no. 1 (1991).

DeVorkin, David. AIP oral history interview with Bart Bok, May 17, 1978, http://www.aip.org/history-programs/niels-bohr-library/oral-histories/4518-2.

———. AIP oral history interview with Fred Whipple, April 29, 1977, https://www.aip.org/history-programs/niels-bohr-library/oral-histories/5403.

DeVorkin, David H. *Henry Norris Russell: Dean of American Astronomers.* Princeton, NJ: Princeton University Press, 2000.

Dick, Steven J., and James Edgar Strick. *The Living Universe: NASA and the Development of Astrobiology.* New Brunswick, NJ: Rutgers University Press, 2004.

Donnan, Frederick G. "The Mystery of Life." *Nature* 122, no. 3075 (October 1, 1928).

Dubos, René Jules. "Oswald Theodore Avery, 1877–1955." *Biographical Memoirs of Fellows of the Royal Society* 2 (November 1, 1956).

———. *The Professor, the Institute, and DNA.* New York: Rockefeller University Press, 1976.

———. "Rene Dubos's Memories of Working in Oswald Avery's Laboratory." Symposium Celebrating the Thirty-Fifth Anniversary of the Publication of "Studies on the Chemical Nature of the Substance Inducing Transformation of Pneumococcal Types," 1979, https://profiles.nlm.nih.gov/101584575X343.

"Due Credit." *Nature* 496, no. 7445 (April 18, 2013).

Echols, Harrison G. *Operators and Promoters: The Story of Molecular Biology and Its Creators.* Berkeley: University of California Press, 2001.

Eijkman, Christiaan. "Christiaan Eijkman Nobel Lecture, 1929," NobelPrize.org.

Eiseley, Loren C. *The Immense Journey.* New York: Vintage Books, 1957.

Emsley, John. *Nature's Building Blocks: An A–Z Guide to the Elements.* Oxford, UK: Oxford University Press, 2011.

Erisman, Jan Willem, Mark A. Sutton, et al. "How a Century of Ammonia Synthesis Changed the World." *Nature Geoscience* 1, no. 10 (October 2008).

Eyles, Don. "Tales from the Lunar Module Guidance Computer." Guidance and Control Conference of the American Astronautical Society, Breckenridge, CO, February 6, 2004.

Falkowski, Paul G. *Life's Engines: How Microbes Made Earth Habitable.* Princeton, NJ: Princeton University Press, 2016.

Farrell, John. *The Day without Yesterday: Lemaître, Einstein, and the Birth of Modern Cosmology.* New York: Basic Books, 2005.

Finlay, Mark R. "Science, Promotion, and Scandal: Soil Bacteriology, Legume Inoculation, and the American Campaign for Soil Improvement in the Progressive Era," in *New Perspectives on the History of Life Sciences and Agriculture*, ed. Denise Phillips and Sharon Kingsland. Heidelberg, Germany: Springer, 2015.

Fisher, Arthur. "Birth of the Moon." *Popular Science* 230, no. 1 (January 1987).

Flamholz, Avi, Rob Phillips, and Ron Milo. "The Quantified Cell." *Molecular Biology of the Cell* 25, no. 22 (November 5, 2014).

Frank, A. B. "On the Nutritional Dependence of Certain Trees on Root Symbiosis with Belowground Fungi (an English Translation of A. B. Frank's Classic Paper of 1885)," trans. James Trappe. *Mycorrhiza* 15, no. 4 (June 2005).

Frankenburg, Frances Rachel. *Vitamin Discoveries and Disasters: History, Science, and Controversies.* Santa Barbara: Prager, 2009.

Frenkel, V., and A. Grib. "Einstein, Friedmann, Lemaître: Discovery of the Big Bang," in *Proceedings of the 2nd Alexander Friedmann International Seminar.* St. Petersburg, Russia: Friedmann Laboratory Publishing, 1994.

Galison, Peter L. "Marietta Blau: Between Nazis and Nuclei." *Physics Today* 50, no. 11 (November 1997).

Gbur, Greg. "Paris: City of Lights and Cosmic Rays." *Scientific American* Blog, July 4, 2011, https://blogs.scientificamerican.com/guest-blog/paris-city-of-lights -and-cosmic-rays.

Gest, Howard. "A 'Misplaced Chapter' in the History of Photosynthesis Research: The Second Publication (1796) on Plant Processes by Dr. Jan Ingen-Housz, MD, Discoverer of Photosynthesis." *Photosynthesis Research* 53, no. 1 (July 1, 1997).

Gibson, Timothy M., Patrick M. Shih, et al. "Precise Age of *Bangiomorpha pubescens* Dates the Origin of Eukaryotic Photosynthesis." *Geology* 46, no. 2 (February 2018).

Gilbert, G. Nigel, and Michael Mulkay. *Opening Pandora's Box: A Sociological Analysis of Scientists' Discourse.* Cambridge, UK: Cambridge University Press, 1984.

Gingerich, Owen. AIP oral history interview with Cecilia Payne-Gaposchkin, March 5, 1968, https://www.aip.org/history-programs/niels-bohr-library/oral -histories/4620.

———. "The Most Brilliant Ph.D. Thesis Ever Written in Astronomy," in *The Starry Universe: The Cecilia Payne-Gaposchkin Centenary: Proceedings of a Symposium Held at the Harvard-Smithsonian Center for Astrophysics, Cambridge, Massachusetts, October 26–27, 2000.* Schenectady, NY: L. Davis Press, 2001.

Gitschier, Jane. "Meeting a Fork in the Road: An Interview with Tom Cech." *PLOS Genetics* 1, no. 6 (December 2005).

Glashow, Sheldon. "Book Review of *Strange Beauty: Murray Gell-Mann and the Revolution in Twentieth-Century Physics*." *American Journal of Physics* 68, no. 6 (June 2000).

Godart, O. "The Scientific Work of Georges Lemaître," in *The Big Bang and Georges Lemaître: Proceedings of a Symposium in Honour of G. Lemaître Fifty Years after His Initiation of Big-Bang Cosmology, Louvain-La-Neuve, Belgium, 10–13 October 1983,* ed. A. Berger. Heidelberg, Germany: Springer, 2012.

Goldscheider, Eric. "Evolution Revolution." *On Wisconsin,* Fall 2009.

Gompel, Claude. *Le destin extraordinaire d'Albert Claude (1898–1983): Découvreur de la cellule, Rénovateur de l'institut Bordet, Prix Nobel de Médecine 1974.* Île-de-France: Connaissances et Savoirs, 2012.

Govindjee and David W. Krogmann. "A List of Personal Perspectives with Selected Quotations, along with Lists of Tributes, Historical Notes, Nobel and Kettering Awards Related to Photosynthesis." *Photosynthesis Research* 73, no. 1 (July 2002).

Govindjee, Dmitriy Shevela, and Lars Olof Björn. "Evolution of the Z-Scheme of Photosynthesis: A Perspective." *Photosynthesis Research* 133, no. 1 (September 2017).

Graham, Loren R. *Science in Russia and the Soviet Union: A Short History.* Cambridge, UK: Cambridge University Press, 1993.

———. *Science, Philosophy, and Human Behavior in the Soviet Union.* New York: Columbia University Press, 1987.

Gratzer, Walter. *Terrors of the Table: The Curious History of Nutrition.* Oxford, UK: Oxford University Press, 2007.

Gregory, Jane. *Fred Hoyle's Universe.* Oxford, UK: Oxford University Press, 2005.

Gribbin, John. *The Scientists: A History of Science Told through the Lives of Its Greatest Inventors.* New York: Random House, 2003.

Gribbin, John, and Mary Gribbin. *Stardust: Supernovae and Life—the Cosmic Connection.* New Haven, CT: Yale University Press, 2001.

Hagmann, Michael. "Between a Rock and a Hard Place." *Science* 295, no. 5562 (March 15, 2002).

Haldane, J.B.S. *Possible Worlds.* London: Chatto and Windus, 1927.

Hammond, Allen L. *A Passion to Know: 20 Profiles in Science.* New York: Scribner's, 1984.

Hanson, Norwood Russell. "Discovering the Positron (I)." *British Journal for the Philosophy of Science* 12, no. 47 (November 1961).

Harder, Ben. "Water for the Rock." *Science News* 161, no. 12 (March 23, 2002).

Hargittai, Balazs, and Istvan Hargittai. *Candid Science V: Conversations with Famous Scientists.* London: Imperial College Press, 2005.

Harold, Franklin M. *To Make the World Intelligible.* Altona, Manitoba, Canada: FriesenPress, 2017.

Hart, Matthew. *Gold: The Race for the World's Most Seductive Metal.* New York: Simon & Schuster, 2013.

Harvie, David I. *Limeys: The True Story of One Man's War against Ignorance, the Establishment and the Deadly Scurvy.* Stroud, Gloustershire, UK: Sutton Publishing, 2002.

Hawkes, Peter W. "Ernst Ruska." *Physics Today* 43, no. 7 (July 1990).

Hazen, Robert M. *Genesis: The Scientific Quest for Life's Origin.* Washington, DC: National Academies Press, 2005.

———. *The Story of Earth: The First 4.5 Billion Years, from Stardust to Living Planet.* New York: Penguin Books, 2013.

Heap, Sir Brian, and Gregory Gregoriadis. "Alec Douglas Bangham, 10 November 1921–9 March 2010." *Biographical Memoirs of Fellows of the Royal Society* 57 (December 1, 2011).

Hedesan, Georgiana D. "The Influence of Louvain Teaching on Jan Baptist Van Helmont's Adoption of Paracelsianism and Alchemy." *Ambix* 68, no. 2–3 (2021).

Helmholtz, Hermann von. *Science and Culture: Popular and Philosophical Essays.* Chicago: University of Chicago Press, 1995.

Henahan, Sean. "From Primordial Soup to the Prebiotic Beach: An Interview with the Exobiology Pioneer Dr. Stanley L. Miller." National Health Museum, Accessexcellence.org, October 1996.

Herculano-Houzel, Suzana. "The Human Brain in Numbers: A Linearly Scaled-Up Primate Brain." *Frontiers in Human Neuroscience* 3 (November 2009).

"Hermann Hellriegel." *Nature* 53, no. 1358 (November 7, 1895).

Hockey, Thomas, Virginia Trimble, et al., eds. "Gilbert, William," in *Biographical Encyclopedia of Astronomers.* New York: Springer, 2014.

Hoffman, Paul. "Recent and Ancient Algal Stromatolites," in *Evolving Concepts in Sedimentology*, ed. Robert N. Ginsburg. Baltimore: Johns Hopkins University Press, 1973.

Hoffmann, Peter M. *Life's Ratchet: How Molecular Machines Extract Order from Chaos.* New York: Basic Books, 2012.

Hofmann, August Wilhelm von. *The Life-Work of Liebig.* London: Macmillan, 1876.

Hom, Jennifer, and Shey-Shing Sheu. "Morphological Dynamics of Mitochondria: A Special Emphasis on Cardiac Muscle Cells." *Journal of Molecular and Cellular Cardiology* 46, no. 6 (June 2009).

Honegger, Rosmarie. "Simon Schwendener (1829–1919) and the Dual Hypothesis of Lichens." *The Bryologist* 103, no. 2 (2000).

Hopkins, Frederick Gowland. *Newer Aspects of the Nutrition Problem.* New York: Columbia University Press, 1922.

Horgan, John. "Francis H. C. Crick: The Mephistopheles of Neurobiology." *Scientific American* 266, no. 2 (1992).

———. "From My Archives: Quark Inventor Murray Gell-Mann Doubts Science Will Discover 'Something Else.'" *Scientific American* Blog, December 17, 2013. https://blogs.scientificamerican.com/cross-check/from-my-archives-quark -inventor-murray-gell-mann-doubts-science-will-discover-e2809csomething -elsee2809d.

———. "Remembering Big Bang Basher Fred Hoyle." *Scientific American* Blog, April 7, 2020, https://blogs.scientificamerican.com/cross-check/remembering -big-bang-basher-fred-hoyle/.

Houston, Peterson. *A Treasury of the World's Great Speeches.* New York: Simon & Schuster, 1954.

Howard Hughes Medical Institute. *The Discovery of Ribozymes*, HHMI BioInteractive video interview with Thomas Cech, 1995, https://www.biointeractive.org /classroom-resources/discovery-ribozymes.

Hoyle, Fred. *Home Is Where the Wind Blows: Chapters from a Cosmologist's Life.* Mill Valley, CA: University Science Books, 1994.

———. *The Small World of Fred Hoyle: An Autobiography.* London: Michael Joseph, 1986.

Hughes, David. "Fred L. Whipple 1906–2004." *Astronomy & Geophysics* 45, no. 6 (December 1, 2004).

Hunter, Graeme. *Vital Forces: The Discovery of the Molecular Basis of Life*. San Diego: Academic Press, 2000.

Ingenhousz, Jan. *An Essay on the Food of Plants and the Renovation of Soils*. London: Bulmer and Co., 1796.

———. *Experiments upon Vegetables: Discovering their great Power of purifying the Common Air in the Sun-shine and of Injuring it in the shade and at Night, to which is joined a new Method of examining the accurate Degree of Salubrity of the Atmosphere*. London: Elmsly and Payne, 1779.

Interview with Albert Claude. Rockefeller Institute Archive Center, RAC FA1444 (Box 1, Folder 5), 1976.

Jess, Allison, and Will Kendrew. "Murchison Meteorite Continues to Dazzle Scientists." ABC News, Goulburn Murray, Australia, December 28, 2016, https://www.abc.net.au/news/2016-12-29/murchison-meteorite/8113520.

Jewitt, David, and Edward Young. "Oceans from the Skies." *Scientific American* 312, no. 3 (March 2015).

Johnson, George. *Strange Beauty: Murray Gell-Mann and the Revolution in Twentieth-Century Physics*, 1st ed. New York: Knopf, 1999.

Johnston, Harold S. *A Bridge Not Attacked: Chemical Warfare Civilian Research during World War II*. Singapore: World Scientific, 2003.

Judson, Horace Freeland. *The Eighth Day of Creation: Makers of the Revolution in Biology*. New York: Simon & Schuster, 1979.

Kaharl, Victoria A. *Water Baby: The Story of Alvin*. New York: Oxford University Press, 1990.

Kahn, Sherry. "How Much Oxygen Does a Person Consume in a Day?" HowStuffWorks, May 11, 2021, https://health.howstuffworks.com/human-body/systems/respiratory/question98.htm.

Kamen, Martin D. "A Cupful of Luck, a Pinch of Sagacity." *Annual Review of Biochemistry* 55, no. 1 (1986).

———. "Early History of Carbon-14." *Science* 140, no. 3567 (May 10, 1963).

———. "Onward into a Fabulous Half-Century." *Photosynthesis Research* 21, no. 3 (September 1, 1989).

———. *Radiant Science, Dark Politics: A Memoir of the Nuclear Age*. Berkeley: University of California Press, 1985.

Kellogg, John Harvey. *The New Dietetics: What to Eat and How: A Guide to Scientific Feeding in Health and Disease*. Battle Creek, MI: Modern Medicine Publishing Company, 1921.

Kelly, Cynthia. "John Earl Haynes's Interview." Atomic Heritage Foundation, Voices of the Manhattan Project, Oak Ridge, TN, February 6, 2017, https://www.manhattanprojectvoices.org/oral-histories/john-earl-hayness-interview.

Kelvin, William Thomson. *Popular Lectures and Addresses*, vol. 2, *Geology and General Physics*. London: Macmillan, 1894.

King, Elbert. *Moon Trip: A Personal Account of the Apollo Program and Its Science.* Houston: University of Houston, 1989.

Kirschvink, Joseph, and Benjamin Weiss. "Mars, Panspermia, and the Origin of Life: Where Did It All Begin?" *Palaeontologia Electronica* 4, no. 2 (2001), https://palaeo-electronica.org/2001_2/editor/mars.htm.

Klickstein, Herbert S. "Charles Caldwell and the Controversy in America over Liebig's 'Animal Chemistry.'" *Chymia* 4 (1953).

Knoll, Andrew H. *A Brief History of Earth: Four Billion Years in Eight Chapters.* New York: HarperCollins, 2021.

Kragh, Helge. "'The Wildest Speculation of All': Lemaître and the Primeval-Atom Universe," in *Georges Lemaître: Life, Science and Legacy*, ed. Rodney D. Holder and Simon Mitton. Heidelberg, Germany: Springer, 2012.

Kraus, John. "A Strange Radiation from Above." North American AstroPhysical Observatory, *Cosmic Search* 2, no. 1 (Winter 1980).

Krulwich, Robert. "Born Wet, Human Babies Are 75 Percent Water: Then Comes the Drying." *Krulwich Wonders*, National Public Radio, November 26, 2013.

Kunzig, Robert. *Mapping the Deep: The Extraordinary Story of Ocean Science.* New York: Norton, 2000.

Kursanov, A. L. "Sketches to a Portrait of A. I. Oparin," in *Evolutionary Biochemistry and Related Areas of Physicochemical Biology: Dedicated to the Memory of Academician A. I. Oparin.* Moscow: Bach Institute of Biochemistry, Russian Academy of Sciences, 1995.

Kusek, Kristen. "Through the Porthole 30 Years Ago." *Oceanography* 20, no. 1 (March 1, 2007).

LaCapra, Véronique. "Bird, Plane, Bacteria? Microbes Thrive in Storm Clouds." *Morning Edition*, National Public Radio, January 29, 2013.

Lambert, Dominique. *The Atom of the Universe: The Life and Work of Georges Lemaître.* Krakow: Copernicus Center Press, 2016.

———. "Einstein and Lemaître: Two Friends, Two Cosmologies." Interdisciplinary Encyclopedia of Religion & Science (Inters.org).

———. "Georges Lemaître: The Priest Who Invented the Big Bang," in *Georges Lemaître: Life, Science and Legacy*, ed. Rodney D. Holder and Simon Mitton. Heidelberg, Germany: Springer, 2012.

Lamm, Ehud, Oren Harman, and Sophie Juliane Veigl. "Before Watson and Crick in 1953 Came Friedrich Miescher in 1869." *Genetics* 215, no. 2 (June 1, 2020).

Lane, Nick. *Life Ascending: The Ten Great Inventions of Evolution.* London: Profile Books, 2010.

———. *Power, Sex, Suicide: Mitochondria and the Meaning of Life*, 2nd ed. Oxford, UK: Oxford University Press, 2018.

———. *The Vital Question: Why Is Life the Way It Is?* London: Profile Books, 2015.

———. "Why Is Life the Way It Is?" *Molecular Frontiers Journal* 3, no. 1 (2019).

Larson, Clarence. Interview with Martin Kamen, Pioneers in Science and Technology

Series, Center for Oak Ridge Oral History, March 24, 1986, http://cdm16107. contentdm.oclc.org/cdm/ref/collection/p15388coll1/id/523.

Laskar, Jacques, and Mickael Gastineau. "Existence of Collisional Trajectories of Mercury, Mars and Venus with the Earth." *Nature* 459, no. 7248 (June 2009).

Lazcano, Antonio. "Alexandr I. Oparin and the Origin of Life: A Historical Reassessment of the Heterotrophic Theory." *Journal of Molecular Evolution* 83, no. 5 (December 2016).

Lazcano, Antonio, and Jeffrey L. Bada. "Stanley L. Miller (1930–2007): Reflections and Remembrances." *Origins of Life and Evolution of Biospheres* 38, no. 5 (October 2008).

Lemaître, Georges. "Contributions to a British Association Discussion on the Evolution of the Universe." *Nature* 128 (October 24, 1931).

———. "My Encounters with A. Einstein," 1958, Interdisciplinary Encyclopedia of Religion & Science, https://www.inters.org/lemaitre-einsten.

———. *The Primeval Atom: An Essay on Cosmogony.* New York: Van Nostrand, 1950.

Levy, David H. *David Levy's Guide to Observing and Discovering Comets.* Cambridge, UK: Cambridge University Press, 2003.

———. *The Quest for Comets: An Explosive Trail of Beauty and Danger.* New York: Plenum Press, 1994.

Lieberman, Daniel. *The Story of the Human Body: Evolution, Health, and Disease.* New York: Vintage Books, 2014.

Liebig, Justus. "Justus Von Liebig: An Autobiographical Sketch," trans. J. C. Brown. *Popular Science Monthly* 40 (March 1892).

Liebig, Justus Freiherr von. *Animal Chemistry: Or Organic Chemistry in Its Application to Physiology and Pathology*, 2nd ed., William Gregory with additional notes and corrections by Dr. Gregory and others. Cambridge, MA: John Owen, 1843.

Lind, James. *A Treatise on the Scurvy, in Three Parts: Containing an Inquiry into the Nature, Causes, and Cure of That Disease, Together with a Critical and Chronological View of What Has Been Published on the Subject.* London: Printed for S. Crowder, D. Wilson and G. Nicholls, T. Cadell, T. Becket and Co., G. Pearch, and W. Woodfall, 1772.

Livio, Mario. *Brilliant Blunders: From Darwin to Einstein—Colossal Mistakes by Great Scientists That Changed Our Understanding of Life and the Universe.* New York: Simon & Schuster, 2013.

Lovelock, James E. "Hands Up for the Gaia Hypothesis." *Nature* 344, no. 6262 (March 1990).

Lucentini, Jack. "Darkness Before the Dawn—of Biology." *The Scientist* 17, no. 23 (December 1, 2003).

MacFarlane, Thos. "The Transmutation of Nitrogen." *Ottawa Naturalist* 8 (1895).

MacLeod, Colin. "Obituary Notice: Oswald Theodore Avery, 1877–1955." *Microbiology* 17, no. 3 (1957).

Maddox, Brenda. *Rosalind Franklin: The Dark Lady of DNA.* London: HarperCollins, 2002.

Magiels, Geerdt. "Dr. Jan IngenHousz, or Why Don't We Know Who Discovered Photosynthesis?" First Conference of the European Philosophy of Science Association, Madrid, November 15–17, 2007.

———. *From Sunlight to Insight: Jan IngenHousz, the Discovery of Photosynthesis & Science in the Light of Ecology.* Brussels: Brussels University Press, 2010.

Maltz, Alesia. "Casimer Funk, Nonconformist Nomenclature, and Networks Surrounding the Discovery of Vitamins." *Journal of Nutrition* 143, no. 7 (July 2013).

Mancuso, Stefano, and Alessandra Viola. *Brilliant Green: The Surprising History and Science of Plant Intelligence.* Washington, DC: Island Press, 2015.

Margulis, Lynn. "Mixing It Up," in *Curious Minds: How a Child Becomes a Scientist,* ed. John Brockman. London: Vintage, 2005.

Margulis, Lynn, and Dorion Sagan. *Microcosmos: Four Billion Years of Evolution from Our Microbial Ancestors.* New York: Summit Books, 1986.

———. *What Is Life?* New York: Simon & Schuster, 1995.

Markel, Howard. *The Secret of Life: Rosalind Franklin, James Watson, Francis Crick, and the Discovery of DNA's Double Helix.* New York: Norton, 2021.

Markham, James M. "European Spacecraft Grazes Comet." *New York Times,* March 14, 1986.

Marsden, Brian G. "Fred Lawrence Whipple (1906–2004)." *Publications of the Astronomical Society of the Pacific* 117, no. 838 (2005).

Marvin, Ursula B. "Fred L. Whipple," Oral Histories in Meteoritics and Planetary Science 13. *Meteoritics & Planetary Science* 39, no. S8 (August 2004).

———. "Gerald J. Wasserburg," Oral Histories in Meteoritics and Planetary Science 12. *Meteoritics & Planetary Science* 39, no. S8 (2004).

McCarty, Maclyn. *The Transforming Principle: Discovering That Genes Are Made of DNA.* New York: Norton, 1986.

McCosh, Frederick William James. *Boussingault: Chemist and Agriculturist.* Dordrecht, Netherlands: D. Reidel, 2012.

Meiklejohn, Arnold Peter. "The Curious Obscurity of Dr. James Lind." *Journal of the History of Medicine and Allied Sciences* 9, no. 3 (July 1954).

Menzel, Donald H. "Blast of Giant Atom Created Our Universe." *Modern Mechanix,* December 1932.

Mesler, Bill, and H. James Cleaves II. *A Brief History of Creation: Science and the Search for the Origin of Life.* New York: Norton, 2016.

Meteoritical Society. "Murchison." *Meteoritical Bulletin,* https://www.lpi.usra.edu/meteor/metbull.php?code=16875.

Meuron-Landolt, Monique de. "Johannes Friedrich Miescher: sa personnalité et l'importance de son œuvre." *Bulletin der Schweizerischen Akademie der Medizinischen Wissenschaften* 25, no. 1–2 (January 1970).

Mikhailov, V. M. *Put' k istinye [The Path to the Truth].* Moscow, Sovetskaia Rossiia, 1984.

Miklós, Vincze. "Seriously Scary Radioactive Products from the 20th Century."

Gizmodo, May 9, 2013, https://gizmodo.com/seriously-scary-radioactive-con sumer-products-from-the-498044380.

Miller, Stanley. "The First Laboratory Synthesis of Organic Compounds under Primitive Earth Conditions," in *The Heritage of Copernicus: Theories "Pleasing to the Mind,"* ed. Jerzy Neyman. Cambridge, MA: MIT Press, 1974.

Miller, Stanley L., and Jeffrey L. Bada. "Submarine Hot Springs and the Origin of Life." *Nature* 334, no. 6183 (August 1988).

Milo, Ron, and Rob Phillips. *Cell Biology by the Numbers.* New York: Garland Science, 2015.

Mirsky, Alfred E. "The Discovery of DNA." *Scientific American* 218, no. 6 (1968).

Mitton, Simon. "The Expanding Universe of Georges Lemaître." *Astronomy & Geophysics* 58, no. 2 (April 1, 2017).

———. *Fred Hoyle: A Life in Science.* New York: Cambridge University Press, 2011.

———. "Georges Lemaître and the Foundations of Big Bang Cosmology." *Antiquarian Astronomer,* July 18, 2020.

Moberg, Carol L. *Entering an Unseen World: A Founding Laboratory and Origins of Modern Cell Biology, 1910–1974.* New York: Rockefeller University Press, 2012.

Monroe, Linda. "2 Dispute Popular Theory on Life Origin." *Los Angeles Times,* August 18, 1988.

Moore, Donovan. *What Stars Are Made Of: The Life of Cecilia Payne-Gaposchkin.* Cambridge, MA: Harvard University Press, 2020.

Morbidelli, A., J. Chambers, et al. "Source Regions and Timescales for the Delivery of Water to the Earth." *Meteoritics & Planetary Science* 35, no. 6 (2000).

Morris, Peter J. T. *The Matter Factory: A History of the Chemistry Laboratory.* London: Reaktion Books, 2015.

Moses, Vivian, and Sheila Moses. "Interview with Al Bassham," in *The Calvin Lab: Oral History Transcript 1945–1963*, chapter 7. Bancroft Library, Regional Oral History Office, Lawrence Berkeley Laboratory, University of California–Berkeley, 2000.

———. "Interview with Rod Quayle," in *The Calvin Lab: Oral History Transcript 1945–1963*, vol. 1, chapter 3. Bancroft Library, Regional Oral History Office, Lawrence Berkeley Laboratory, University of California–Berkeley, 2000.

Mulder, Gerardus. *Liebig's Question to Mulder Tested by Morality and Science.* London and Edinburgh: William Blackwood and Sons, 1846.

National Geographic Channel. "Birth of the Oceans." *Naked Science* series, March 2009.

New York Times. "Finds Spiral Nebulae Are Stellar Systems: Dr. Hubbell [sic] Confirms View That They Are 'Island Universes' Similar to Our Own," November 23, 1924.

———. "Scientists Find Indication of a Vitamin Which Prevents Softening of the Brain," April 10, 1931.

Nicholson, Wayne L., Nobuo Munakata, et al. "Resistance of Bacillus Endospores to Extreme Terrestrial and Extraterrestrial Environments." *Microbiology and Molecular Biology Reviews* 64, no. 3 (September 1, 2000).

Nobel Lectures Physics: Including Presentation Speeches and Laureates' Biographies, 1922–1941. Amsterdam: Elsevier, 1965.

Nutman, P. S. "Centenary Lecture." *Philosophical Transactions of the Royal Society of London*, Series B, *Biological Sciences* 317, no. 1184 (1987).

Olby, Robert. "Cell Chemistry in Miescher's Day." *Medical History* 13, no. 4 (October 1969).

———. *The Path to the Double Helix: The Discovery of DNA.* Seattle: University of Washington Press, 1974.

Oliveira, Patrick Luiz Sullivan De. "Martyrs Made in the Sky: The *Zénith* Balloon Tragedy and the Construction of the French Third Republic's First Scientific Heroes." *Notes and Records: The Royal Society Journal of the History of Science* 74, no. 3 (September 18, 2019).

Oparin, Aleksandr. *The Origin of Life*, trans. Sergius Morgulis, 2nd ed. New York: Dover, 1952.

O'Raifeartaigh, Cormac, and Simon Mitton. "Interrogating the Legend of Einstein's 'Biggest Blunder.'" *Physics in Perspective* 20 (December 2018).

Orgel, Leslie E. "Are You Serious, Dr. Mitchell?" *Nature* 402, no. 6757 (November 4, 1999).

Otis, Laura. *Rethinking Thought: Inside the Minds of Creative Scientists and Artists.* New York: Oxford University Press, 2015.

Pagel, Walter. *Joan Baptista Van Helmont: Reformer of Science and Medicine.* Cambridge, UK: Cambridge University Press, 1982.

Pais, Abraham. *Inward Bound: Of Matter and Forces in the Physical World.* Oxford, UK: Clarendon Press, 1988.

Palade, George E. "Albert Claude and the Beginnings of Biological Electron Microscopy." *Journal of Cell Biology* 50, no. 1 (July 1971).

Patel, Bhavesh H., Claudia Percivalle, et al. "Common Origins of RNA, Protein and Lipid Precursors in a Cyanosulfidic Protometabolism." *Nature Chemistry* 7, no. 4 (April 2015).

"Paul C. Aebersold Interview." *Longines Chronoscope*, CBS, 1953. https://www.youtube.com/watch?v=RFcxsX1UO44.

Payne-Gaposchkin, Cecilia. *Cecilia Payne-Gaposchkin: An Autobiography and Other Recollections.* Cambridge, UK: Cambridge University Press, 1996.

Peretó, Juli, Jeffrey L. Bada, and Antonio Lazcano. "Charles Darwin and the Origin of Life." *Origins of Life and Evolution of the Biosphere* 39, no. 5 (October 2009).

Perutz, M. F. "Co-Chairman's Remarks: Before the Double Helix." *Gene* 135, no. 1–2 (December 15, 1993).

Petterson, Roger. "The Chemical Composition of Wood," in *The Chemistry of Solid Wood: Advances in Chemistry*, vol. 207. American Chemical Society, 1984.

Phillips, J. P. "Liebig and Kolbe, Critical Editors." *Chymia* 11 (January 1966).

Plitt, Charles C. "A Short History of Lichenology." *The Bryologist* 22, no. 6 (1919).

Plumb, Robert. "Brookhaven Cosmotron Achieves the Miracle of Changing Energy Back into Matter." *New York Times*, December 21, 1952.

Portree, David. "The *Eagle* Has Crashed (1966)." *Wired*, May 15, 2012.

Poundstone, William. *Carl Sagan: A Life in the Cosmos*. New York: Henry Holt, 2000.

Powell, James. "To a Rocky Moon," in *Four Revolutions in the Earth Sciences: From Heresy to Truth*. New York: Columbia University Press, 2014.

Prebble, John. "Peter Mitchell and the Ox Phos Wars." *Trends in Biochemical Sciences* 27, no. 4 (April 2002).

———. "The Philosophical Origins of Mitchell's Chemiosmotic Concepts." *Journal of the History of Biology* 34 (2001).

Prebble, John, and Bruce Weber. *Wandering in the Gardens of the Mind: Peter Mitchell and the Making of Glynn*. New York: Oxford University Press, 2003.

Price, Catherine. *Vitamania: How Vitamins Revolutionized the Way We Think about Food*. New York: Penguin Books, 2016.

Prothero, Donald R. *The Story of Life in 25 Fossils: Tales of Intrepid Fossil Hunters and the Wonders of Evolution*. New York: Columbia University Press, 2015.

Quammen, David. *The Tangled Tree: A Radical New History of Life*. New York: Simon & Schuster, 2018.

Racker, Efraim. "Reconstitution, Mechanism of Action and Control of Ion Pumps." *Biochemical Society Transactions* 3, no. 6 (December 1, 1975).

Radetsky, Peter. "How Did Life Start?" *Discover*, November 1992.

Ralfs, John. "The Lichens of West Cornwall," in *Transactions of the Penzance Natural History and Antiquarian Society*, vol. 1. Plymouth, 1880.

Reinhard, Christopher T., Noah J. Planavsky, et al. "Evolution of the Global Phosphorus Cycle." *Nature* 541, no. 7637 (January 19, 2017).

Rentetzi, Maria. AIP oral history interview with Leopold Halpern, March 10, 1999, https://www.aip.org/history-programs/niels-bohr-library/oral-histories/32406.

———. "Blau, Marietta," in *Complete Dictionary of Scientific Biography*, vol. 19. Detroit: Charles Scribner's Sons, 2008.

———. *Trafficking Materials and Gendered Experimental Practices: Radium Research in Early 20th Century Vienna*. New York: Columbia University Press, 2008.

Rheinberger, Hans-Jörg. "Claude, Albert," in *Complete Dictionary of Scientific Biography*, vol. 20. Detroit: Charles Scribner's Sons, 2008.

Rhodes, Richard. *The Making of the Atomic Bomb*. New York: Simon & Schuster, 1986.

Righter, Kevin, John Jones, et al. "Michael J. Drake (1946–2011)." *Geochemical Society News*, October 1, 2011.

Riordan, Michael. *The Hunting of the Quark: A True Story of Modern Physics*. New York: Simon & Schuster, 1987.

Roddis, Louis Harry. *James Lind, Founder of Nautical Medicine*. New York: Henry Schuman, 1950.

Rosenfeld, Louis. "The Last Alchemist—the First Biochemist: J. B. van Helmont (1577–1644)." *Clinical Chemistry* 31, no. 10 (October 1985).

Roskoski, Robert. "Wandering in the Gardens of the Mind: Peter Mitchell and the Making of Glynn." *Biochemistry and Molecular Biology Education* 32, no. 1 (2004).

Rosner, Robert W., and Brigitte Strohmaier. *Marietta Blau, Stars of Disintegration: Biography of a Pioneer of Particle Physics.* Riverside, CA: Ariadne Press, 2006.

Russell, Percy, and Anita Williams. *The Nutrition and Health Dictionary.* New York: Chapman and Hall, 1995.

Ryan, Frank. *Darwin's Blind Spot: Evolution Beyond Natural Selection.* Boston: Houghton Mifflin Harcourt, 2002.

Sagan, Carl. *Conversations with Carl Sagan,* ed. Tom Head. Jackson: University Press of Mississippi, 2006.

Sagan, Dorion. *Lynn Margulis: The Life and Legacy of a Scientific Rebel.* White River Junction, VT: Chelsea Green, 2012.

Saier, Milton H., Jr. "Peter Mitchell and the Life Force," https://petermitchellbiography.wordpress.com/.

Sapp, Jan. *Evolution by Association: A History of Symbiosis.* New York: Oxford University Press, 1994.

Schmidt, Albert. "Essai d'une biologie de l'holophyte des Lichens." *Mémoires du Muséum national d'histoire naturelle, Série B, Botanique* 3 (1953).

Schopf, William. *Cradle of Life: The Discovery of Earth's Earliest Fossils.* Princeton, NJ: Princeton University Press, 1999.

———. *Life in Deep Time: Darwin's "Missing" Fossil Record.* Boca Raton, FL: CRC Press, 2018.

———. *Major Events in the History of Life.* Boston: Jones & Bartlett Learning, 1992.

Schuchert, Charles. "Charles Doolittle Walcott, (1850–1927)." *Proceedings of the American Academy of Arts and Sciences* 62, no. 9 (1928).

Segré, Daniel, and Doron Lancet. "Theoretical and Computational Approaches to the Study of the Origin of Life" in *Origins: Genesis, Evolution and Diversity of Life,* ed. Joseph Seckbach. Dordrecht, Netherlands: Springer, 2005.

Sender, Ron, Shai Fuchs, and Ron Milo. "Revised Estimates for the Number of Human and Bacteria Cells in the Body." *PLOS Biology* 14, no. 8 (August 19, 2016).

Sender, Ron, and Ron Milo. "The Distribution of Cellular Turnover in the Human Body." *Nature Medicine* 27, no. 1 (January 2021).

Seward, Albert Charles. *Plant Life through the Ages: A Geological and Botanical Retrospect,* 2nd ed. New York: Hafner, 1959.

Sharkey, Thomas D. "Discovery of the Canonical Calvin-Benson Cycle." *Photosynthesis Research* 140, no. 2 (May 1, 2019).

Shaw, Bernard. *Annajanska, the Bolshevik Empress: A Revolutionary Romancelet,* in *Selected One Act Plays.* Harmondsworth: Penguin, 1976.

Shindell, Matthew. *The Life and Science of Harold C. Urey.* Chicago: University of Chicago Press, 2019.

Sime, Ruth Lewin. "Marietta Blau: Pioneer of Photographic Nuclear Emulsions and Particle Physics." *Physics in Perspective* 15 (2013).

Smith, Annie Lorrain. *Lichens.* Cambridge, UK: Cambridge University Press, 1921.

Stager, Curt. *Your Atomic Self: The Invisible Elements That Connect You to Everything Else in the Universe.* New York: Thomas Dunne Books, 2014.

Steinmaurer, Rudolf. "Erinnerungen an V.F. Hess, Den Entdecker der Kosmischen Strahlung, und an Die ersten Jahre des Betriebes des Hafelekar-Labors." *Early History of Cosmic Ray Studies* 118 (1985).

Step, Edward. *Plant-Life: Popular Papers on the Phenomena of Botany.* London: Marshall Japp, 1881.

Stevens, Charles. "The Neuron." *Scientific American* 241, no. 3 (September 1979).

St. Louis Post-Dispatch, "Is Vitamine Starvation the True Cause of Cancer?" October 27, 1924.

Sullivan, Walter. "Subatomic Tests Suggest a New Layer of Matter." *New York Times,* April 25, 1971.

———. *We Are Not Alone: The Search for Intelligent Life on Other Worlds,* rev. ed. New York: Dutton, 1993.

Sundermier, Ali. "The Particle Physics of You." *Symmetry* magazine, November 3, 2015, https://www.symmetrymagazine.org/article/the-particle-physics-of-you.

Tegmark, Max. "Solid. Liquid. Consciousness." *New Scientist* 222, no. 2964 (April 12, 2014).

Telegraph, The (London). "Lynn Margulis," December 13, 2011.

Tera, Fouad, Dimitri A. Papanastassiou, and Gerald J. Wasserburg. "A Lunar Cataclysm at ~3.95 AE and the Structure of the Lunar Crust," in *Lunar Science* IV (1973).

Thoreau, Henry David. *Walden.* Boston: Ticknor & Fields, 1854; Beacon Press, 2004.

Thorpe, Thomas Edward. *Essays in Historical Chemistry.* London: Macmillan, 1902.

Times, The (London). "The British Association: Evolution of the Universe," September 30, 1931.

Tobey, Ronald C. *Saving the Prairies: The Life Cycle of the Founding School of American Plant Ecology, 1895–1955.* Berkeley: University of California Press, 1981.

Townes, Charles H. "The Discovery of Interstellar Water Vapor and Ammonia at the Hat Creek Radio Observatory," in *Revealing the Molecular Universe: One Antenna Is Never Enough,* Proceedings of a Symposium Held at University of California, Berkeley, California, USA, September 9–10, 2005. Astronomical Society of the Pacific.

———. *How the Laser Happened: Adventures of a Scientist.* New York: Oxford University Press, 2002.

———. "Microwave and Radio-Frequency Resonance Lines of Interest to Radio Astronomy," in *International Astronomical Union Symposium,* no. 4, *Radio Astronomy.* Cambridge, UK: Cambridge University Press, 1957.

"Tracing the Lost Railway Lines of Indonesia: The Forgotten Steamtram of Bat-

avia," https://indonesialostrailways.blogspot.com/p/the-forgotten-steamtram-of
-batavia.html.

Trewavas, Anthony. "The Foundations of Plant Intelligence." *Interface Focus* 7, no.
3 (June 6, 2017).

———. "Mindless Mastery." *Nature* 415, no. 6874 (February 21, 2002).

Trewavas, Anthony, and František Baluška. "The Ubiquity of Consciousness." European Molecular Biology Organization, *EMBO Reports* 12, no. 12 (December 1, 2011).

Turner, R. Steven. "Justus Liebig versus Prussian Chemistry: Reflections on Early Institute-Building in Germany." *Historical Studies in the Physical Sciences* 13, no. 1 (1982).

USDA FoodData Central website. "Bananas, Ripe and Slightly Ripe, Raw," April 1, 2020, https://fdc.nal.usda.gov/fdc-app.html#/food-details/1105314 /nutrients.

Valley, John W. "A Cool Early Earth?" *Scientific American* 293, no. 4 (October 2005).

Van Klooster, H. S. "Jan Ingenhousz." *Journal of Chemical Education* 29, no. 7 (July 1, 1952).

Vedder, Edward Bright. *Beriberi*. New York: William Wood, 1913.

Vernadsky, Vladimir I. *The Biosphere*, ed. Mark Mcmenamin, trans. David Langmuir. New York: Copernicus, 1998.

Von Braun, Wernher, Fred L. Whipple, and Willy Ley. *Conquest of the Moon*, ed. Cornelius Ryan. New York: Viking Press, 1953.

Wächtershäuser, Günter. "Before Enzymes and Templates: Theory of Surface Metabolism." *Microbiological Reviews* 52, no. 4 (December 1988).

———. "The Origin of Life and Its Methodological Challenge." *Journal of Theoretical Biology* 187, no. 4 (August 21, 1997).

Wade, Nicholas. "Heart Muscle Renewed over Lifetime, Study Finds." *New York Times*, April 2, 2009.

———. "Making Sense of the Chemistry That Led to Life on Earth." *New York Times*, May 4, 2015.

———. "Meet Luca, the Ancestor of All Living Things." *New York Times*, July 25, 2016.

———. "Stanley Miller, Who Examined Origins of Life, Dies at 77." *New York Times*, May 23, 2007.

———. "Your Body Is Younger Than You Think." *New York Times*, August 2, 2005.

Wagener, Leon. *One Giant Leap: Neil Armstrong's Stellar American Journey*. Brooklyn, NY: Forge Books, 2004.

Walcott, Charles Doolittle. *Pre-Cambrian Fossiliferous Formations*. Rochester, NY: Geological Society of America, 1899.

———. "Pre-Carboniferous Strata in the Grand Canyon of the Colorado, Arizona." *American Journal of Science* 26 (December 1883).

———. "Report of Mr. Charles D. Walcott, July 2," in *Fourth Annual Report of the*

Director of the United States Geological Survey. Washington, DC: US Government Printing Office, 1885.

Wald, George. Nobel Banquet Speech, Nobel Prize in Physiology or Medicine 1967, Stockholm, December 10, 1967.

Walker, Gabrielle. *Snowball Earth: The Story of the Great Global Catastrophe That Spawned Life as We Know It*. New York: Crown, 2003.

Walker, John. *Fuel of Life*, video recording of Nobel Laureate Lecture, 2018, https://www.royalacademy.dk/en/ENG_Foredrag/ENG_Walker.

Walker, Timothy. *Plants: A Very Short Introduction*. Oxford, UK: Oxford University Press, 2012.

Walter, Michael. "From the Discovery of Radioactivity to the First Accelerator Experiments," in *From Ultra Rays to Astroparticles: A Historical Introduction to Astroparticle Physics*, ed. Brigitte Falkenburg and Wolfgang Rhode. Dordrecht, Netherlands: Springer, 2012.

Watson, James D., and Andrew Berry. *DNA: The Secret of Life*. New York: Knopf, 2003.

Watson, James D., Alexander Gann, and Jan Witkowski. *The Annotated and Illustrated Double Helix*. New York: Simon & Schuster, 2012.

Weaver, Kenneth. "What the Moon Rocks Tell Us." *National Geographic*, December 1969.

Web of Stories. Interview with Francis Crick, "Molecular Biology in the Late 1940s," 1993, https://www.webofstories.com/people/francis.crick/33?o=SH.

Web of Stories. Interview with James Watson, "Complementarity and My Place in History," 2010, https://www.webofstories.com/people/james.watson/29?o=SH.

Webb, Richard. "Listening for Gravitational Waves from the Birth of the Universe." *New Scientist*, March 16, 2016.

Weiner, Charles. AIP oral history interview with William Fowler, February 6, 1973, https://www.aip.org/history-programs/niels-bohr-library/oral-histories/4608-4.

Weiss, Benjamin P., Joseph L. Kirschvink, et al. "A Low Temperature Transfer of ALH84001 from Mars to Earth." *Science* 290, no. 5492 (October 27, 2020).

West, Bert. "Moon Rocks Go to Experts on Friday." *Newsday*, September 10, 1969.

Wetherill, George W. "Contemplation of Things Past." *Annual Review of Earth and Planetary Sciences* 26, no. 1 (1998).

———. "The Formation of the Earth from Planetesimals." *Scientific American* 244, no. 6 (June 1981).

Whipple, Fred L. "Of Comets and Meteors." *Science* 289, no. 5480 (August 4, 2000).

Wilford, John Noble. "Moon Rocks Go to Houston; Studies to Begin Today: Lunar Rocks and Soil Are Flown to Houston Lab." *New York Times*, July 26, 1969.

Wilkins, Maurice. *Maurice Wilkins: The Third Man of the Double Helix: An Autobiography*. Oxford, UK: Oxford University Press, 2005.

Williams, Gareth. *Unravelling the Double Helix: The Lost Heroes of DNA*. London: Weidenfeld & Nicolson, 2019.

Wills, Christopher, and Jeffrey Bada. *The Spark of Life: Darwin and the Primeval Soup*. Oxford, UK: Oxford University Press, 2000.

Wilson, Charles Morrow. *Roots: Miracles Below*. New York: Doubleday, 1968.

Wolchover, Natalie. "Geological Explorers Discover a Passage to Earth's Dark Age." *Quanta Magazine*, December 22, 2016.

Woodard, Helen Q., and David R. White. "The Composition of Body Tissues." *British Journal of Radiology* 59, no. 708 (December 1986).

Yarris, Lynn. "Ernest Lawrence's Cyclotron: Invention for the Ages." Lawrence Berkeley National Laboratory, Science Articles Archive, https://www2.lbl.gov/Science-Articles/Archive/early-years.html.

Yochelson, Ellis Leon. *Charles Doolittle Walcott, Paleontologist*. Kent, OH: Kent State University Press, 1998.

Yong, Ed. "Trees Have Their Own Internet." *The Atlantic*, April 14, 2016.

Zahnle, Kevin, Laura Schaefer, and Bruce Fegley. "Earth's Earliest Atmospheres." *Cold Spring Harbor Perspectives in Biology* 2, no. 10 (October 2010).

"The *Zenith* Tragedy: The Dangers of Hypoxia." Those Magnificent Men in Their Flying Machines, https://www.thosemagnificentmen.co.uk/balloons/zenith.html.

Ziegler, Charles A. "Technology and the Process of Scientific Discovery: The Case of Cosmic Rays." *Technology and Culture* 30, no. 4 (October 1989).

Zimmer, Carl. "Vitamins' Old, Old Edge." *New York Times*, December 9, 2013.

Zimmermann, Albrecht. "Nachrufe: Simon Schwendener." *Berichte der Deutschen Botanischen Gesellschaft* 40 (1922).

Zweig, George. "Origin of the Quark Model," in *Proceedings of the Fourth International Conference on Baryon Resonances*, Toronto, July 14–16, 1980.

INDEX

ABOUT THE AUTHOR

DAN LEVITT spent over twenty-five years writing, producing, and directing award-winning documentaries for National Geographic, Discovery, Science, History, PBS, and the Howard Hughes Medical Institute. His film topics included how Galileo, Newton, Einstein, and Hawking made their greatest discoveries; the archeology of Custer's Last Stand; a new theory on dinosaur evolution; and the scientific search for alien life. Dan began his career as a Peace Corps volunteer in Kenya, teaching high school physics and biology. He lives in Cambridge with his wife, two children, and their dog.